Quantum Mechanics of the Diatomic Molecule (Second Edition)

Online at: https://doi.org/10.1088/978-0-7503-6204-7

IOP Series in Coherent Sources, Quantum Fundamentals, and Applications

About the Editor
F J Duarte is a laser physicist based in Western New York, USA. His career has covered three continents while contributing within the academic, industrial, and defense sectors. Duarte is editor/author of 15 laser optics books and sole author of three books: *Tunable Laser Optics, Quantum Optics for Engineers*, and *Fundamentals of Quantum Entanglement*. Duarte has made original contributions in the fields of coherent imaging, directed energy, high-power tunable lasers, laser metrology, liquid and solid-state organic gain media, narrow-linewidth tunable laser oscillators, organic semiconductor coherent emission, *N*-slit quantum interferometry, polarization rotation, quantum entanglement, and space-to-space secure interferometric communications. He is also the author of the generalized multiple-prism grating dispersion theory and pioneered the use of Dirac's quantum notation in *N*-slit interferometry and classical optics. His contributions have found applications in numerous fields, including astronomical instrumentation, dispersive optics, femtosecond laser microscopy, geodesics, gravitational lensing, heat transfer, laser isotope separation, laser medicine, laser pulse compression, laser spectroscopy, mathematical transforms, nonlinear optics, polarization optics, and tunable diode-laser design. Duarte was elected Fellow of the Australian Institute of Physics in 1987 and Fellow of the Optical Society of America in 1993. He has received various recognitions, including the *Paul F Foreman Engineering Excellence Award* and the *David Richardson Medal* from the Optical Society.

Coherent Sources, Quantum Fundamentals, and Applications
Since its discovery the laser has found innumerable applications from astronomy to zoology. Subsequently, we have also become familiar with additional sources of coherent radiation such as the free electron laser, optical parametric oscillators, and coherent interferometric emitters. The aim of this book Series in Coherent Sources, Quantum Fundamentals, and Applications is to explore and explain the physics and technology of widely applied sources of coherent radiation and to match them with utilitarian and cutting-edge scientific applications. Coherent sources of interest are those that offer advantages in particular emission characteristics areas such as broad tunability, high spectral coherence, high energy, or high power. An additional area of inclusion are the coherent sources capable of high performance in the miniaturized realm. Understanding of quantum fundamentals can lead to new and better coherent sources and unimagined scientific and technological applications. Application areas of interest include the industrial, commercial, and medical sectors. Also, particular attention is given to scientific applications with a bright future such as coherent spectroscopy, astronomy, biophotonics, space communications, space interferometry, quantum entanglement, and quantum interference.

Publishing benefits
Authors are encouraged to take advantage of the features made possible by electronic publication to enhance the reader experience through the use of color, animation and video, and incorporating supplementary files in their work.

Do you have an idea of a book that you'd like to explore?
For further information and details of submitting book proposals, see iopscience.org/books or contact Ashley Gasque at ashley.gasque@iop.org.

A full list of titles published in this series can be found here: https://iopscience.iop.org/bookListInfo/series-in-coherent-sources-and-applications.

Quantum Mechanics of the Diatomic Molecule (Second Edition)

Christian G Parigger
Former address:
Physics and Astronomy Department, University of Tennessee, University of Tennessee Space Institute, Center for Laser Applications, Tullahoma, TN, USA

Current address:
CGP Consulting, Manchester, TN, USA

James O Hornkohl
Hornkohl Consulting, Tullahoma, TN, USA

IOP Publishing, Bristol, UK

ISBN 978-0-7503-6204-7 (ebook)
ISBN 978-0-7503-6202-3 (print)
ISBN 978-0-7503-6205-4 (myPrint)
ISBN 978-0-7503-6203-0 (mobi)

DOI 10.1088/978-0-7503-6204-7

Version: 20241001

IOP ebooks

British Library Cataloguing-in-Publication Data: A catalogue record for this book is available from the British Library.

Published by IOP Publishing, wholly owned by The Institute of Physics, London

IOP Publishing, No.2 The Distillery, Glassfields, Avon Street, Bristol, BS2 0GR, UK

US Office: IOP Publishing, Inc., 190 North Independence Mall West, Suite 601, Philadelphia, PA 19106, USA

To John, Anna, and Melissa

To Justin, Jason, and Jola

Contents

Preface

0.1 First edition

The book notes from J O Hornkohl and extensive scientific discussions and research engagements in my work at the University of Tennessee Space Institute, Center for Laser Applications, motivate completion of this ebook. Communication exchanges occurred since the spring of 1987, and continued regularly until winter 2017 [1]. Over the years, several colleagues and postgraduate MSc and PhD students have contributed to applications of fundamental insights in the physics of the diatomic molecule. Thanks go to David Plemmons, Guoming Guan, Ying-Ling Chen, Wenhong Qin, Ivan Dors, Alexander Woods, David Surmick, Michael Witte, Ghaneshwar Gautam, and Christopher Helstern.

Significant emphasis has been placed on the application of the diatomic spectroscopy predictions in analysis of experimental data. For this reason, this ebook includes several chapters on applications in studies of diatomic molecules, especially important molecules such as cyanide (CN), aluminum monoxide (AlO), diatomic carbon (C_2), titanium monoxide (TiO), hydroxyl (OH), but also selected work on other diatomic molecules.

This text introduces insights that are essential in utilizing the inherent symmetries associated with diatomic molecules. Consequently, line positions and strengths associated with transitions from lower and upper state-manifolds are determined without invoking approximations that separate vibrations and rotations of diatomic nuclei from electron motion based on mass. The approach utilized in this work makes use of the separation of angular coordinates from electronic vibrational coordinates. Consequently, the volley of selection rules for diatomic spectroscopy is no longer required, including methodologies that rely on so-called reversed angular momentum techniques.

This work summarizes well over 30 years of quantitative analysis of temporally and spatially resolved experimental records, almost all of the experiments discussed in this ebook were conducted at the Center for Laser Applications (CLA) at the University of Tennessee Space Institute. Applications include understanding on nonequilibrium fluid and plasma physics and interpretation of stellar astrophysics spectra. In several cases of laser-induced plasma investigations, both atomic and molecular signatures or superposed spectral characteristics from molecules and atoms can be identified. Analysis of such superposition spectra requires accurate knowledge of wavelength positions and transition strengths. The revival and replacement of electrical-spark spectroscopy with laser-spark or laser-plasma spectroscopy for quantitative elemental composition analysis since the mid-1990s, viz. laser-induced breakdown spectroscopy (LIBS), extends into increased interests in molecular LIBS since (give-or-take) the mid-2000s. From an analytical and practical point of view, the requirements can be reduced to the availability of a set of diatomic line-strengths in tabular form along with programs that are designed to appropriately read the records. However, this ebook provides a reasonable account of the quantum mechanics of the diatomic molecule, along with selected applications that were important for motivating a consistent approach and for analyzing recorded data sets from various experiments in the CLA laboratories.

The challenge of this work has been the prediction of spectra with a focus on diatomic spectroscopy. The aim of the lifetime work of Jim Hornkohl is the design of an algorithm to predict and fit computed and measured molecular spectra to provide inferences on parameters such as excitation temperature. The means to accomplish goals for various diatomic molecules are the consistent application of standard quantum theory of angular momentum. During his career, Jim engaged in efforts to overcome techniques such as Van Vleck's reversed angular momentum approach based on angular momentum commutators. The apparent difficulties included the battles with the established practice to predict and compute spectra and design programs despite the mathematical inconsistencies associated with the reversed angular momentum practice. The experimental investigations, and again the stimulating discussions, motivated refinements such as enlarging the data sets for the CN, C_2, or TiO diatomic molecules. In turn, the discussed applications in this book are intended to alleviate analysis of diatomic spectra composed of superpositions of a significant amount of transition lines within typical resolution for laser-plasma emission spectroscopy, to name but one example.

Christian Parigger
August 2019

0.2 Second edition

The second edition includes 10 additional chapters, one on the fundamentals and nine on the applications parts. Three additional appendices are included, namely: communication of NMT and BESP scripts for computation of diatomic spectra, Abel inversion scripts with one specific example, and an appendix on select recent publications that include C.G.P. as author. The additions primarily address communication of spatial profiles analyses, including Abel inversion and communication of scripts for diatomic spectroscopy and Abel inversions. However, comparisons with other existing databases clearly reveal the significance of the line strengths for the selected electronic transitions of diatomic molecules. The comparisons also include a section of C_2 laser-induced fluorescence. The existing databases comparisons include PGOPHER, LIFBASE, and ExoMol databases that are compared with line-strength data of diatomic molecules of interest, particularly for laser-induced plasma that is generated in gases and gas mixtures.

- New chapter 2 addresses the foundations of quantum mechanics and the mathematical implementation of specific symmetries. Application of the correspondence principle, relating classical and quantum mechanics, leads to the occurrence of the infamous sign-reversal. This chapter addresses formal treatment of symmetries in quantum mechanics. Quantum theory contraindicates sign changes of the fundamental angular momentum algebra.

Reversed angular momentum sign changes are of a heuristic nature and are actually undesirable in the analysis of diatomic spectra.

- New chapter 15 communicates line-strength data and associated scripts for the computation and spectroscopic fitting of selected transitions of diatomic molecules. The scripts for data analysis are designed for inclusion in various software packages or program languages. Selected results demonstrate the applicability of the program for data analysis in laser-induced optical breakdown spectroscopy, primarily at the University of Tennessee Space Institute, Center for Laser Applications. Representative spectra are calculated and referenced to measured data records. Comparisons of experiment data with predictions from other tabulated diatomic molecular databases confirm the accuracy of the communicated line-strength data.
- New chapter 17 discusses cavity ring-down spectroscopy of methylidyne in a chemiluminescent plasma that is produced in a microwave cavity. Of interest are the rotational lines of selected vibrational transitions for the A–X and B–X bands. This chapter also includes recent analysis that shows excellent agreement of measured and computed data, and it communicates CH line-strength data. The CH radical is an important diatomic molecule in hydrocarbon combustion diagnosis and analysis of stellar plasma emissions, to name just two examples for analytical plasma chemistry.
- New chapter 19 discusses diatomic molecular spectroscopy of laser-induced plasma and analysis of data records, specifically signatures of cyanide (CN). Line-strength data from various databases are compared for simulation of the cyanide spectra. Of interest are recent predictions using an astrophysical database, i.e., ExoMol, a laser-induced fluorescence database, i.e., LIFBASE, and a program for simulating rotational, vibrational, and electronic spectra, i.e., PGOPHER.
- New chapter 21 presents analysis of carbon Swan bands laser-plasma emission records using line-strength data and the ExoMol database. The temperature inferences are elaborated when using nonlinear fitting with both databases. The line-strength data are also utilized for analysis of laser-induced fluorescence experiments that employ a spectral resolution of the order of 5 pm. Accurate diatomic carbon databases show many applications in laboratory diagnosis and interpretation of astrophysical plasma records.
- New chapter 23 elaborates on analysis of aluminum monoxide (AlO), laser-plasma emission records using line-strength data, and the ExoMol astrophysical database. A nonlinear fitting program computes comparisons of measured and simulated diatomic molecular spectra. This work also presents a comparison of the AlO line strength and of ExoMol data for the AlO diatomic molecule. Accurate AlO databases show a volley of applications in laboratory and astrophysical plasma diagnosis.
- New chapter 25 applies NMT and BESP scripts for the fitting of recorded experimental hydroxyl data. The fitting program also incorporates a slight, overall wavelength offset. The ExoMol and line-strength data yield close to identical temperature with a slightly different linear background.

The databases for specific hydroxyl transition yield similar predictions of the recorded laser-plasma spectra for time delays of the order of one hundred microseconds after optical breakdown initiation.

- New chapter 26 communicates measurement and analysis of diatomic molecular hydroxyl spectra after generation of laser-induced plasma, and it also shows details of the expanding plasma including associations of shadowgraphs with spectroscopy. Formation of OH is clearly discernible at time delays of several dozen microseconds after plasma initiation. Optical emissions are dispersed by a Czerny–Turner spectrometer and an intensified charge-coupled device records the data along the wavelength and slit dimensions.
- New chapter 29 combines time-resolved emission spectroscopy with Abel integral inversion techniques to obtain radial electron density values in laser-induced plasma. This chapter also includes details of the Abel transforms Hydrogen beta line profiles are recorded following optical breakdown in ultra-high-pure hydrogen gas. Asymmetric Abel inversion techniques are utilized in the analysis of collected, time-resolved data. The averaged, line-of-sight electron densities are found to be in of the order of one hundredth of an amagat for time delays close to one-half microseconds. The electron densities indicate variations across the laser-induced plasma.
- New chapter 30 elucidates the connection of measured shadowgraphs from optically induced air breakdown with emission spectroscopy in selected gas mixtures. Spectroscopic analysis explores well-above hypersonic expansion dynamics using primarily diatomic molecule cyanide and atomic hydrogen emission spectroscopy. Analysis of the air breakdown and selected gas breakdown events permits the use of Abel inversion for inference of the expanding species distribution. Typically, species are prevalent at higher density near the hypersonically expanding shock wave, measured by tracing cyanide and a specific carbon atomic line.
- New appendix J presents NMT and BESP MATLAB-scripts for computation of diatomic spectra.
- New appendix K presents Abel inversion MATLAB-scripts with one specific example.
- New appendix L summarizes select recent publications that include C.G.P. as author.

Christian Parigger
February 2024

Reference

[1] Parigger C G and Nemes L 2017 *Int. J. Mol. Theor. Phys.* **1** 00105

Acknowledgements

CGP and JOH appreciate the support by the Center for Laser Applications (CLA) at the University of Tennessee Space Institute during over 30 years of research engagement in spectroscopy. The research led to this book and its second edition that describe how quantum mechanics can be used to predict diatomic molecule spectra. The book and its second edition provide a comprehensive overview on diatomic molecule fundamentals and emphasize the applications of spectroscopy predictions in analysis of experimental data. Both CGP and JOH are happy to report interests and contributions by postgraduate students. Moreover, CGP is delighted about publication of the first and second editions in memoriam to JOH, a long time collaborator in analysis of laser-induced plasma spectra recorded at the CLA. Last but not least, CGP thanks researchers and colleagues at the CLA and international collaborators throughout the world for extensive discussions and motivation towards completion of both book editions.

Author biographies

Christian G. Parigger

Christian G. Parigger (Courtesy UTSI photo albums: L Horton.)

The research interests of Dr Christian Parigger include fundamental and applied spectroscopy, nonlinear optics, quantum optics, ultrafast phenomena, ultrasensitive diagnostics, lasers, combustion and plasma physics, optical diagnostics, biomedical applications, and, in general, atomic and molecular and optical physics. His Mag. rer. nat. degree shows work on optical bistability at the University of Innsbruck, Austria, with guidance by Dr Peter Zoller. His PhD degree studies are on the subject of polarization spectroscopy and magnetically induced switching at the University of Otago, Dunedin, New Zealand, with guidance by Drs Wes Sandle and Rob Ballagh. He also holds the Dr rer. nat. degree in Physics from the University of Innsbruck, Austria. Since 1987, his work encompasses experimental, theoretical and computational research, together with teaching, service, and outreach at the Center for Laser Applications at The University of Tennessee Space Institute, Tullahoma, Tennessee, USA.

James O and Jeri Hornkohl

James O and Jeri Hornkohl (Courtesy UTSI photo albums: L Horton.)

The contributions of James Hornkohl, or 'Jim', encompass the spectroscopy of diatomic molecules, and the application of such spectroscopy in diagnosis of combustion, plasmas, rocket propulsion, and related problems. His support of student theses and dissertations has been especially significant, including the application of numerical methods in analysis of experiments. Moreover, his help has been greatly appreciated by his collaborators in the design of computational and experimental methods to record digital data. During the last 30 years prior to his death on February 7, 2017, Jim had been strongly engaged in the description of the very details on diatomic spectroscopy. The challenge to Jim's work has been the prediction of spectra with a focus on diatomic spectroscopy. The aim of his lifetime work was the design of an algorithm for the prediction and fitting of computed to measured molecular spectra and to provide inferences of parameters such as excitation temperature. The means to accomplish goals for various transitions of diatomic molecules are the consistent application of standard quantum theory of angular momentum.

Part I

Fundamentals of the diatomic molecule

IOP Publishing

Quantum Mechanics of the Diatomic Molecule (Second Edition)

Christian G Parigger and James O Hornkohl

Chapter 1

Primer on diatomic spectroscopy

1.1 Overview

This book describes how one uses quantum mechanics to predict the spectra of diatomic molecules in their gaseous state. The two most important attributes of a spectral line are its position in the electromagnetic spectrum and the strength with which the molecule can interact with the radiation field to produce spectral lines. Thus, a book that discusses the calculation of positions and intensities of spectral lines of a diatomic molecule equally communicates the application of quantum theory to the diatomic molecule.

The theoretically convenient measure of spectral line position is its vacuum wave number $\tilde{\nu}_{u\ell}$, which is the difference between the upper term T_u (i.e., upper energy eigenvalue expressed in the units of cm^{-1}) and the lower term T_ℓ,

$$\tilde{\nu}_{u\ell} = T_u - T_\ell. \tag{1.1}$$

In the optical region, the term difference corresponds to a specific color. However, experiments usually measure the wavelength positions in a laboratory setting at standard ambient temperature and pressure. For typical laser spectroscopy investigations of, say, optical emission spectroscopy subsequent to generation of a laser spark, spectral resolutions of the instrument spectrometer and detector amount to 0.1–0.01 nm, rarely to 0.001 nm or 1 pm. At the wavelength, λ, of 400 nm, a spectral resolution, $\Delta\lambda$, of better than 1 pm corresponds to a resolving power, R,

$$R = \lambda/\Delta\lambda \geqslant 400\,000, \tag{1.2}$$

or a wave number resolution of better than 0.05 cm^{-1}. The spectral resolution of diatomic molecular data computed in this book is better than 0.05 cm^{-1}. For laser-induced optical breakdown experiments, which is a recent application of diatomic molecular spectroscopy, resolving powers are of the order of 4000–10 000. For high-resolution, absorption measurements of stellar astrophysical objects, resolving powers of the order of 40 000 are quite common.

doi:10.1088/978-0-7503-6204-7ch1

The theoretically most convenient measure of a molecule's ability to interact with electromagnetic radiation is its Condon and Shortley [1] line strength, $S_{u\ell}$, which describes transitions between an upper, u, and a lower level, ℓ. The line strength represents a summation over individual states that comprise upper and lower levels. Both the vacuum wave number $\tilde{\nu}_{u\ell} = \tilde{\nu}_{\ell u}$ and the line strength $S_{u\ell} = S_{\ell u}$ are symmetric with regard to the upper and lower levels. In addition, the symbols u and ℓ represent a collection of quantum numbers. In diatomic spectroscopy, upper state quantum numbers are normally denoted with a single prime, while lower states are denoted with the absence of a prime or a double prime. The absence of a double prime has become the standard way of denoting a lower state diatomic quantum number.

1.2 Reversed angular momentum

Historically, the reversed-angular-momentum (RAM) methodology has successfully predicted diatomic spectra without the use of modern digital computers. The RAM method establishes a reduced set of basis states; in other words, works with an a priori approximation. Sets of rules are introduced when applying a transformation to a molecular-fixed from the laboratory-fixed coordinate system. These rules utilize a supposed reversal of sign in the application of quantum mechanical angular momentum algebra. This section provides a brief historic account of the challenges associated with the RAM method.

The reversed-angular momentum approach is mentioned first in an article on the quantization question of the asymmetric top [2]. Klein writes in the introduction that the paper might be of interest for methods of quantization. The reversed sign is introduced for the equations of the components of angular momentum in the molecular-fixed coordinate system in order to obtain agreement with the well-established classical equations for the symmetric top. Conversely, the application of the standard, laboratory-fixed angular momentum equations would lead to the wrong classical result. This article also makes reference to canonical conjugate Euler angles that are interpreted as references to dual space.

The RAM methodology is embraced by Van Vleck in his work on the coupling of angular momentum vectors in molecules [3]. Notably, Sir Harold Kroto communicates in his acceptance lecture for the 1996 Nobel Prize in Chemistry, 'Symmetry, Space, Stars and C_{60}' [4], the importance of 'Symmetry, the Key to the Theory of Everything'. With reference to the RAM work, Sir Kroto quotes Van Vleck: '*Practically every-one (!) knows that the components of total angular momentum (NB the angular momentum operator is usually denoted by* J *and the associated quantum number by* j*) of the molecule relative to the axes [x, y, z] fixed in space satisfy the commutation relation of the form*

$$J_x J_y - J_y J_x = iJ_z \tag{1.3}$$

Klein discovered the rather surprising fact that when total angular momentum is referred to axes mounted in the molecule which we will denote by [x', y', z'] the sign of i in the commutation relation is reversed i.e.

$$J_{x'}J_{y'} - J_{y'}J_{x'} = -iJ_{z'} \tag{1.4}$$

Sir Kroto goes on to say: *Does practically everyone know this?—I wondered whether to check this claim out by asking everyone on the main street in Brighton whether they did. I hardly knew—or more accurately—really understood the first relation, let alone the second. However I did know that angular momentum was quantised and governed by the fundamental relations*

$$\langle j|J^2|\rangle = \hbar j(j+1) \tag{1.5}$$

$$M_J = -j \ldots +j \tag{1.6}$$

which means that J has $2j + 1$ *possible orientations, and*

$$\Delta j = 0, \pm 1 \tag{1.7}$$

which indicates that when a transition occurs, j may only change by one unit or on occasion remain unchanged.' Previously, in 1975 and then in 1992, Sir Kroto discussed the molecule-fixed angular momentum following Van Vleck [3], leading to the reversed-angular momentum equations in his Nobel laureate lecture [4] and in his book on molecular rotation spectra [5].

However, an accurate review shows that there is no reversal of the sign when moving from a laboratory-fixed to a molecule-fixed coordinate system; in other words, there is no mathematical support of the reversed sign. Sustenance of the angular momentum equations can be explained as follows. In terms of classical mechanics, reversal of motion occurs as one goes from a rotating system to a fixed system, or vice versa. For example, motion reversal can be experienced by looking at the surroundings while on a rotating merry-go-round versus observing the rotation in the fixed reference frame. The quantum mechanical implementation of motion reversal or time reversal changes the sign and takes the conjugate complex, leading to the preservation of the sign. Reference to dual space would confuse things because clearly the standard angular momentum operator equations are not affected by a transformation from laboratory-fixed to molecule-fixed coordinates (see appendix A).

A reasonably concise treatment shows preservation of the commutator relations under a unitary transformation. Consider the operators A, B, and C which satisfy the commutation formula

$$AB - BA = iC \tag{1.8}$$

and subject these three operators to the unitary transformation U; that is,

$$A' = U^{\dagger}\, A\, U \tag{1.9a}$$

$$A = U\, A'\, U^{\dagger} \tag{1.9b}$$

with similar equations holding for B and C. Then,

$$AB - BA = U\, A'\, U^{\dagger}\, U\, B'\, U^{\dagger} - U\, B'\, U^{\dagger}\, U\, A'\, U^{\dagger} \tag{1.10a}$$

$$= U\ A'\ B'\ U^{\dagger} \tag{1.10b}$$

$$= iC \tag{1.10c}$$

$$iU^{\dagger}\ C\ U = A'B' \tag{1.10d}$$

$$iC' = A'B' - B'A'. \tag{1.10e}$$

The above result, e.g., see Davydov [6], holds for all commutators, including those for angular momentum. Thus,

$$J_{x'}J_{y'} - J_{y'}J_{x'} = i\ J_{z'} \tag{1.11a}$$

$$J_{y'}J_{z'} - J_{z'}J_{y'} = i\ J_{x'} \tag{1.11b}$$

$$J_{z'}J_{x'} - J_{x'}J_{z'} = i\ J_{y'} \tag{1.11c}$$

In summary, the RAM method is not utilized in this book for the computation of diatomic molecular spectra. RAM is avoided due, in part, to not needing approximations thanks to the availability of modern digital computers and due in part to the mathematical inconsistency of the supposed change of sign, as implied by the 'reversed-angular momentum' descriptive nomenclature.

1.3 Exact diatomic eigenfunction

An exact expression of the diatomic eigenfunction is essential for prediction of spectra. The major difference between this book and other treatments of the diatomic molecule is the use of the Wigner–Witmer diatomic eigenfunction [7] in place of invoking the Born–Oppenheimer approximation [8] from the very beginning of a theory description. In the Wigner–Witmer approach, angular coordinates are exactly separated from the electronic–vibrational coordinates. In this book, the Wigner–Witmer eigenfunction is employed for computation of the vacuum wave numbers and the rotational line strengths. If one were to instead adopt the Born–Oppenheimer approximation, then the rotational line strengths would be labeled as Hönl–London factors. The Born–Oppenheimer approximation breaks the electronic–vibrational strength into electronic and vibrational parts that correspond to r-centroids and Franck–Condon factors, and both may be functions of the total angular momentum in the upper and lower levels.

The expression *spectroscopic accuracy* refers to the accuracy with which line position measurements can be performed. Whereas wavelength measurements having an accuracy of 1 part per million are routinely performed, achieving an accuracy of 1 part per hundred in the measurement of relative intensities of a group of spectral lines is fully adequate for many purposes. Thus, one may elect to directly use the Born–Oppenheimer approximation for many practical calculations of molecular line intensity; namely, approximating the diatomic eigenfunction as a product of electronic, vibrational, and rotational factors. However, the Born–Oppenheimer approximation cannot produce diatomic term values with spectroscopic accuracy without generalization. To achieve spectroscopic accuracy within the Born–Oppenheimer approximation, one must include sums over the many electronic states of the molecule and sums over the many vibrational states of each electronic state. Van Vleck transformations [9] or other mathematical procedures reduce the dimension of the Hamiltonian matrix prior to numerically diagonalization [10–15].

In this book, only one diatomic selection rule is used. A spectral line, i.e., a term difference, is allowed if the angular momentum part of its line strength is nonvanishing. However, a modification of the line strength computation is required if the diatomic molecule in question is homonuclear, i.e., the two nuclei are identical. An unresolved hyperfine structure in the spectrum of a homonuclear molecule causes states of positive parity and negative parity to have different nuclear spin statistical weights, g_+ and g_-. If the nuclear spin is zero, then either g_+ or g_- will be zero. Thus, exchange symmetry, the symmetry associated with the exchange of identical particles, rigorously forbids certain spectral lines, even when the rotational line strength is nonzero. However, if the rotational line strength factor vanishes, then the spectral line is rigorously forbidden.

1.4 Computation of diatomic spectra

The required steps for computation of spectra can be summarized as follows:

- An angular momentum momentum coupling model must be chosen because angular momentum theory does not tell us how the total angular momentum is formed from the orbital and spin momenta.
- The eigenfunctions for everything in the system except the total angular momentum are computed.
- With the eigenfunctions obtained in the previous step and the chosen angular momentum coupling model, upper and lower Hamiltonians are computed and diagonalized.
- From the orthogonal matrices that diagonalize the upper and lower Hamiltonians, the line strengths are computed for various possible types of transitions, e.g., electric dipole, magnetic dipole, electric quadrupole, etc. Typically, one knows precisely what type of transition dominates in the spectrum, but this is not invariably the case.
- The nonvanishing of the rotational angular momentum part of the line strength selects the subset of allowed spectral lines from the computed term differences.

Consequently, the minimal information required for computation of a spectrum includes selected term differences $\tilde{\nu}_{u\ell}$ and the computed line strengths $S_{u\ell}$. A description of a diatomic molecule having N electrons and residing in field free space requires $3N + 6$ spatial or angular coordinates, the time t, N electronic spin variables, and two nuclear spin variables. In the case of the diatomic molecule, the only exactly separable variables are the time t, the coordinates of the total mass, and three Euler angles which describe the total angular momentum. The Wigner–Witmer diatomic eigenfunction provides the exact separation of three Euler angles, but $3N$ internal spatial coordinates and the numerous spins remain. Unless the number of electrons N is very small, the diatomic problem remains unsolvable with spectroscopic accuracy because there are $3N$ independent variables that cannot be treated with mathematical exactness.

Despite the challenges mentioned in the previous paragraph, one can, with two stringent caveats, apply the above algorithm to the diatomic molecule. The first caveat is that one must have extensive experimentally recorded wave number tables, $\tilde{\nu}_{u\ell}^{\text{exp}}(J', J)$, versus upper and lower total angular momenta, J' and J, respectively, for many vibrational bands in the spectrum of a molecule of interest. The second caveat is associated with using trial values of semiempirical molecular parameters for each vibrational level, v, such as Bv, Dv, Av, λ_v, γ_v, and so on. One computes term differences, $\tilde{\nu}_{u\ell}(J', J)$, from numerically diagonalized upper and lower Hamiltonians, calculates corrections to the trial values of the parameters from differences $\tilde{\nu}_{u\ell}(J', J) - \tilde{\nu}_{u\ell}^{\text{exp}}(J', J)$, and iterates the computations until the errors in the computed line positions are comparable to the estimated errors in the experimental line positions. When successful, this procedure yields working models for the upper and lower Hamiltonians and sets of molecular parameters that predict the measured line positions.

The practical significance of molecular parameters was their appearance in term value equations, semiempirical equations with which one can compute the upper T_u and lower T_ℓ terms, and thereby the vacuum wave number $\tilde{\nu}_{u\ell}$. Herzberg [16] gives many examples of term value equations, but note that when Herzberg wrote his book the numerical diagonalization of thousands of matrices was impractical. The current significance of the molecular parameters is that they can be used to compute diatomic Hamiltonian matrix representations in one of the Hund's bases.

In this book the computation of $\tilde{\nu}_{u\ell}(J', J)$ and $S_{u\ell}(J', J)$ is based upon the Wigner–Witmer diatomic eigenfunction instead of the eigenfunction associated with the Born–Oppenheimer approximation, but computations of the electronic–vibrational strengths utilize separation of electronic from vibrational contributions familiar from the Born–Oppenheimer approximation.

References

[1] Condon E U and Shortley G 1953 *The Theory of Atomic Spectra* (Cambridge: Cambridge University Press)

[2] Klein O 1929 Zur Frage der Quantelung des asymmetrischen Kreisels *Z. Phys.* **58** 730

[3] Van Vleck J H 1951 The coupling of angular momentum vectors in molecules *Rev. Mod. Phys.* **23** 213

[4] Kroto H W 1996 Symmetry, space, stars and C_{60}, 1996 (accessed 19 January 2016). https://www.nobelprize.org/prizes/chemistry/1996/kroto/facts/

[5] Kroto H W 1992 *Molecular Rotation Spectra* (New York: Dover)

[6] Davydov A S 1965 *Quantum Mechanics* (Oxford: Pergamon)

[7] Wigner E and Witmer E E 1928 *Z. Phys.* **51** 859
Hettema H 2000 *Quantum Chemistry: Classic Scientific Papers* p 287 (Singapore: World Scientific)

[8] Born M and Oppenheimer R 1927 *Ann. Phys.* **84** 457
Hettema H 2000 *Quantum Chemistry: Classic Scientific Papers* p 1 (Singapore: World Scientific)
[9] Kemble E C 2005 *The Fundamental Principles of Quantum Mechanics* (Mineola, NY: Dover)
[10] Zare R N, Schmeltekopf A L, Harrop W J and Albritton D L 1973 *J. Mol. Spectrosc.* **46** 37
[11] Bunker P R and Jensen P 1998 *Molecular Symmetry and Spectroscopy* 2nd edn (Ottawa: NRC)
[12] Brown J M and Carrington A 2003 *Rotational Spectroscopy of Diatomic Molecules* (Cambridge: Cambridge University Press)
[13] Lefebvre-Brion H and Field R W 2004 *The Spectra and Dynamics of Diatomic Molecules* (New York: Elsevier/Academic)
[14] James K G 2005 Watson. Different forms of effective hamiltonians *J. Mol. Spectrosc.* **103** 3283
[15] Field R W, Baraban J H, Lipoff S H and Annelise R B 2011 Effective hamiltonians for electronic fine structure and polyatomic molecules *Handbook of High-Resolution Spectroscopy* ed M Quack and F Merkt (New York: Wiley)
[16] Herzberg G 1950 Molecular Spectra and Molecular Structure I *Spectra of Diatomic Molecules* 2nd edn (New York: Van Nostrand Reinhold)

IOP Publishing

Quantum Mechanics of the Diatomic Molecule (Second Edition)

Christian G Parigger and James O Hornkohl

Chapter 2

Formal quantum mechanics of diatomic molecular spectroscopy

2.1 Introduction

The interpretation of optical spectra requires thorough comprehension of quantum mechanics, especially an understanding of the concept of angular momentum operators. Suppose now that a transformation from laboratory-fixed to molecule-attached coordinates, by invoking the correspondence principle, induces reversed angular momentum (RAM) operator identities. However, the foundations of quantum mechanics and the mathematical implementation of specific symmetries assert that reversal of motion or time reversal includes complex conjugation as part of anti-unitary operation. Quantum theory contraindicates sign changes of the fundamental angular momentum algebra. RAM sign changes are of a heuristic nature and are actually not needed in analysis of diatomic spectra. This chapter addresses sustenance of usual angular momentum theory [1], including presentation of straightforward proofs leading to falsification of the occurrence of RAM identities. This chapter also summarizes aspects of a consistent implementation of quantum mechanics for spectroscopy with selected diatomic molecules of interest in astrophysics and in engineering applications.

Identification of diatomic molecular spectra necessitates a clear description of angular momentum in order to demarcate the various features that comprise optical fingerprints. Quantum mechanics theory (QMT) asserts that not all three components of angular momentum can be measured simultaneously, usually the total angular momentum and one projection of the total angular momentum describe upper and lower states of molecular transitions.

Classical mechanics description and associated quantization of the asymmetric top [2] suggests occurrence of commutator relations with different signs when computing momenta with respect to the principal axes of inertia. In other words, a laboratory-fixed system shows standard angular momentum commutators, but with

doi:10.1088/978-0-7503-6204-7ch2

respect to the molecule-attached coordinate system there is a sign change that carries the name ‘reversed’ internal angular momentum [3]. The derivation by Klein in 1929 [2] is based on the correspondence principle, which in essence emphasizes that QMT reproduces classical physics in the limit of large quantum numbers. From a classical mechanics point of view, reversal of motion occurs when transforming from a laboratory-fixed to a molecule-attached coordinate system, akin to experience of motion reversal when jumping onto a moving merry-go-around. However, reversal of motion in quantum mechanics is described by an anti-unitary transformation, requiring sign change and complex conjugation. The reversed internal angular momentum concept [3] and applications were actually communicated and applied in analysis of molecular spectra by Van Vleck in 1951 in his review article on coupling angular momenta, i.e., angular momentum, referring to axes mounted on the molecule, which adheres to opposite-sign commutator algebra. This evolved into so-called RAM concepts for prediction of molecular spectra.

However, orthodox or classic quantum mechanics abides by strict mathematical rules associated with the theory. The use of RAM techniques is contraindicated, especially because the Nöther-type symmetry transformation [4, 5] sustains the standard commutator relations, viz. reversal of motion is an anti-unitary transformation, just like in Schrödinger’s wave equation that is invariant with respect to motion reversal or time reversal due to anti-unitary operation, as expected. It is important to recognize that a transformation from laboratory-fixed to molecular-attached coordinates within standard quantum mechanics does not condone anomalous angular momentum operator identities [6].

This chapter communicates proofs that the quantum-mechanic angular momentum equations remain the same in a transition from laboratory-fixed to molecular-attached coordinates. Methods that invoke RAM for the prediction of molecular spectra are misleading. Application of standard quantum mechanics establishes within the concept of line strengths [7] consistent computation of diatomic spectra [8]; examples include hydroxyl, cyanide (CN), and diatomic carbon spectra [9]. First, Oscar Klein’s paper [2] is discussed showing his original argumentation. This is followed by presenting proofs consistent with QMT opposing RAM concepts and the occurrence of a minus sign in unitary and anti-unitary transformations. The ‘new’ aspect of this review is the emphasis of invoking mathematics consistent with QMT.

2.2 Theory details

The premise of this chapter is Klein’s work [2] ‘Zur Frage der Quantelung des asymmetrischen Kreisels’ or ‘On the question of the quantization of the asymmetric top.’ This particular work is in German without an available translation; the essential contents are in the Einleitung, viz. the introduction, and on the page following the introduction. Klein’s paper reflects the initial argumentation of the RAM method, and essential aspects of this paper are discussed below, up to equation (2.6).

The purpose of the 1929 work is, as Klein writes, to reduce quantization of the asymmetric top to simple algebra for the components of the angular momentum '*that were developed by Dirac [10] and as well by Born, Heisenberg and Jordan [11].*' For a solid body, the main moments of inertia are labeled as A, B, and C, the angular momenta are labeled P, Q, R, and one finds the classical mechanics energy of rotation, E,

$$E = \frac{1}{2}\left(\frac{P^2}{A} + \frac{Q^2}{B} + \frac{R^2}{C}\right), \tag{2.1}$$

or perhaps with convenient notation, using for operators $\tilde{J}_1 = P$, $\tilde{J}_2 = Q$, $\tilde{J}_3 = R$, where the tilde-symbol indicates that angular momenta (that would be angular momentum operators in quantum mechanics) are referred to the main axis of the ellipse of inertia (or in molecules, referred to molecular-fixed coordinates), and for moments of inertia $I_1 = A$, $I_2 = B$, $I_3 = C$,

$$E = \frac{1}{2}\sum_{k=1}^{k=3}\frac{1}{I_k}\tilde{J}_k. \tag{2.2}$$

Subsequently, Klein writes that P, Q, R can be understood to describe matrices satisfying quantum mechanics equations of motion, with $i = \sqrt{-1}$ and using the standard $\hbar$ for Planck's constant divided by 2π,

$$\frac{dP}{dt} = \frac{i}{\hbar}(EP - PE), \qquad \frac{dQ}{dt} = \frac{i}{\hbar}(EQ - QE), \qquad \frac{dR}{dt} = \frac{i}{\hbar}(ER - RE). \tag{2.3}$$

In terms of operators, using the Hamilton operator $\mathcal{H}$ instead of E and writing the equation in the Heisenberg picture for an abstract observable (operator), $\mathcal{O}$, without explicit time-dependence of the observable, i.e., $\frac{\partial \mathcal{O}}{\partial t} = 0$, and using the commutator $[\mathcal{H}, \mathcal{O}] = \mathcal{H}\mathcal{O} - \mathcal{O}\mathcal{H}$, gives

$$\frac{d\mathcal{O}}{dt} = \frac{i}{\hbar}[\mathcal{H}, \mathcal{O}] + \frac{\partial \mathcal{O}}{\partial t}. \tag{2.4}$$

Klein's hypothesis comprises the requirement of utilizing equation (2.3) in equation (2.1). Consequently, Klein assumes commutator relations for P, Q, R,

$$i\hbar P = RQ - QR, \qquad i\hbar Q = PR - RP, \qquad i\hbar R = QP - PQ, \tag{2.5}$$

or using abbreviated nomenclature and the Levi-Civita symbol, with $\varepsilon_{klm} = 1$ for even permutations, and $\varepsilon_{klm} = -1$ for odd ones; otherwise, $\varepsilon_{klm} = 0$ for identical indices, k, l, $m = 1, 2, 3$,

$$[\tilde{J}_k, \tilde{J}_l] = -i\hbar\varepsilon_{klm}\tilde{J}_m. \tag{2.6}$$

With the commutator relations in equation (2.5), the correspondence principle leads to the equations of motion, and as Klein writes, 'as we overlook occurrence of the action-quant,' viz. overlook $\hbar$. Furthermore, Klein remarks that equation (2.5) differs only by the sign of i from the well-known quantum-mechanical commutators for a laboratory-fixed system. In summary, Klein's work concludes that a minus sign is required for consistency with classical mechanics and a result of the application of the correspondence principle.

Clearly, writing equation (2.5) in the compact form of equation (2.6) highlights the minus sign that differs from the standard equations of angular momentum operators J_k, $k = 1, 2, 3$,

$$[J_k, J_l] = i\hbar\varepsilon_{klm}J_m. \tag{2.7}$$

The minus sign in equation (2.6) is labeled 'anomalous' by some authors, e.g., J. Van Vleck [3], but there is no justification for the anomalous minus sign to occur within QMT. Usually, one considers right-hand systems, so equation (2.7) is termed as the standard quantum-mechanic angular momentum operator identity. Sustenance of RAM concepts may appear convenient, even calling the negative sign an 'anomaly' but without QMT support. In the scientific approach, and in spite of the initial success in explaining spectra within various approximations, one usually avoids starting with an 'anomaly' and/or inaccurate presuppositions that are readily falsified [12]. However, several textbooks and works continue to support RAM in the theory of molecular spectra [13–25], in spite of obvious falsification by QMT. This work emphasizes that there is no need to resort to RAM 'cook book' [25] methods.

The methods in this work utilize standard QMT [26, 27] and standard mathematical methods [28], showing that there is no sign change of the standard commutator relations when transforming from a laboratory-fixed to a molecule-attached coordinate system. Consistent application of standard angular momentum algebra in the establishment of computed spectra yield a nice agreement with laboratory experimental results [8] and agreement in analysis of astrophysical C_2 Swan data from the white dwarf Procyon B [8], including agreement in comparisons with computed spectra that are obtained with other molecular fitting programs, such as PGOPHER [29].

Methods for measurement of optical emission signals from diatomic molecules are comprised of standard molecular spectroscopy experimental arrangements, such as in laser-induced plasma or breakdown spectroscopy [30–37], which are also encountered in stellar plasma physics or astrophysics (to name other areas of interest). Particular interests in astrophysics include 'cool' stars, brown dwarfs, and extra-solar planets, and the associated need for accurate theoretical models for ab initio calculations of diatomic molecular spectra, which were nicely reviewed in [38].

2.3 Results

2.3.1 Angular momentum commutators

The invariance of standard QMT commutator relations (see equation (2.7)) is communicated in this section.

2.3.1.1 Invariance for unitary transformations

Application of unitary transformation, viz. transforming from one coordinate system to another, leaves the angular momentum commutator relations invariant [39]. A unitary transformation operator, U, acting on an operator $\mathcal{O} \longrightarrow \mathcal{O}'$, with $U^\dagger = U^{-1}$, is defined by

$$\mathcal{O}' = U\mathcal{O}U^\dagger \quad \text{or} \quad \mathcal{O} = U^\dagger \mathcal{O}' U. \tag{2.8}$$

The invariance of the angular momentum commutators with respect to a unitary transformation, equation (2.8),

$$[J_k, J_l] = i\varepsilon_{klm} J_m \quad \longrightarrow \quad [J_k', J_l'] = i\varepsilon_{klm} J_m', \tag{2.9}$$

can be derived by inserting $J_k = U^\dagger J_k' U$ and $J_l = U^\dagger J_l' U$ in equation (2.9) to obtain the intermediate step,

$$U^\dagger J_k' U U^\dagger J_l' U - U^\dagger J_l' U U^\dagger J_k' U = U^\dagger J_k' J_l' U - U^\dagger J_l' J_k' U = i\varepsilon_{klm} U^\dagger J_m' U. \tag{2.10}$$

Multiplying from left with U and from right with U^{-1} yields the transformed identity in equation (2.9). In other words, a unitary transformation preserves the quantum-mechanic angular momentum commutators. For example, the Euler rotation matrix is easily demonstrated to be unitary [9]. In other words, there is no anomaly when going from a laboratory-fixed to a molecule-attached coordinate system.

2.3.1.2 Invariance for time reversal or reversal of motion

Time reversal or reversal of motion in QMT requires sign changes of the operators and complex conjugation, leaving the QMT commutators invariant,

$$[J_k, J_l] = i\varepsilon_{klm} J_m \quad \longleftrightarrow \quad [(-J_k), (-J_l)] = (-i)\varepsilon_{klm}(-J_m). \tag{2.11}$$

Classical mechanics would indicate a reversal of motion when going from a laboratory-fixed to a molecular-fixed coordinate system; however, reversal of motion requires complex conjugation due to the anti-unitary requirement. In other words, the sign is preserved. QMT so-to-speak opposes the hypothesis by Klein.

The invariance regarding time reversal or reversal of motion of course also would apply to the abstract form of the time-dependent Schrödinger equation,

$$i\hbar\frac{\partial}{\partial t}\psi = \mathcal{H}\psi \quad \longleftrightarrow \quad (-i)\hbar\frac{\partial}{\partial(-t)}\psi = \mathcal{H}\psi, \tag{2.12}$$

where ψ describes an abstract vector in Hilbert space, and $\mathcal{H}$ is a Hamiltonian. Changing time $t \longrightarrow -t$ and applying conjugate complex of i preserves the left-hand side of the equation. For example, for a free particle of mass m and momentum $\mathcal{P}$, the Hamiltonian is $\mathcal{H} = \mathcal{P}^2/2m$, and the form of Schrödinger's equation is preserved.

Equally, the operator equation in the Heisenberg picture, see equation (2.4), preserves form under time reversal or reversal of motion,

$$\frac{d\mathcal{O}}{dt} = \frac{i}{\hbar}[\mathcal{H}, \mathcal{O}] + \frac{\partial \mathcal{O}}{\partial t} \quad \longleftrightarrow \quad \frac{d(-\mathcal{O})}{d(-t)} = \frac{(-i)}{\hbar}[\mathcal{H}, (-\mathcal{O})] + \frac{\partial(-\mathcal{O})}{\partial(-t)}. \quad (2.13)$$

A change of sign for the operators and complex conjugation leaves the equation invariant. The mentioned symmetry can also be associated with usual Nöther symmetries [4, 5].

2.3.2 Diatomic wave function

For diatomic molecules, symmetry properties allow one to invoke simplifications when evaluating the laboratory wave-function in terms of rotated coordinates [8]. For internuclear geometry, the spherical polar coordinates are r, ϕ, and θ, and one (arbitrary) electron is described by cylindrical coordinates ρ, χ, ζ. For coordinate rotation, one uses Euler angles α, β, γ, and without loss of generality one can choose $\alpha = \phi$, $\beta = \theta$, $\chi = \gamma$ [8]. The result is the Wigner–Witmer eigenfunction for diatomic molecules [40, 41],

$$\langle \rho, \zeta, \chi, \mathbf{r}_2, \ldots, \mathbf{r}_N, r, \theta, \phi \,|nvJM\rangle = \sum_{\Omega=-J}^{J} \langle \rho, \zeta, \mathbf{r}'_2, \ldots, \mathbf{r}'_N, r \,|nv\rangle \, D^{J*}_{M\Omega}(\phi, \theta, \chi). \quad (2.14)$$

The usual total angular momentum quantum numbers are J and M, and the electronic–vibrational eigenfunction is explicitly written by extracting v from the collection of quantum numbers, n. The Wigner–Witmer eigenfunction exactly separates ϕ, θ, χ. The quantum numbers J, M, Ω refer to the total angular momentum. The sum over Ω in equation (2.14) originates from the usual abstract transformation,

$$|JM\rangle = \sum_{\Omega=-J}^{J} |J\Omega\rangle \, \langle J\Omega \,|JM\rangle, \quad (2.15)$$

where Ω is the magnetic quantum number along the rotated, or new, z'-axis. The sum in equation (2.15) ensures that the quantum numbers for total angular momentum are J and M. In Hund's case *a* [42], Ω describes the projection of the total angular momentum, within L–S coupling. Hund's case *a* eigenfunctions form a basis; therefore, from a computational point of view, these eigenfunctions form a complete (sufficient) set. In various approximate descriptions and for specific diatomic molecules, it may be desirable to use other Hund cases.

From the rotation operator $\mathcal{R}(\alpha, \beta, \gamma)$, with the Euler angles α, β, γ, one finds for D-matrix elements,

$$D^{J*}_{M\Omega}(\alpha, \beta, \gamma) = \langle JM|\, \mathcal{R}(\alpha, \beta, \gamma) \,|J\Omega\rangle^*. \quad (2.16)$$

D-matrices are the usual mathematical tool for transformation from one basis to another, but the D-matrix cannot represent an eigenfunction due to the presence of two magnetic quantum numbers, M and Ω, so the sum over Ω is needed in the transformed coordinates.

Diatomic spectra composed of line positions and line strengths are based on Wigner–Witmer eigenfunctions instead of the eigenfunctions used for the Born–Oppenheimer approximation. Extensive experimental studies confirm agreement of computed spectra with measured emission spectra from laser-induced optical plasma, see the application sections of this book.

2.3.3 Selected diatomic spectra

Various reported studies of plasma spectra, including astrophysics plasma, and of molecular laser-induced breakdown spectroscopy (LIBS) [43–46] illustrate nice comparisons of recorded and of computed diatomic spectra. The published line strength data [43, 44] are derived consistent with standard quantum mechanics; in other words, without anomalous commutators and without states that have two magnetic quantum numbers associated with angular momentum. The published program package [43] also includes a worked high-temperature CN example, the Boltzmann equilibrium spectrum program for computation of equilibrium spectra, and the Nelder–Mead temperature (NMT) routine that utilizes a nonlinear fitting algorithm. The OH line strength data have been made available recently [44].

In LIBS, plasma generated by focusing coherent radiation is analyzed primarily in visible/optical or in near-ultraviolet to near-infrared regions. After initiation of optical breakdown with typically 10 ns, 100 mJ laser pulses focused in standard ambient temperature and pressure air or in gas mixtures [45], molecule formation including, for example, OH in air, C_2 in carbon monoxide, and CN in 1:1 molar N_2: CO_2 mixture, leads to recombination radiation that is typically measured using time-resolved optical emission laser spectroscopy. When using a metallic target, other diatomic molecules can be investigated, e.g., TiO or AlO, and molecular spectra can be computed from line strength data [43].

2.4 Summary

Angular momentum operators are well defined in QMT, including the fact that there is an inherent limit in the measurement of its components. Another way of formulating this could be that there are only two quantum numbers needed for description of AM, usually the total angular momentum and its projection onto a quantization axis. The use of the correspondence principle to ensure compatibility with classical mechanics equations of motion brings about an ad hoc hypothesis of a negative sign for the commutators, as originally communicated by Oscar Klein in 1929. Subsequent application of RAM coupling continues to find support in analytic description of molecules that also includes modeling of quantum-mechanic vector-operators as vectors.

However, QMT already ensures how to mathematically describe angular momentum, not supporting heuristic conclusions involving RAM concepts nor the occurrence of more than two quantum numbers for the total angular momentum of diatomic molecules. This review emphasizes that there is no mathematical justification of RAM algebra, and it also discusses applications in diatomic molecular spectroscopy. Consistent application of standard QMT is preferred, including avoidance of a priori use of separating electronic, vibrational, and rotational wave functions. Subsequent to the implementation of diatomic molecular symmetries, line strengths for selected diatomic molecules that contain effects of spin splitting and lambda-doubling as function of wavelength are in agreement with results from optical emission spectroscopy. The computed and fitted diatomic spectra nicely match within reasonable error bars, but without invoking heuristic selection rules that may be affected by initial approximations or by spurious use of reversal of angular momentum.

References

[1] Parigger C G 2021 *Foundations* **1** 208
[2] Klein O 1929 *Z. Phys.* **58** 730
[3] Van Vleck J H 1951 *Rev. Mod. Phys.* **23** 213
[4] Nöther E, Nachr , König D, Gesellsch and Wiss D 1918 *Nachr. D. König. Gesellsch. D. Wiss. Göttingen* **918** 235
[5] Nöther E 1971 *Transp. Theory Statist. Phys.* **1** 183
[6] Parigger C G 2021 *Foundations* **1** 208
[7] Condon E U and Shortley G 1953 *The Theory of Atomic Spectra* (Cambridge: Cambridge University Press)
[8] Parigger C G and Hornkohl J O 2020 *Quantum Mechanics of the Diatomic Molecule with Applications* (Bristol: IOP Publishing)
[9] Parigger C G and Hornkohl J O 2010 *Int. Rev. At. Mol. Phys.* **1** 25
[10] Dirac P A M 1926 *Proc. Roy. Soc. Lond. A* 111 281
[11] Heisenberg W, Born M and Jordan P 1926 *Z. Phys.* **35** 557
[12] Popper K 1992 The logic and evolution of scientific theory *All Life Is Problem Solving* (London: Routledge)
[13] Brown J and Carrington A 2003 *Rotational Spectroscopy of Diatomic Molecules* (Cambridge: Cambridge University Press)
[14] Lefebvre-Brion H and Field R W 2004 *The Spectra and Dynamics of Diatomic Molecules* (Amsterdam: Elsevier)
[15] Jensen P and Bunker P R 2005 *Fundamentals of Molecular Spectroscopy* (Bristol: IOP Publishing)
[16] Gottfried K 1989 *Quantum Mechanics* (Reading: Addison-Wesley)
[17] Baym G 1969 *Lectures on Quantum Mechanics* (Reading: Benjamin/Cummings)
[18] Shore B W and Menzel D H 1968 *Principles of Atomic Spectra* (Reading: Addison-Wesley)
[19] Judd B R 1975 *Angular Momentum Theory for Diatomic Molecules* (New York: Academic)
[20] Mizushima M 1975 *The theory of rotating diatomic molecules*
[21] Kovacs I 1969 *Rotational Structure in the Spectra of Diatomic Molecules* (New York: Elsevier)
[22] Hougen J T 1970 *The Calculation of Rotational Energy Levels and Rotational Line Intensities in Diatomic Molecules* **volume NBS Monograph 115** (Washington, DC: U.S. Government Printing Office)
[23] Kroto H W 1992 *Molecular Rotation Spectra* (New York: Dover)
[24] Carrington A, Levy D H and Miller T A 1970 *Adv. Chem. Phys.* **18** 149
[25] Freed K F 1966 *J. Chem. Phys.* **45** 4214

[26] Cohen-Tannoudji C, Diu B and Laloe F 2019 *Quantum Mechanics, Volume 1: Basic Concepts, Tools, and Application* 2nd edn (Weinheim: Wiley-VCH)
[27] Cohen-Tannoudji C, Diu B and Laloe F 2019 *Quantum Mechanics, Volume 2: Angular Momentum, Spin, and Approximation Methods* 2nd edn (Weinheim: Wiley-VCH)
[28] Arfken G B, Weber H J and Harris F E 2012 *Mathematical Methods for Physicists, A Comprehensive Guide* 7th edn (New York: Academic)
[29] Western C M 2017 *J. Quant. Spectrosc. Radiat. Transf.* **186** 221
[30] Kunze H-J 2009 *Introduction to Plasma Spectroscopy* (Heidelberg: Springer)
[31] Demtröder W 2014 *Laser Spectroscopy 1: Basic Principles* 5th edn (Heidelberg: Springer)
[32] Demtröder W 2015 *Laser Spectroscopy 2: Experimental Techniques* 5th edn (Heidelberg: Springer)
[33] Hertel I V and Schulz C-P 2015 *Atoms, Molecules and Optical Physics 1, Atoms and Spectroscopy.* (Heidelberg: Springer)
[34] Hertel I V and Schulz C-P 2015 *Atoms, Molecules and Optical Physics 2, Molecules and Photons–Spectroscopy and Collisions* (Heidelberg: Springer)
[35] Radziemski L J and Cremers D E 2006 *Handbook of Laser-Induced Breakdown Spectroscopy* (New York: Wiley)
[36] Miziolek A W, Palleschi V and Schechter I 2006 *Laser Induced Breakdown Spectroscopy* (New York: Cambridge University Press)
[37] Singh J P and Thakur S N (ed) 2020 *Laser-Induced Breakdown Spectroscopy* 2nd edn (New York: Elsevier)
[38] Tennyson J, Lodi L, McKemmish L K and Yurchenko S N 2016 *aeXiv* p 1605.023 01v1
[39] Davydov A S 1965 *Quantum Mechanics* (Oxford: Pergamon)
[40] Wigmer E and Witmer E E 1928 *Z. Phys.* **51** 859
[41] Hettema E H 2000 *On the Structure of the Spectra of Two-Atomic Molecules According to Quantum Mechanics. In: Quantum Chemistry: Classic Scientific Papers* (Singapore: World Scientific)
[42] Bransden B H and Joachain C J 2003 *Physics of Atoms and Molecules* 2nd edn (Essex: Prentice-Hall)
[43] Parigger C G, Woods A C, Surmick D M, Gautam G, Witte M J and Hornkohl J O 2015 *Spectrochim. Acta* B **107** 132
[44] Parigger C G, Jordan H B S, Surmick D M and Splinter R 2020 *Molecules* **25** 988
[45] Parigger C G, Surmick D M, Helstern C M, Gautam G, Bol'shakov A A and Russo R 2020 *Molecular Laser-Induced Breakdown Spectroscopy. In: Laser Induced Breakdown Spectroscopy* ed J P Singh and S N Thakur 2nd edn (New York: Elsevier)
[46] Parigger C G, Helstern C M, Jordan B S, Surmick D M and Splinter R 2020 *Molecules* **25** 615

IOP Publishing

Quantum Mechanics of the Diatomic Molecule (Second Edition)

Christian G Parigger and James O Hornkohl

Chapter 3

Line strength computations

3.1 Introduction

The quantum mechanics theory description of optical transitions in diatomic molecules uses the Condon and Shortley [1] line strength, $S_{u\ell}$,

$$S_{u\ell} = \sum_{m_u}\sum_{m_\ell}\sum_{k=-q}^{q} \left| \langle \ell | T_k^{(q)} | u \rangle \right|^2, \tag{3.1}$$

where $T_k^{(q)}$ is the kth component of the tensor operator $\mathbf{T}^{(q)}$ of degree q, e.g., electric dipole, magnetic dipole, electric quadrupole, etc. The tensor operator describes the type of interaction between the molecule and the radiation field. The basic idea behind summations in the line strength formula is to include all transitions that produce the same spectral line. In field free space, the summations are over the upper m_u and lower m_ℓ magnetic quantum numbers, and the k components of the tensor operator.

The line position is described by the vacuum wave number $\tilde{\nu}_{u\ell}$ which is the difference between the upper term T_u, i.e., upper energy eigenvalue expressed in the units of cm^{-1}, and the lower term T_ℓ,

$$\tilde{\nu}_{u\ell} = T_u - T_\ell. \tag{3.2}$$

The vacuum wave number $\tilde{\nu}_{u\ell} = \tilde{\nu}_{\ell u}$ and the line strength $S_{u\ell} = S_{\ell u}$ are symmetric with regard to the upper and lower levels. In addition, the symbols u and ℓ represent a collection of quantum numbers. For a diatomic molecule the symbol u represents the collection n', v', J', and M', while ℓ represents n, v, J, and M in which v is the vibrational quantum number, J is the total angular momentum quantum number, M is the quantum number for the z-component of $\mathbf{J}$, and n represents the collection of all other required quantum numbers. In diatomic spectroscopy, upper state quantum numbers are normally denoted with a single prime while lower states are denoted

doi:10.1088/978-0-7503-6204-7ch3

with the absence of a prime or a double prime. The absence of a double prime has become the standard way of denoting a lower state diatomic quantum number.

The line position is described by the vacuum wave number $\tilde{\nu}_{u\ell}$, which is the difference between the upper term T_u, i.e., upper energy eigenvalue expressed in the units of cm^{-1}, and the lower term T_ℓ,

$$\tilde{\nu}_{u\ell} = T_u - T_\ell. \tag{3.3}$$

In concise summary, this book is about how to compute the vacuum wave numbers,

$$\tilde{\nu}(n'v'J', nvJ) = T(n'v'J') - T(nvJ) \tag{3.4}$$

and line strengths,

$$S(n'v'J', nvJ) = \sum_{M'=-J'}^{J'} \sum_{M=-J}^{J} \sum_{k=-q}^{q} \left| \langle nvJM | T_k^{(q)} | n'v'J'M' \rangle \right|^2, \tag{3.5}$$

for diatomic molecules. The upper and lower states are assumed to have a degeneracy (statistical weight) of $2J' + 1$ and $2J + 1$, respectively. This implies that the molecule is free from any static fields. This also explains why M' and M do not appear in the equation for the vacuum wave number $\tilde{\nu}$, and why there are sums over M' and M in the equation for the line strength $S(n'v'J', nvJ)$. Note that the Wigner–Eckart theorem separates the transition matrix element into two parts, one of which depends entirely on angular momentum,

$$\langle nvJM | T_k^{(q)} | n'v'J'M' \rangle = \frac{C_{J'M'qk}^{JM}}{\sqrt{2J+1}} \langle nvJ || \mathbf{T}^{(q)} || n'v'J' \rangle. \tag{3.6}$$

The exact Wigner–Witmer [2] diatomic eigenfunction,

$$\langle \rho, \zeta, \chi, \mathbf{r}_2, \ldots, \mathbf{r}_N, r, \theta, \phi | \mathcal{R}(\alpha, \beta, \gamma) | nJM \rangle = \sum_{\Omega=-J}^{J} \langle \rho, \zeta, \mathbf{r}'_2, \ldots, \mathbf{r}'_N, r | nv \rangle D_{M\Omega}^{J*}(\phi, \theta, \chi), \tag{3.7}$$

which is further elaborated and derived in chapter 5, provides the primary theory model used in this book as opposed to the Born–Oppenheimer [3] approximation, and this is the major difference between the contents in this book and other treatments of the diatomic molecule. In the Wigner–Witmer eigenfunction, the angular coordinates ϕ, θ, and χ are exactly separated from the electronic–vibrational coordinates $\rho, \zeta, \mathbf{r}'_2, \ldots, \mathbf{r}'_N, r$. Coordinates r, θ, and ϕ are the spherical coordinates of the internuclear vector $\mathbf{r}(r, \theta, \phi)$ with the coordinate origin at the center of mass of the two nuclei. Coordinates ρ, ζ, and χ are the cylindrical coordinates $\mathbf{r}'_1(\rho, \zeta, \chi)$ of one of the N electrons of the mutual nuclei and electron configuration. Rotation of the mutual configuration is accomplished by arbitrarily choosing electron 1 in rotated, primed coordinates obtained from unprimed coordinates through coordinate rotations ϕ, θ, and χ. A primed quantum number denotes the upper state. A primed coordinate denotes a rotated coordinate.

In this book, the Wigner–Witmer eigenfunction is used for computation of the vacuum wave number and the unitless rotational line strength factor, $S(J', J)$, the angular momentum part of the diatomic line strength, *viz.* Hönl–London factor in descriptions that utilize the Born–Oppenheimer approximation from the very start of computing diatomic spectra,

$$S(n'v'J', nvJ) = S_{ev}(n'v'J', nvJ)\, S(J', J), \tag{3.8}$$

in which $S_{ev}(n'v'J', nvJ)$ is the electronic–vibrational strength of the diatomic molecule and which carries the units of line strength (C^2 m^2 for electric dipole transitions). The electronic–vibrational strength $S_{ev}(n'v'J', nvJ)$ is mostly controlled by the electronic and vibrational eigenfunctions and is a weaker function of the angular momentum quantum numbers J' and J. Typically, if the diatomic molecule is not a hydride, the variation of electronic–vibrational strength $S_{ev}(n'v'J', nvJ)$ can often be ignored and it is denoted by $S_{ev}(n'v', nv)$. Subsequently, the reduced Born–Oppenheimer approximation breaks the electronic–vibrational strength into electronic and vibrational parts

$$S_{ev}(n'v', nv) \approx |a_0 + a_1\bar{r}_1(v', v) + a_2\bar{r}_2(v', v) + \cdots|^2\, q(v', v) \tag{3.9}$$

in which

$$\bar{r}_n(v', v) = \int_0^\infty \psi_{v'}(r)\, r^n\, \psi_v(r)\, dr \;\Big/ \int_0^\infty \psi_{v'}(r)\, \psi_v(r)\, dr, \tag{3.10}$$

$$q(v', v) = \left| \int_0^\infty \psi_{v'}(r)\, \psi_v(r)\, dr \right|^2 \tag{3.11}$$

The upper and lower vibrational eigenfunctions $\psi_{v'}(r)$ and $\psi_v(r)$ are not elements of the same vector space (e.g., $\langle v'|v\rangle \neq \delta_{v',v}$) because they are eigenfunctions for different potential energy wells. For those cases for which the variation of $S_{ev}(n'v'J', nvJ)$ with J' and J cannot be ignored, the vibrational eigenfunctions become functions of angular momentum, and the Franck–Condon factors $q(v', v)$ and r-centroids $\bar{r}_n(v', v)$ acquire angular momentum dependence to become $q(v'J', vJ)$ and $\bar{r}_n(v'J', vJ)$.

The expression *spectroscopic accuracy* refers to the accuracy with which line position measurements can be performed. Whereas wavelength measurements having an accuracy of one part per million are routinely performed, achieving an accuracy of one part per hundred in the measurement of relative intensities of For measurement accuracies of relative intensities of the order of one part per hundred, standard Born–Oppenheimer approximation,

$$\Psi_{nvJM} \approx \psi_n\, \psi_v\, \psi_{JM}, \tag{3.12}$$

is adequate for many practical calculations of molecular spectra. However, the Born–Oppenheimer approximation (3.12) cannot produce diatomic term values with spectroscopic accuracy. To achieve spectroscopic accuracy using the Born–Oppenheimer approximation, one would have to replace equation (3.12) with

$$\Psi_{nvJM} \approx \sum_{n}\sum_{v_n} \psi_n \, \psi_{v_n} \, \psi_{JM}, \tag{3.13}$$

in which the n sum is over the many electronic states of the molecule and the v_n sums are over the many vibrational states of each electronic state.

3.2 Idealized computation of spectra

What if mathematics were no impediment and we could actually perform the procedures described in quantum mechanics textbooks required for the solution of a quantum system having bound states? For example, what if we were given a complete basis $|a\rangle$?. Subsequently, we would compute a matrix representation of the Hamiltonian H having matrix elements

$$H_{ij} = \langle a_i | H | a_j \rangle. \tag{3.14}$$

In principle, the Hamiltonian matrix is diagonal with dimensions $(2J + 1) \times (2J + 1)$, where J is the total angular momentum quantum number. Angular momentum theory tells us that the total angular momentum $\mathbf{J}$ is the quantity associated with the empirical observation that field free space is isotropic, but there are two kinds of angular momenta, orbital $\mathbf{N}$ and spin $\mathbf{S}$, and angular momentum theory does not tell us how the orbital and spin momenta are coupled to form the total $\mathbf{J}$. We must form a mathematically complete basis for an assumed coupling scheme, e.g., ℓs or jj for atoms or one of Hund's models for molecules. The physical state $|nJM\rangle$ must be expressed as a sum of our basis states $|a_i\rangle$,

$$|nJM\rangle = \sum_i |a_i\rangle \, \langle a_i | nJM \rangle. \tag{3.15}$$

The result is still mathematically exact because the basis $|a_i\rangle$ is complete, but two practical problems are caused by using basis functions instead of the eigenfunctions for the system. First, the Hamiltonian matrix computed using the basis functions is no longer diagonal. Second, the Hamiltonian matrix has become larger. For example, if $J = 10$ and 10 basis functions are required, then the Hamiltonian matrix dimensions are now $(210) \times (210)$, i.e., $(2J + 1) \times 10 = 210$. The quantum mechanics of diatomic molecules played an important part in the early development of quantum theory, but at the time numerical or analytical diagonalization of a 4×4 was a serious challenge, and this remained true until roughly 1960. Today, numerical diagonalization of much larger matrices can be accomplished routinely.

Numerical diagonalization of the upper H_u and lower H_ℓ Hamiltonians gives the upper and lower energy eigenvalues, i.e., upper and lower terms T_u and T_ℓ,

$$T_u = \tilde{U}_u \, H_u \, U_u \tag{3.16a}$$

$$T_\ell = \tilde{U}_\ell \, H_\ell \, U_\ell \tag{3.16b}$$

and from equation (3.2) we find the line position $\tilde{\nu}_{u\ell}$. The tilde on a matrix (unrelated to the tilde on the vacuum wave number $\tilde{\nu}$) denotes the transpose of that matrix.

The Hamiltonians are assumed to be real symmetric matrices. Thus, the matrices U_u and U_ℓ are real unitary matrices. That is, U_u and U_ℓ are orthogonal matrices,

$$U_u^{-1} = \tilde{U}_u, \tag{3.17a}$$

$$U_\ell^{-1} = \tilde{U}_\ell. \tag{3.17b}$$

The elements of U_ℓ are the coefficients from equation (3.15),

$$U_{ij}^{(u)} = \langle a_i | (n'J'M')_j \rangle, \tag{3.18a}$$

$$U_{nm}^{(\ell)} = \langle a_n | (nJM)_m \rangle. \tag{3.18b}$$

In software, a two-dimensional array has two indices, say i and j. In the above equation, the replacement of quantum numbers nJM with the single software index j is an example of something that often happens in software pertaining to applied quantum mechanics.

The line strength $S_{u\ell}$ can be computed from the matrices U_u and U_ℓ that diagonalized the upper and lower Hamiltonians,

$$S_{nj}^{(u\ell)} = \sum_{k=-q}^{q} \sum_{m=1}^{\mathcal{N}_\ell} \sum_{i=1}^{\mathcal{N}_u} \tilde{U}_{nm}^{(\ell)} \langle m | T_k^{(q)} | i \rangle \; U_{ij}^{(u)} \tag{3.19a}$$

where $\mathcal{N}_u$ is the dimension of the square upper Hamiltonian matrix H_u, and $\mathcal{N}_\ell$ is the dimension of H_ℓ.

In equations (3.16), the matrices T_u and T_ℓ of the upper and lower terms are diagonal square matrices of dimensions $\mathcal{N}_u$ and $\mathcal{N}_\ell$ (in software, one-dimensional arrays). The total number of term differences, i.e., potential spectral lines, is $\mathcal{N}_u \times \mathcal{N}_\ell$, but this number of term differences is much larger than the number of actual spectral lines. Selection rules determine which of the term differences represents a real spectral line. In this book, only one diatomic selection rule is used. A spectral line, i.e., a term difference, is allowed if the angular momentum part of its line strength, $S(J', J)$ in equation (3.8), is nonvanishing.

References

[1] Condon E U and Shortley G 1953 *The Theory of Atomic Spectra* (Cambridge: Cambridge University Press)

[2] Wigner E and Witmer E E 1928 *Z. Phys.* **51** 859
Hettema H 2000 *Quantum Chemistry: Classic Scientific Papers* p 287 (Singapore: World Scientific)

[3] Born M and Oppenheimer R 1927 *Ann. Phys.* **84** 457
Hettema H 2000 *Quantum Chemistry: Classic Scientific Papers* p 1 (Singapore: World Scientific)

IOP Publishing

Quantum Mechanics of the Diatomic Molecule (Second Edition)

Christian G Parigger and James O Hornkohl

Chapter 4

Framework of the Wigner–Witmer eigenfunction

In this book, the eigenfunction $\Psi_{nJM}(\mathbf{r}_1, \mathbf{r}_2, \ldots, \mathbf{r}_\mathcal{N})$ is viewed as the projection of the abstract state $|nJM\rangle$ into coordinate space,

$$\Psi_{nJM}(\mathbf{r}_1, \mathbf{r}_2, \ldots, \mathbf{r}_\mathcal{N}) \equiv \langle \mathbf{r}_1, \mathbf{r}_2, \ldots, \mathbf{r}_\mathcal{N} \,|nJM\rangle. \tag{4.1}$$

Quantum mechanics theory postulates the existence of the state $|nJM\rangle$. Schrödinger's eigenfunction reveals how this state manifests itself in the world of calculus, differential equations, and matrix algebra. The quantum numbers here are J for the total angular momentum $\mathbf{J}$, M for the z component of $\mathbf{J}$, and the symbol n representing all other required quantum numbers.

It is the nature of angular momentum that the total $\mathbf{J}$ [actually, the total squared whose eigenvalue is $J(J + 1)$] and *only one component*, conventionally taken to be the z component, are constants of the motion. That is, $\mathbf{J}^2$ and J_z commute with the Hamiltonian operator and other operators composing the complete set of commuting observables (CSCO), but no other components of $\mathbf{J}$ are constants of the motion.

Angular-momentum theory gives us the standard angular-momentum representation [1], $|JM\rangle$, in which

$$\mathbf{J}^2|JM\rangle = J(J + 1)\,|JM\rangle \tag{4.2a}$$

$$J_z|JM\rangle = M\,|JM\rangle \tag{4.2b}$$

$$J_\pm = J_x \pm i\,J_y \tag{4.2c}$$

$$J_\pm|JM\rangle = \sqrt{J(J + 1) - M(M \pm 1)}\,|J, M \pm 1\rangle \tag{4.2d}$$

$$= \sqrt{(J \mp M)(J \pm M + 1)}\,|J, M \pm 1\rangle \tag{4.2e}$$

$$= C_{\pm}(JM)\,|J,\, M \pm 1\rangle \tag{4.2f}$$

The homogeneity of time and space and the isotropy of space permit one to perform coordinate translations and rotations in an attempt to simplify the mathematics of the problem by separating some of the variables, i.e., by breaking the eigenfunction into a product of independent functions. Each coordinate rotation changes the direction of the z-axis giving a new operator $J_{z'}$ requiring one to work with a new $|J\Omega\rangle$ standard representation of angular momentum,

$$J_{z'}|J\Omega\rangle = \Omega|J\Omega\rangle \tag{4.3a}$$

$$J'_{\pm}|J\Omega\rangle = C_{\pm}(J\Omega)\,|J,\, \Omega \pm 1\rangle \tag{4.3b}$$

in which Ω is the new magnetic quantum number for the new z'-axis. Of particular importance is the ability to move between two standard representations,

$$|J\Omega\rangle = \sum_{M=-J}^{J} |JM\rangle\,\langle JM\,|J\Omega\rangle, \tag{4.4}$$

$$|JM\rangle = \sum_{\Omega=-J}^{J} |J\Omega\rangle\,\langle J\Omega\,|JM\rangle. \tag{4.5}$$

Application of the rotation operator [2], $\mathcal{R}(\alpha,\, \beta,\, \gamma)$, on the eigenfunction Ψ_{nJM} gives a new a new mathematical function $\Phi_{nJ\Omega}$,

$$\mathcal{R}(\alpha,\, \beta,\, \gamma)\,\Psi_{nJM}(\mathbf{r}_1,\, \mathbf{r}_2,\, \dots,\, \mathbf{r}_N) = \Phi_{nJ\Omega}(\mathbf{r}'_1,\, \mathbf{r}'_2,\, \dots,\, \mathbf{r}'_N) \tag{4.6}$$

where primes denote rotated coordinates. This equation describes coordinate rotation, not mechanical rotation. The Euler angles α, β, and γ are parameters of coordinate rotation, not angles describing mechanical rotation of mass and charge. The eigenfunctions $\Psi_{nJM}(\mathbf{r}_1,\, \mathbf{r}_2,\, \dots,\, \mathbf{r}_N)$ and $\Phi_{nJ\Omega}(\mathbf{r}'_1,\, \mathbf{r}'_2,\, \dots,\, \mathbf{r}'_N)$ describe the exact same physical state and differ only because they are expressed in different coordinate systems.

Using Dirac notation, we apply $\mathcal{R}\,\mathcal{R}^{\dagger} = 1$ (because $\mathcal{R}$ is unitary) to the eigenfunction $\Psi_{nJM}(\mathbf{r}_1,\, \mathbf{r}_2,\, \dots,\, \mathbf{r}_N)$,

$$\begin{aligned}\langle \mathbf{r}_1,\, \mathbf{r}_2,\, \dots,\, \mathbf{r}_N\,|nJM\rangle &= \langle \mathbf{r}_1,\, \mathbf{r}_2,\, \dots,\, \mathbf{r}_N\,|\mathcal{R}(\alpha,\, \beta,\, \gamma)\,\mathcal{R}^{\dagger}(\alpha,\, \beta,\, \gamma)|JM\rangle \\ &= \sum_{\Omega=-J}^{J} \langle \mathbf{r}_1,\, \mathbf{r}_2,\, \dots,\, \mathbf{r}_N\,|\mathcal{R}(\alpha,\, \beta,\, \gamma)|nJ\Omega\rangle\,\langle J\Omega\,|\mathcal{R}^{\dagger}(\alpha,\, \beta,\, \gamma)|JM\rangle \\ &= \sum_{\Omega=-J}^{J} \mathbf{r}'_1,\, \mathbf{r}'_2,\, \dots,\, \mathbf{r}'_N\,|nJ\Omega\rangle\,\langle JM\;\mathcal{R}(\alpha,\, \beta,\, \gamma)\,|J\Omega\rangle^*,\end{aligned} \tag{4.7}$$

to yield

$$\langle \mathbf{r}_1,\, \mathbf{r}_2,\, \dots,\, \mathbf{r}_N\,|nJM\rangle = \sum_{\Omega=-J}^{J} \mathbf{r}'_1,\, \mathbf{r}'_2,\, \dots,\, \mathbf{r}'_N\,|nJ\Omega\rangle\; D^{J*}_{M\Omega}(\alpha,\, \beta,\, \gamma). \tag{4.8}$$

The above result shows that the rotation matrix element or Wigner D-function is the mathematical tool one uses to perform transformations between two standard angular-momentum representations which differ only because of coordinate rotation.

Equation (4.8) strictly describes coordinate rotation, not mechanical rotation. However, the simple geometry of the diatomic molecule gives one a way to use (4.8) to describe the physical rotation of the diatomic molecule. This is possible because the mechanical rotations ϕ, θ, and χ of the molecule duplicate the coordinate rotations α, β, and γ.

The eigenfunction describing the internal motions in a diatomic molecule having N electrons will require the electronic coordinates $\mathbf{r}_1, \mathbf{r}_2, \ldots, \mathbf{r}_N$ and the internuclear vector $\mathbf{r}$ with the origin at the center of mass of the two nuclei. The internuclear vector $\mathbf{r}$ is expressed in terms of its polar angle θ, azimuthal angle ϕ, and internuclear distance r,

$$x = r \sin\theta \cos\phi \tag{4.9a}$$

$$y = r \sin\theta \sin\phi \tag{4.9b}$$

$$z = r \cos\theta. \tag{4.9c}$$

The coordinate vector of one of the electrons, arbitrary chosen to be electron 1, is expressed in cylindrical coordinates

$$x_1 = \rho \cos\chi, \tag{4.10a}$$

$$y_1 = \rho \sin\chi, \tag{4.10b}$$

$$z_1 = \zeta, \tag{4.10c}$$

where ρ is the distance of the chosen electron from the internuclear vector, ζ is the distance of that electron above or below the plane perpendicular to the internuclear vector and passing through the center of mass of the two nuclei, and χ is mechanical rotation of the chosen electron about the internuclear vector. With these variables, one can write the internal diatomic eigenfunction as

$$\langle \mathbf{r}_1, \mathbf{r}_2, \ldots, \mathbf{r}_N, \mathbf{r} \,|nvJM\rangle = \langle \rho, \zeta, \chi, \mathbf{r}_2, \ldots, \mathbf{r}_N, r, \theta, \phi \,|nvJM\rangle. \tag{4.11}$$

The vibrational quantum number v has been extracted from the collection n.

In accord with equation (4.8), coordinate rotation of the right-hand side of equation (4.11) gives

$$\begin{aligned} &\langle \rho, \zeta, \chi, \mathbf{r}_2, \ldots, \mathbf{r}_N, r, \theta, \phi \,|nvJM\rangle \\ &= \sum_{\Omega=-J}^{J} \langle \rho, \zeta, \chi', \mathbf{r'}_2, \ldots, \mathbf{r'}_N, r, \theta', \phi' \,|nvJ\Omega\rangle\, D^{J*}_{M\Omega}(\alpha, \beta, \gamma) \end{aligned} \tag{4.12}$$

with

$$\phi' = \phi - \alpha, \tag{4.13a}$$

$$\theta' = \theta - \beta, \tag{4.13b}$$

$$\chi' = \chi - \gamma. \tag{4.13c}$$

The rotationally invariant scalars ρ, ζ, and r remain unprimed. Equations (4.13) hold because mechanical rotation ϕ and coordinate rotation α are about the z-axis, mechanical rotation θ and coordinate rotation β are about the first intermediate y-axis of the full coordinate rotation, and mechanical rotation χ and coordinate rotation γ are about the z'-axis. One is at liberty to choose the angles of coordinate rotation. The choices

$$\alpha = \phi, \tag{4.14a}$$

$$\beta = \theta, \tag{4.14b}$$

$$\gamma = \chi, \tag{4.14c}$$

remove all angular dependence from

$$\langle \rho, \zeta, \chi' = 0, \mathbf{r}'_2, \dots, \mathbf{r}'_N, r, \theta' = 0, \phi' = 0 \,|nJ\Omega\rangle \tag{4.15}$$

and the result is the Wigner–Witmer diatomic eigenfunction [3],

$$\begin{aligned} &\langle \rho, \zeta, \chi, \mathbf{r}_2, \dots, \mathbf{r}_N, r, \theta, \phi \,|nvJM\rangle \\ &= \sum_{\Omega=-J}^{J} \langle \rho, \zeta, \mathbf{r}'_2, \dots, \mathbf{r}'_N, r \,|nv\rangle\, D^{J*}_{M\Omega}(\phi, \theta, \chi). \end{aligned} \tag{4.16}$$

The electronic–vibrational eigenfunction $\langle \rho, \zeta, \mathbf{r}'_2, \dots, \mathbf{r}'_N, r \,|nv\rangle$ is influenced by angular momentum (e.g., by centrifugal stretching) but it is not the total angular-momentum eigenfunction. The variables ϕ, θ, and χ are exactly separated in the Wigner–Witmer diatomic eigenfunction. Because the general equation for coordinate rotation (4.8) holds only if J, M, and Ω refer to the total angular momentum, all orbital and spin momenta are included in $D^{J*}_{M\Omega}(\phi, \theta, \chi)$.

References

[1] Arfken G B, Weber H J and Harris F E 2012 *Mathematical Methods for Physicists: A Comprehensive Guide* 7th edn (New York: Academic)

[2] Wolf A A 1969 *Am. J. Phys.* **37** 531

[3] Wigner E and Witmer E E 1928 *Z. Phys.* **51** 859
Hettema H 2000 *Quantum Chemistry: Classic Scientific Papers* p 287 (Singapore: World Scientific)

IOP Publishing

Quantum Mechanics of the Diatomic Molecule (Second Edition)

Christian G Parigger and James O Hornkohl

Chapter 5

Derivation of the Wigner–Witmer eigenfunction

This chapter gives a derivation of the Wigner–Witmer diatomic eigenfunction based on symmetries associated with conservation of energy, linear momentum, and angular momentum. An experimental observation recorded at a given time, location, and orientation should be reproducible at all other times, locations, and orientations. Time and space are homogeneous, and space is isotropic. The Wigner–Witmer diatomic eigenfunction [1], i.e., the diatomic eigenfunction in which time, the coordinates of the center of mass, and the coordinates of the total angular momentum have been exactly separated, follows directly from consideration of translational symmetry of time and space and rotational symmetry of space [2]. Application of the time translation operator $\mathcal{U}$, spatial translation operator $\mathcal{T}$, and rotation operator $\mathcal{R}$ to the total eigenfunction yields a separation of time, spatial coordinates of the total linear momentum, and angular coordinates of the total angular momentum.

The system under consideration in this book is a diatomic molecule in the radiation field. This system is conservative. Total energy, total linear momentum, and total angular momentum are constants of the motion. In this chapter, we assume the molecule is in its ground state and the radiation field is empty, i.e., in this chapter, we treat a diatomic molecule as a conservative system.

5.1 Outline of the derivation

The eigenfunction for a diatomic molecule having N electrons and, of course, two nuclei depends on the N spatial coordinates $\mathbf{R}_1, \mathbf{R}_2, \ldots, \mathbf{R}_N$ of the electrons, the spatial coordinates $\mathbf{R}_a$ and $\mathbf{R}_b$ of the nuclei, and time,

$$\Psi_{nvJM}(\mathbf{R}_1, \mathbf{R}_2, \ldots, \mathbf{R}_N, \mathbf{R}_a, \mathbf{R}_b, t) \equiv \langle \mathbf{R}_1, \mathbf{R}_2, \ldots, \mathbf{R}_N, \mathbf{R}_a, \mathbf{R}_b, t | nvJM \rangle. \quad (5.1)$$

J and M again refer to the total angular momentum including all spins, and n refers to all other quantum numbers except the vibrational quantum number v.

doi:10.1088/978-0-7503-6204-7ch5

The following derivation consists of replacing the coordinates in equation (5.1) with coordinates ρ, ζ, χ, $\mathbf{r}_2$, ... , $\mathbf{r}_N$, r, ϕ, θ, $\mathbf{R}_{\text{CM}}$, τ. The new coordinates are introduced for the purpose of making symmetry transformations. At first, the symmetry transformations merely change the origins of the coordinates (independent variables). These coordinate transformations give new mathematical pictures of a physically unchanged system. The independent variables displayed in equation (5.1) eigenfunctions are each referenced to an arbitrarily chosen origin 0, but there is no absolute origin for time, Cartesian coordinates, or angular coordinates. Time and space are homogeneous, and space is isotropic. If the new origins of the independent variables are referenced to the internal dynamics of the molecule, the new independent variables become dynamical variables describing physical motion independent of fictitious absolute origins for time, space, or orientation. The homogeneity of time and space, and the isotropy of space allow one to choose new origins for the independent variables. Certain numerical values, e.g., the total energy, total linear momentum, total angular momentum, and the mathematical forms of certain important equations, e.g., the Schrödinger equation and the angular momentum commutation formulas, are invariant under the symmetry transformations. If the origins of the new variables are connected to the electrons and nuclei composing a conservative system, then the invariance of the total energy, linear momentum, and angular momentum under symmetry transformations is converted to their conservation during physical motion. The approach taken here is to minimize initial approximations in the treatment of diatomic molecules in favor of using the tools of modern quantum physics.

5.2 Time translation symmetry

Our discussion of time translation symmetry begins with the time-dependent Schrödinger equation,

$$\frac{\partial \Psi(t)}{\partial t} = -\frac{i}{\hbar} H \, \Psi(t). \tag{5.2}$$

For a conservative system, the Hamiltonian H is not a function of time. For a system having bound states, the Hamiltonian H has eigenvalues E,

$$H \, \Psi(t) = E \, \Psi(t). \tag{5.3}$$

The time translation symmetry operator $\mathcal{U}(t, t_0)$ is defined by the effect on the eigenfunction $\Psi(t)$ caused by changing the origin of the time coordinate t from $t = 0$ to some new origin t_0,

$$\mathcal{U}(t, t_0) \, \Psi(t) = \Psi(t - t_0). \tag{5.4}$$

The Taylor series of $\Psi(t - t_0)$,

$$\Psi(t - t_0) = \Psi(t) - t_0 \frac{\partial \Psi(t)}{\partial t} + \frac{t_0^2}{2!} \frac{\partial^2 \Psi(t)}{\partial t^2} - \frac{t_0^3}{3!} \frac{\partial^3 \Psi(t)}{\partial t^3} + \frac{t_0^4}{4!} \frac{\partial^4 \Psi(t)}{\partial t^4} + \cdots \tag{5.5}$$

in combination with the result

$$\frac{\partial^k \Psi}{\partial t^k} = \left(\frac{i\,E}{\hbar}\right)^k \Psi(t) \quad k \text{ even} \qquad = -\left(\frac{i\,E}{\hbar}\right)^k \Psi(t) \quad k \text{ odd} \tag{5.6}$$

from the time-dependent Schrödinger equation (5.2) and equation (5.3) gives

$$\begin{aligned}\Psi(t - t_0) = \Psi(t) + \frac{i\,t_0\,E}{\hbar}\Psi(t) + \frac{1}{2!}\left(\frac{i\,t_0\,E}{\hbar}\right)^2 \Psi(t) + \frac{1}{3!}\left(\frac{i\,t_0\,E}{\hbar}\right)^3 \Psi(t) \\ + \frac{1}{4!}\left(\frac{i\,t_0\,E}{\hbar}\right)^4 \Psi(t) + \cdots\end{aligned} \tag{5.7}$$

yielding the time translation operator,

$$\mathcal{U}(t,\, t_0) = \exp\left(\frac{i\,t_0\,E}{\hbar}\right). \tag{5.8}$$

The time translation operator is unitary,

$$\mathcal{U}^{-1}(t,\, t_0) = \mathcal{U}^{\dagger}(t,\, t_0) \tag{5.9a}$$

$$= \exp\left(-\frac{i\,t_0\,E}{\hbar}\right). \tag{5.9b}$$

Equation (5.4) can be rewritten as

$$\Psi(t) = \mathcal{U}^{\dagger}(t,\, t_0)\Psi(t - t_0) \tag{5.10a}$$

$$= \exp\left(-\frac{i\,t_0\,E}{\hbar}\right)\Psi(t - t_0). \tag{5.10b}$$

If the new time origin is not referenced to a physical event in the system, then the new origin t_0 becomes just another independent variable. If t_0 is equated to the time at which some event occurred in the system, then $\mathcal{U}^{\dagger}(t,\, t_0)$ describes the temporal evolution of the system from that event. Equating t_0 to the time at which some event occurred in the system is identical to imposing a boundary condition on the solution of the time-dependent Schrödinger equation. The time-dependent Schrödinger equation is easily solved when the Hamiltonian H is not a function of time and has the eigenvalue E. The time-dependent Schrödinger equation becomes the simple differential equation

$$\frac{d\Psi(t)}{\Psi(t)} = -\frac{i}{\hbar}E\;dt \tag{5.11}$$

whose solution is

$$\Psi(t) = \text{const}\;\exp\left(-\frac{i\,E\,t}{\hbar}\right). \tag{5.12}$$

One might use the boundary condition $\Psi(t = 0) = \Psi(0)$, which gives

$$\Psi(t) = \exp\left(-\frac{i\,E\,t}{\hbar}\right)\Psi(0). \tag{5.13}$$

Note that this same result is obtained when t_0 is set equal to t in equation (5.10).

The alternative boundary condition $\Psi(t = t_0) = \Psi(t_0)$ gives

$$\Psi(t) = \exp\left[-\frac{i(t - t_0)E}{\hbar}\right]\Psi(t_0) \tag{5.14}$$

which agrees with the previous result equation (5.13) when $t_0 = 0$. Equation (5.14) can be rewritten as

$$\Psi(t) = \exp\left(-\frac{itE}{\hbar}\right)\exp\left(\frac{it_0E}{\hbar}\right)\Psi(t_0) \tag{5.15a}$$

$$= \exp\left(-\frac{itE}{\hbar}\right)U(t,\, t_0)\,\Psi(t_0) \tag{5.15b}$$

$$= \exp\left(-\frac{itE}{\hbar}\right)\Psi(t_0 - t_0) \tag{5.15c}$$

$$= \exp\left(-\frac{itE}{\hbar}\right)\Psi(0) \tag{5.15d}$$

If some physical event occurred in the system at $t = 0$, then equation (5.13) gives the temporal evolution of the system from $t = 0$, and time does not appear in $\Psi(0)$ on the right-hand side of equation (5.13). Similarly, if t_0 in equation (5.14) is the time at which some event occurred in the system, then equation (5.14) gives the temporal evolution of the system after than event, and $\Psi(t_0)$ is not a function of time variable t. However, if the new time origin that one chooses is unrelated to any internal event in the system, then the new time origin simply becomes a new independent variable, and separation of the time variable has not been accomplished. We will see below that these same results are obtained with the time displacement operator $\mathcal{T}(\mathbf{R}_0)$ if the new coordinate origin $\mathbf{R}_0$ is just some arbitrary new origin not referenced to the locations of the electrons and nuclei composing the system.

5.3 Spatial translation symmetry

The spatial translation symmetry operator $\mathcal{T}$ is defined by the effect on the eigenfunction $\Psi(\mathbf{R}_1, \mathbf{R}_2, \ldots, \mathbf{R}_N)$ caused by changing the origin of spatial coordinates from $\mathbf{R} = 0$ to some new origin $\mathbf{R}_0(X_0, Y_0, Z_0)$,

$$\mathcal{T}(\mathbf{R}_1, \mathbf{R}_2, \ldots, \mathbf{R}_N, \mathbf{R}_0)\,\Psi(\mathbf{R}_1, \mathbf{R}_2, \ldots, \mathbf{R}_N) = \Psi(\mathbf{r}_1, \mathbf{r}_2, \ldots, \mathbf{r}_N, \mathbf{R}_0) \tag{5.16}$$

where

$$\mathbf{r}_i = \mathbf{R}_i - \mathbf{R}_0, \qquad i = 1, 2, \ldots, N. \tag{5.17}$$

The finite translation operator $\mathcal{T}$ will be constructed as an infinite sequence of infinitesimal translations. The translation operator $\Delta\mathcal{T}_1$ for the infinitesimal translation along the X-axis $\Delta X_1 = X_1 - \Delta X_0$ is given by

$$\Delta\mathcal{T}_1 \Psi(X_1) = \Psi(X_1 - \Delta X_0) \tag{5.18a}$$

$$\approx 1 - \Delta X_0 \frac{\partial\Psi}{\partial X_1} + \cdots \tag{5.18b}$$

Similar equations hold for the Y_1 and Z_1 components of $\mathbf{R}_1$ and for components of $\mathbf{R}_2$, $\mathbf{R}_3$, … , $\mathbf{R}_N$. Spatial derivatives of the eigenfunction are given by

$$\frac{\partial\Psi(X)}{\partial X} = \frac{i}{\hbar} P_X \Psi(X) \tag{5.19a}$$

$$\frac{\partial\Psi(Y)}{\partial Y} = \frac{i}{\hbar} P_Y \Psi(Y) \tag{5.19b}$$

$$\frac{\partial\Psi(Z)}{\partial Z} = \frac{i}{\hbar} P_Z \Psi(Z). \tag{5.19c}$$

Note the sign difference between the right-hand side of equation (5.2) and the above equations. Insertion of equation (5.19) into equation (5.18) gives

$$\Delta\mathcal{T}_1 \Psi(X_1) = 1 - \frac{i}{\hbar} \Delta X_0 \, P_{X_1} \Psi(X_1) \tag{5.20}$$

The three-dimensional version of the above equation reads

$$\Delta\mathcal{T}_1 = 1 - \frac{i}{\hbar} \Delta\mathbf{R}_0 . \mathbf{P}_1 \, \Psi(\mathbf{R}_1) \tag{5.21}$$

The finite translation operator $\mathcal{T}_1$ is the product of infinitesimal translations in which

$$\Delta\mathbf{R}_1 = \lim_{n\to\infty} \frac{\mathbf{R}_0}{n}, \tag{5.22}$$

$$\mathcal{T}_1 = \lim_{n\to\infty} \prod_{k=1}^{n} \left(1 - \frac{i}{\hbar}\frac{\mathbf{R}_0}{n} . \mathbf{P}_1\right) = \lim_{n\to\infty}\left(1 - \frac{i}{\hbar}\frac{\mathbf{R}_0}{n} . \mathbf{P}_1\right)^n \tag{5.23a}$$

$$= \exp\left(-\frac{i}{\hbar}\, \mathbf{R}_0 . \mathbf{P}_1\right) \tag{5.23b}$$

The total translation operator $\mathcal{T}(\mathbf{R}_1, \mathbf{R}_2, \ldots, \mathbf{R}_N, \mathbf{R}_0)$ is the product of the individual translations,

$$\begin{aligned} \mathcal{T}(\mathbf{R}_1, \mathbf{R}_2, \ldots, \mathbf{R}_N, \mathbf{R}_0) &= \mathcal{T}_1(\mathbf{R}_1, \mathbf{R}_0)\, \mathcal{T}_2(\mathbf{R}_2, \mathbf{R}_0) \ldots \mathcal{T}_N(\mathbf{R}_N, \mathbf{R}_0) \\ &= \exp\left(-\frac{i}{\hbar}\, \mathbf{R}_0 . \mathbf{P}_1\right) \exp\left(-\frac{i}{\hbar}\, \mathbf{R}_0 . \mathbf{P}_2\right) \ldots \exp\left(-\frac{i}{\hbar}\, \mathbf{R}_0 . \mathbf{P}_N\right) \end{aligned} \tag{5.24}$$

Because the linear momentum operators for the different electrons and nuclei commute, the product of the individual translation operators can be written in any order. The first two terms of the above product can be written

$$\begin{aligned} \mathcal{T}_1(\mathbf{R}_1, \mathbf{R}_0)\, \mathcal{T}_2(\mathbf{R}_2, \mathbf{R}_0) &= \exp\left(-\frac{i}{\hbar}\, \mathbf{R}_0 . \mathbf{P}_1\right) \exp\left(-\frac{i}{\hbar}\, \mathbf{R}_0 . \mathbf{P}_2\right) \\ &= \exp\left[-\frac{i}{\hbar}\, \mathbf{R}_0 .(\mathbf{P}_1 + \mathbf{P}_2)\right] \exp\, [\mathbf{R}_0 . \mathbf{P}_1,\, \mathbf{R}_0 . \mathbf{P}_2]/2 \end{aligned} \tag{5.25}$$

in which the right-most term is the commutator of $\mathbf{R}_0.\mathbf{P}_1$ and $\mathbf{R}_0.\mathbf{P}_2$. The new origin $\mathbf{R}_0$ is a fixed vector, not an operator, and it commutes with the linear momentum operators $\mathbf{P}_i$,

$$\exp\, [\mathbf{R}_0 . \mathbf{P}_i,\, \mathbf{R}_0 . \mathbf{P}_i]/2 = 1, \tag{5.26}$$

and the final result for the translation operator is

$$\mathcal{T}(\mathbf{R}_1, \mathbf{R}_2, \ldots, \mathbf{R}_N, \mathbf{R}_0) = \exp\left[-\frac{i}{\hbar}\, \mathbf{R}_0 . \mathbf{P}\right] \tag{5.27a}$$

$$= \exp\left[-\frac{i}{\hbar}\, \mathbf{P} . \mathbf{R}_0\right] \tag{5.27b}$$

in which $\mathbf{P}$ is the total linear momentum operator,

$$\mathbf{P} = \sum_{i=1}^{N} \mathbf{P}_i. \tag{5.28}$$

If the new origin of coordinates $\mathbf{R}_0$ is not referenced to the locations of the mass points from which the atomic or molecular system is composed, then the Cartesian components of $\mathbf{R}_0(X_0, Y_0, Z_0)$ add three independent variables to the problem. If the new origin is referenced to the internal spatial coordinates of the system, *and* if a two-body reduction is applied to two of the mass points, *and* if the coordinates of the center of mass $\mathbf{R}_{CM}(X_{CM}, Y_{CM}, Z_{CM})$ are introduced as variables of the system, then application of the translation operator introduces no new independent variables and separates the three spatial components of the center of mass. An example of this will be seen below when the translation operator $\mathcal{T}(\mathbf{R}, \mathbf{R}_0)$ is applied to the diatomic (5.1) eigenfunction.

The time translation operator $\mathcal{U}(t, t_0)$ always involves two coordinate systems, the original time axis and the translated time axes. The spatial translation operator $\mathcal{T}(\mathbf{R}, \mathbf{R}_0)$ always involves two Cartesian coordinate systems, the original system whose origin lies at 0 and the translated system whose origin is $\mathbf{R}_0$. The rotation operator $\mathcal{R}$ below will always involve two coordinate systems, the original system and the rotated system. When do coordinates of translated or rotated coordinate systems become coordinates describing physical motion? As seen above, coordinates of the new coordinate system become internal coordinates describing physical motion when the time origin is referenced to an event in the system and the coordinate origin is referenced to locations of the mass points composing the system. Translation of coordinates to new origins having no physical connection to the system under study provide a new mathematical viewpoint of the system, but are excluded as coordinates describing physical motion. When the new origins of coordinate transformation are physically referenced to the system, the invariance of values and equations under symmetry transformations becomes conservation of energy, linear momentum, and (as shown below) angular momentum during physical motion.

5.4 Two-body symmetry

The model for an atomic or molecular system contains a minimum of two mass points. The first step in finding suitable internal coordinates for any such system is the selection of two particles for the two-body reduction. This is not historically the first step nor is it the first step in many textbook treatments of atomic and molecular systems, but sooner or later as approximations are dropped in favor of higher degrees of approximation one will select two mass points located at $\mathbf{R}_1$ and $\mathbf{R}_2$ and replace their motion with that of two fictitious particles one located $\mathbf{r} = \mathbf{R}_1 - \mathbf{R}_2$ and having reduced mass μ,

$$\mu = \frac{m_1 m_2}{m_1 + m_2}, \tag{5.29}$$

and a second fictitious particle of mass $\mathcal{M}$,

$$\mathcal{M} = m_1 + m_2, \tag{5.30}$$

located at the center of mass of the two particles,

$$\mathbf{R}_0 = \frac{m_1\mathbf{R}_1 + m_2\mathbf{R}_2}{m_1 + m_2}. \tag{5.31}$$

The change of variables

$$\mathbf{R}_1, \mathbf{R}_2 \rightarrow \mathbf{r}, \mathbf{R}_0 \tag{5.32}$$

in which

$$\mathbf{r} = \mathbf{R}_1 - \mathbf{R}_2 \tag{5.33a}$$

$$= \mathbf{r}_1 - \mathbf{r}_2 \tag{5.33b}$$

and the chain rule, e.g.,

$$\frac{\partial\Psi}{\partial X_1} = \frac{\partial\Psi}{\partial x}\frac{dx}{dX_1} + \frac{\partial\Psi}{\partial X_0}\frac{dX_0}{dX_1} = \frac{\partial\Psi}{\partial x} + \frac{m_1}{\mathcal{M}}\frac{\partial\Psi}{\partial X_0} \tag{5.34a}$$

$$\frac{\partial\Psi}{\partial X_2} = \frac{\partial\Psi}{\partial x}\frac{dx}{dX_2} + \frac{\partial\Psi}{\partial X_0}\frac{dX_0}{dX_2} = -\frac{\partial\Psi}{\partial x} + \frac{m_2}{\mathcal{M}}\frac{\partial\Psi}{\partial X_0}, \tag{5.34b}$$

give the result that the total linear momentum **P** of the two masses in terms the new variables is the linear momentum of the center of mass of the two masses,

$$\mathbf{P} = \mathbf{P}_1 + \mathbf{P}_2 \tag{5.35a}$$

$$= \mathbf{p} + \frac{m_1}{\mathcal{M}}\mathbf{P}_0 - \mathbf{p} + \frac{m_2}{\mathcal{M}}\mathbf{P}_0 \tag{5.35b}$$

$$= \mathbf{P}_0 \tag{5.35c}$$

with

$$\mathbf{p} = \frac{\hbar}{i}\left[\frac{\partial}{\partial x} + \frac{\partial}{\partial y} + \frac{\partial}{\partial z}\right]. \tag{5.36}$$

The inverse change of variables,

$$\mathbf{R}_1 = \mathbf{R}_0 + \frac{m_2}{\mathcal{M}}\mathbf{r} \tag{5.37a}$$

$$\mathbf{R}_2 = \mathbf{R}_0 - \frac{m_1}{\mathcal{M}}\mathbf{r}, \tag{5.37b}$$

and the chain rule give the internal (relative) linear momentum **p** in terms of the original coordinates $\mathbf{R}_1$ and $\mathbf{R}_2$,

$$\mathbf{p} = \frac{m_2\mathbf{P}_1 - m_1\mathbf{P}_2}{m_1 + m_2}. \tag{5.38}$$

When the system in question contains only two mass points, the new coordinate origin expressed in the original coordinate system $\mathbf{R}_0$ is the the center of total mass $\mathbf{R}_{CM}$. Two-body symmetry separates the linear momentum of the two particles from their relative motion. Separation of the total linear momentum in a two particle system is accomplished when the coordinate transformation

$$\mathbf{R}_1, \mathbf{R}_2 \rightarrow \mathbf{r}, \mathbf{R}_{CM} \tag{5.39}$$

is performed. Application of the translation operator $\mathcal{T}(\mathbf{R}_1, \mathbf{R}_2, \mathbf{R}_{CM})$ to the eigenfunction $\Psi(\mathbf{R}_1, \mathbf{R}_2)$ produces this transformation (change of variables).

In a generic system consisting of more than two mass points, application of the two-body reduction still dictates that the new coordinate origin be at the center of mass of two mass points $\mathbf{R}_0$, and $\mathbf{R}_0$ no longer coincides with center of total mass $\mathbf{R}_{CM}$. In this case the coordinate transformation is

$$\mathbf{R}_1, \mathbf{R}_2, \ldots, \mathbf{R}_N \rightarrow \mathbf{r}_1, \mathbf{r}_2, \ldots, \mathbf{r}_{N-1}, \mathbf{R}_{CM} \tag{5.40}$$

5.5 Time and spatial translations together

The symmetry operators $\mathcal{U}$ and $\mathcal{T}$ are unitary

$$\mathcal{T}^{\dagger}\mathcal{T}\mathcal{U}^{\dagger}\mathcal{U}\Psi = \Psi. \tag{5.41}$$

Applied to generic eigenfunction $\Psi(\mathbf{R}_1, \mathbf{R}_2, \ldots, \mathbf{R}_N, t)$ in which N is the number of mass points,

$$\begin{aligned}&\mathcal{T}^{\dagger}\mathcal{T}\mathcal{U}^{\dagger}\mathcal{U}\Psi(\mathbf{R}_1, \mathbf{R}_2, \ldots, \mathbf{R}_N, t) = \mathcal{T}^{\dagger}\mathcal{U}^{\dagger}\Phi(\mathbf{r}_1, \mathbf{r}_2, \ldots, \mathbf{r}_{N-2}, \mathbf{r})\\ &= \exp\left(\frac{i}{\hbar}\mathbf{R}_0.\mathbf{P}\right)\exp\left[-\frac{i}{\hbar}t\ E\right]\Phi(\mathbf{r}_1, \mathbf{r}_2, \ldots, \mathbf{r}_{N-2}, \mathbf{r}).\end{aligned} \tag{5.42}$$

A superposition of the exponentials

$$\Psi_{\mathrm{CM}}(\mathbf{R}_0, t) = \exp\left[i\left(\frac{1}{\hbar}\mathbf{R}_0.\mathbf{P} - \omega t\right)\right] \tag{5.43}$$

with $\omega = E/\hbar$ yielding a wave packet describing linear motion the molecule as a whole.

5.6 Rotational symmetry

The eigenfunction $\Phi(\mathbf{r}_1, \mathbf{r}_2, \ldots, \mathbf{r}_{N-2}, \mathbf{r})$ is expressed in generic coordinates and quantum numbers. Rewritten for the diatomic molecule,

$$\langle \mathbf{r}_1, \mathbf{r}_2, \ldots, \mathbf{r}_N, \mathbf{r}\ |nvJM\rangle = \langle \rho, \zeta, \chi, \mathbf{r}_2, \ldots, \mathbf{r}_N, r, \theta, \phi\ |nvJM\rangle, \tag{5.44}$$

this eigenfunction is identical to that presented in the framework for the Wigner–Witmer eigenfunction (WWE), see equation (4.11). Although the Euler angles α, β, and γ and the rotation operator $\mathcal{R}(\alpha, \beta, \gamma)$ are not properly introduced until chapter 9, they have already been used in chapter 4, see equations (4.6)–(4.16). Analogous to the presentation of the framework for WWE, the Wigner–Witmer diatomic eigenfunction for the diatomic molecule is written as

$$\begin{aligned}&\langle \rho, \zeta, \chi, \mathbf{r}_2, \ldots, \mathbf{r}_N, r, \theta, \phi\ |nvJM\rangle\\ &= \sum_{\Omega=-J}^{J} \langle \rho, \zeta, \mathbf{r}'_2, \ldots, \mathbf{r}'_N, r\ |nv\rangle\ D^{J*}_{M\Omega}(\phi, \theta, \chi).\end{aligned} \tag{5.45}$$

Note that the Euler angles in the Wigner–Witmer diatomic eigenfunction (5.45) are both angles of coordinate rotation and angles of mechanical rotation is extremely atypical. Normally, one must maintain a clear distinction between coordinate (passive) rotation and mechanical (active) rotation. The first two Euler angles ϕ and θ are also exactly separable in the hydrogen atom eigenfunction, and serve both as angles of coordinate rotation and mechanical rotation. The angles ϕ and θ in the hydrogen atom are identical to the same angles in the diatomic molecule. In both cases, θ is the polar angle and ϕ is the azimuthal angle of the vector $\mathbf{r}$ drawn from the origin taken to be the center of mass of two particles, the electron and proton in the hydrogen atom and the two nuclei in the diatomic molecule.

References

[1] Wigner E and Witmer E E 1928 *Z. Phys.* **51** 859
Hettema H 2000 *Quantum Chemistry: Classic Scientific Papers* p 287 (Singapore: World Scientific)
[2] Weissbluth M 1978 *Atoms and Molecules* (New York: Academic)

IOP Publishing

Quantum Mechanics of the Diatomic Molecule (Second Edition)

Christian G Parigger and James O Hornkohl

Chapter 6

Diatomic formula inferred from the Wigner–Witmer eigenfunction

In this chapter,a formula useful in diatomic spectroscopy is derived from the Wigner–Witmer eigenfunction [1] (equation (4.16)).

Equation (4.2d) holds in a coordinate system and equation (4.3b) holds in a second coordinate system obtained from the first by coordinate rotation, but on the right-hand side of equation (4.16) the only thing that $J'_{\pm}$ can operate on is $D^{J*}_{M\Omega}(\phi, \theta, \chi)$. Equation (4.3b) shows what happens when $J'_{\pm}$ is applied to $|J\Omega\rangle$, but what happens when $J'_{\pm}$ is applied to $D^{J*}_{M\Omega}(\phi, \theta, \chi)$? This mathematical question is answered in the following paragraphs.

Angular-momentum theory gives the result of applying the raising and lowering operators to the left-hand side of the Wigner–Witmer equation (4.16), and from this we can find how $J_{\pm}$ operates on $D^{J*}_{M\Omega}(\phi, \theta, \chi)$,

$$\begin{aligned}
&\langle \rho, \zeta, \chi, \mathbf{r}_2, \dots, \mathbf{r}_N, r, \theta, \phi\, | J_{\pm} | nvJM \rangle \\
&= \sqrt{J(J+1) - M(M \pm 1)}\ \langle \rho, \zeta, \chi, \mathbf{r}_2, \dots, \mathbf{r}_N, r, \theta, \phi\, | nvJ, M \pm 1 \rangle \\
&= \sqrt{J(J+1) - M(M \pm 1)} \sum_{\Omega=-J}^{J} \langle \rho, \zeta, \mathbf{r}'_2, \dots, \mathbf{r}'_, r\, | nv \rangle\, D^{J*}_{M\pm 1,\,\Omega}(\phi, \theta, \chi) \\
&= \sum_{\Omega=-J}^{J} \langle \rho, \zeta, \mathbf{r}'_2, \dots, \mathbf{r}'_, r\, | nv \rangle\, J_{\pm}\, D^{J*}_{M\Omega}(\phi, \theta, \chi).
\end{aligned} \tag{6.1}$$

The eigenfunctions $\langle \rho, \zeta, \mathbf{r}'_2, \dots, \mathbf{r}'_, r\, | nv \rangle$ are independent, and the result

$$J_{\pm}\, D^{J*}_{M\Omega}(\phi, \theta, \chi) = \sqrt{J(J+1) - M(M \pm 1)}\ D^{J*}_{M\pm 1,\,\Omega}(\phi, \theta, \chi) \tag{6.2}$$

holds for each term in the above summations, and is intuitively pleasing. The raising and lowering operators raise and lower M on $D^{J*}_{M\pm 1,\,\Omega}(\phi, \theta, \chi)$. However, on taking

doi:10.1088/978-0-7503-6204-7ch6

the complex conjugate of both sides of the above equation (note from (4.2c) that $J_{\pm}^{*} = J_{\mp}$),

$$J_{\mp} D_{M\Omega}^{J}(\phi,\, \theta,\, \chi) = \sqrt{J(J+1) - M(M \pm 1)}\; D_{M \pm 1,\, \Omega}^{J}(\phi,\, \theta,\, \chi). \tag{6.3}$$

and intuition is now faring less well. The raising and lowering operators lower and raise M, respectively, when applied to $D_{M \mp 1,\, \Omega}^{J}(\phi,\, \theta,\, \chi)$. The above is a little more easily read if one swaps all the $\pm$ with $\mp$, and vice versa,

$$J_{\pm} D_{M\Omega}^{J}(\phi,\, \theta,\, \chi) = \sqrt{J(J+1) - M(M \mp 1)}\; D_{M \mp 1,\, \Omega}^{J}(\phi,\, \theta,\, \chi). \tag{6.4}$$

The above describes operation $J_{\pm}$ on the D-functions and its complex conjugate, but we have missed the goal of learning how the rotated raising and lowering operators $J'_{\pm}$ operate on the D-function. To accomplish this goal, we need to turn the Wigner–Witmer equation (4.16) inside out to express the rotated eigenfunction as a sum of unrotated eigenfunctions.

The rotation matrix elements form a $(2J+1) \times (2J+1)$ unitary matrix. By multiplying equation (4.12) by $D_{M'\Omega}^{J}(\alpha,\, \beta,\, \gamma)$ and summing over M' one obtains

$$\begin{aligned} &\langle \rho,\, \zeta,\, \chi',\, \mathbf{r}'_2,\, \dots,\, \mathbf{r}'_N,\, r,\, \theta',\, \phi' \,|nvJ\Omega\rangle \\ &= \sum_{M=-J}^{J} \langle \rho,\, \zeta,\, \chi,\, \mathbf{r}_2,\, \dots,\, \mathbf{r}_N,\, r,\, \theta,\, \phi \,|nvJM\rangle\, D_{M\Omega}^{J}(\alpha,\, \beta,\, \gamma). \end{aligned} \tag{6.5}$$

Note that when equation (4.12) was inverted to yield the above, the Euler angles of mechanical rotation and the Euler angles of coordinate rotation were unrelated. Having inverted equation (4.12), we choose Euler angles of coordinate rotation to remove the angular dependence from $\langle \rho,\, \zeta,\, \chi,\, \mathbf{r}_2,\, \dots,\, \mathbf{r}_N,\, r,\, \theta,\, \phi \,|nvJM\rangle$,

$$\phi' = \phi + \alpha, \tag{6.6a}$$

$$\theta' = \theta + \beta, \tag{6.6b}$$

$$\chi' = \chi + \gamma. \tag{6.6c}$$

Note that α, β, and γ chosen in equations (6.6) are the negative to those in equations (4.13). The inverse of the rotation operator $\mathcal{R}^{-1}(\alpha,\, \beta,\, \gamma)$ is $\mathcal{R}(-\gamma,\, -\beta,\, -\alpha)$ [2]. The choices in equations (6.6) yield

$$\begin{aligned} &\langle \rho,\, \zeta,\, \chi',\, \mathbf{r}'_2,\, \dots,\, \mathbf{r}'_N,\, r,\, \theta',\, \phi' \,|nvJ\Omega\rangle \\ &= \sum_{M=-J}^{J} \langle \rho,\, \zeta,\, \mathbf{r}_2,\, \dots,\, \mathbf{r}_N,\, r \,|nv\rangle\, D_{M\Omega}^{J}(\phi',\, \theta',\, \chi') \end{aligned} \tag{6.7}$$

We can now apply $J'_{\pm}$ on both sides of this equation, and using the standard result (4.3b) mathematically determine the operation of $J'_{\pm}$ on $D_{M\Omega}^{J}(\phi',\, \theta',\, \chi')$. Operation of $J'_{\pm}$ to both sides of the above equation gives

$$
\begin{aligned}
&\sqrt{J(J+1)-\Omega(\Omega\pm 1)}\,\langle \rho, \zeta, \chi', \mathbf{r}'_2, \dots, \mathbf{r}'_N, r, \theta', \phi' \,|nvJ, \Omega \pm 1\rangle \\
&\sqrt{J(J+1)-\Omega(\Omega\pm 1)}\sum_{M=-J}^{J} \langle \rho, \zeta, \mathbf{r}_2, \dots, \mathbf{r}_N, r \,|nv\rangle\, D^{J}_{M,\,\Omega\pm 1}(\phi', \theta', \chi') \\
&= \sum_{M=-J}^{J} \langle \rho, \zeta, \mathbf{r}_2, \dots, \mathbf{r}_N, r \,|nvJM\rangle\, J'_{\pm}\, D^{J}_{M\Omega}(\phi', \theta', \chi').
\end{aligned}
\tag{6.8}
$$

from which, after dropping primes on the Euler angles, one obtains

$$
J'_{\pm}\, D^{J}_{M\Omega}(\phi, \theta, \chi) = \sqrt{J(J+1)-\Omega(\Omega\pm 1)}\, D^{J}_{M,\,\Omega\pm 1}(\phi, \theta, \chi). \tag{6.9}
$$

This result, like equation (6.2), is intuitively acceptable. The rotated raising and lowering operators $J'_{\pm}$ raise and lower Ω on $D^{J}_{M\Omega}(\phi, \theta, \chi)$. On taking the complex conjugate of both sides of the above equation, one finds a less intuitively pleasing result,

$$
J'_{\pm}\, D^{J*}_{M\Omega}(\phi, \theta, \chi) = \sqrt{J(J+1)-\Omega(\Omega\mp 1)}\, D^{J*}_{M,\,\Omega\mp 1}(\phi, \theta, \chi). \tag{6.10}
$$

This equation is a mathematical result, not an accepted 'phase convention.' The mysterious difference between equation (4.3b) and equation (6.10) disappears when one understands that $|J\Omega\rangle$ is an eigenstate of angular momentum but neither $D^{J*}_{M\Omega}(\phi, \theta, \chi)$ nor $D^{J}_{M\Omega}(\phi, \theta, \chi)$ is.

The effects of raising and lowering operators are further discussed in appendix B. The importance of equation (6.10) derives from its use in the calculation of the Hund's case (a) matrix representation of the diatomic Hamiltonian.

References

[1] Wigner E and Witmer E E 1928 *Z. Phys.* **51** 859 H. Hettema, *Quantum Chemistry: Classic Scientific Papers*, p 287, (Singapore: World Scientific) 2000

[2] Arfken G B, Weber H J and Harris F E 2012 *Mathematical Methods for Physicists: A comprehensive Guide* 7th edn (New York: Academic)

IOP Publishing

Quantum Mechanics of the Diatomic Molecule (Second Edition)

Christian G Parigger and James O Hornkohl

Chapter 7

Hund's cases (a) and (b)

7.1 Introduction

In spectroscopic notation, the total angular momentum $\mathbf{F}$ of the diatomic molecule is given by

$$\mathbf{F} = \mathbf{L} + \mathbf{R} + \mathbf{S} + \mathbf{T} \tag{7.1a}$$

$$= \mathbf{J} + \mathbf{T} \tag{7.1b}$$

$$\mathbf{J} = \mathbf{L} + \mathbf{R} + \mathbf{S} \tag{7.1c}$$

$$= \mathbf{N} + \mathbf{S} \tag{7.1d}$$

$$\mathbf{N} = \mathbf{L} + \mathbf{R} \tag{7.1e}$$

$$\mathbf{T} = \mathbf{I}_\mathrm{a} + \mathbf{I}_\mathrm{b} \tag{7.1f}$$

in which $\mathbf{I}_\mathrm{a}$ and $\mathbf{I}_\mathrm{b}$ are the nuclear spins and $\mathbf{T}$ is their sum (the a on $\mathbf{I}_\mathrm{a}$ and b on $\mathbf{I}_\mathrm{b}$ are the labels for the two nuclei and are unrelated to Hund's cases (a) and (b) angular momentum coupling models). The total orbital angular momentum $\mathbf{N}$ is the sum $\mathbf{L}$ of the individual electronic orbital angular momenta ℓ_i,

$$\mathbf{L} = \sum_{i=1}^{N} \ell_i, \tag{7.2}$$

plus the orbital angular momentum $\mathbf{R}$ of the two nuclei. The total electronic spin is the sum of the individual electronic spins,

$$\mathbf{S} = \sum_{i=1} \mathbf{s}_i. \tag{7.3}$$

doi:10.1088/978-0-7503-6204-7ch7

In spectroscopic notation, **J** is the total angular momentum including everything except nuclear spin. In this chapter, nuclear spin is ignored, and the spectroscopic J, M, and Ω quantum numbers are treated as if they represent the true total angular momentum.

Orbital and spin momenta produce magnetic moments. It is obvious from equations (7.1) that even when nuclear spin is ignored there are many ways in which magnetic moments produced by **L**, **R**, their sum $\mathbf{N} = \mathbf{L} + \mathbf{R}$, and **S** could interact. In the Hund's angular momentum coupling models, a particular interaction is assumed to dominate all others. Only Hund's cases (a) and (b) are considered in this book, and they will be used as mathematically complete bases, not mechanical models. The following shows how cases (a) and (b) basis functions are obtained from the Wigner–Witmer eigenfunction.

7.2 Case (b) basis functions

Construction of case (b) basis functions will be familiar to those familiar with ℓs coupling in atoms where one builds $|jm\rangle$ states from orbital angular momentum states $|\ell m\rangle$ and spin states $|sm_s\rangle$ states. In the diatomic case, one builds $|JM\rangle$ states from total orbital angular momentum states $|NM_N\rangle$ and, ignoring nuclear spin, total electronic spin states $|SM_S\rangle$.

In the absence of both electronic and nuclear spin, the Wigner–Witmer eigenfunction is

$$\langle \mathbf{r}_1, \mathbf{r}_2, \ldots, \mathbf{r}_{\mathcal{N}}, \mathbf{r}\, |nvNM_N\rangle = \sum_{\Lambda=-N}^{N} \langle \rho, \zeta, \mathbf{r}_2, \ldots, \mathbf{r}_{\mathcal{N}}, r\, |nv\rangle\, D^{N*}_{M_N\Lambda}(\phi, \theta, \chi). \quad (7.4)$$

We have used $\mathcal{N}$ as the number of electrons to avoid confusion with the quantum number N. Dropping the summation over Λ from the right-hand side and renormalization gives the function

$$\begin{aligned} &\langle \rho, \zeta, \chi, \mathbf{r}'_2, \ldots, \mathbf{r}'_{\mathcal{N}}, r, \theta, \phi\, |nvNM_N\Lambda\rangle \\ &= \sqrt{\frac{2N+1}{8\pi^2}}\langle \rho, \zeta, \mathbf{r}_2, \ldots, \mathbf{r}_{\mathcal{N}}, r\, |nv\rangle\, D^{N*}_{M_N\Lambda}(\phi, \theta, \chi). \end{aligned} \quad (7.5)$$

In case (b), the mechanical coupling of the magnetic moment of the spin state $|SM_S\rangle$ to the magnetic moment of the orbital $|NM_N\rangle$ state is assumed to be negligibly small. This is ideally suited to building total angular momentum states $|JM\rangle$ using Clebsch–Gordan coefficients $\langle NM_N;\, SM_S\, |JM\rangle$,

$$|JM\rangle = \sum_{M_N=-N}^{N} \sum_{M_S=-S}^{S} |NM_N\rangle \otimes |SM_S\rangle\, \langle NM_N;\, SM_S\, |JM\rangle. \quad (7.6)$$

An unfortunate notational convention leaves the use of the angular momentum coupling coefficients most valid when the mechanical coupling is negligible. Other notational features of the above equation deserve explanation. The tensor (or direct or Kronecker) product $|NM_N\rangle \otimes |SM_S\rangle$ is not written as $|NM_N\rangle\, |SM_S\rangle$ because the latter lacks mathematical meaning, e.g., $|NM_N\rangle$ and $|SM_S\rangle$ are normally regarded as column matrices having $(2N + 1)$ rows and $(2S + 1)$ rows, and their product is not commensurate with standard matrix multiplication [1, 2]. In addition, a semicolon appears in the Clebsch–Gordan coefficient $\langle NM_N;\ SM_S\ |JM\rangle$ to remind one that it is merely a number, and not a collection of bras and kets.[1]

The case (b) basis function is obtained when equation (7.5) is inserted into equations(7.6),

$$|b\rangle = \langle \rho,\ \zeta,\ \chi,\ \mathbf{r}_2,\ \dots\ ,\ \mathbf{r}_N,\ r,\ \theta,\ \phi\ |nvN\Lambda S\rangle = \sqrt{\frac{2N+1}{8\pi^2}}\langle \rho,\ \zeta,\ \mathbf{r}_2,\ \dots\ ,\ \mathbf{r}_N,\ r\ |nv\rangle \times \sum_{M_N=-N}^{N}\sum_{M_S=-S}^{S}\langle NM_N;\ SM_S\ |SM_S\rangle\ |SM_S\rangle\ D^{N*}_{M_N\Lambda}(\phi,\ \theta,\ \chi). \tag{7.7}$$

The direct product symbol $\otimes$ should appear between $|SM_S\rangle$ and $D^{N*}_{M_N\Lambda}(\phi,\ \theta,\ \chi)$ in the case (b) basis function, but in practice one never attempts to form the direct product of $|SM_S\rangle$ with the Wigner *D*-function and the symbol $\otimes$ is conventionally dropped.

It is important that one remember that the summation over Λ in equation (7.4) has been dropped. As a result, the case (b) basis functions are not *energy* eigenfunctions, not *angular momentum* eigenfunctions, and not *parity* eigenfunctions. For example, a sum of case (b) basis functions,

$$|JM\rangle = \sum_{\Lambda=-N}^{N}|b\rangle\ \langle b\ |JM\rangle, \tag{7.8}$$

is required to build the angular momentum eigenfunction. The case (b) basis possesses three magnetic quantum numbers, M_N, Λ, and M_S, two more than possessed by a quantum mechanical state of angular momentum.

The Hund's case (b) basis is mathematically complete. In principle, one can successfully perform in the case (b) basis any calculation that can be performed in any other basis or with the exact eigenfunction.

7.3 Case (a) eigenfunctions

When the summation over Ω is dropped from the right-hand side of the Wigner–Witmer eigenfunction (4.16) and the result is renormalized, one obtains the function

$$\begin{aligned}&\langle \rho,\ \zeta,\ \chi,\ \mathbf{r'}_2,\ \dots\ ,\ \mathbf{r'}_N,\ r,\ \theta,\ \phi\ |nvJM\Omega\rangle\\ &= \sqrt{\frac{2J+1}{8\pi^2}}\langle \rho,\ \zeta,\ \mathbf{r'}_2,\ \dots\ ,\ \mathbf{r'}_N,\ r\ |nv\rangle\ D^{J*}_{M\Omega}(\phi,\ \theta,\ \chi).\end{aligned} \tag{7.9}$$

In case (b), the interaction of the orbital states $|NM_N\rangle$ and $|SM_S\rangle$ is assumed negligibly small, and the $|JM\rangle$ state is built from Clebsch–Gordan coefficients. In case (a), one assumes the electronic spin state $|S\Sigma\rangle$ referenced to rotated coordinates

[1] Notations for the Clebsch–Gordon coefficients are so varied that one is almost at liberty to invent new ones. For example, coefficients such as Zare's [3] help one to avoid double subscripts and subscripted superscripts.

is locked to the orbital state $|N\Lambda\rangle$ also referenced to rotated coordinates. The perfect, complete coupling of $|S\Sigma\rangle$ to $|N\Lambda\rangle$ negates the use of Clebsch–Gordon coefficients. The quantum numbers Ω, Λ, and Σ each refer to a z' (internuclear) component of angular momentum. This means that

$$J_{z'} = L_{z'} + S_{z'}, \tag{7.10}$$

$$\Omega = \Lambda + \Sigma. \tag{7.11}$$

Because $\mathbf{J} = \mathbf{L} + \mathbf{S}$, the quantum numbers Λ and Σ are already implied in equation (7.9), but the spin state $|S\Sigma\rangle$ is missing. The case (a) basis function is obtained when the spin ket $|S\Sigma\rangle$ is simply inserted into equation (7.9),

$$\begin{aligned} |a\rangle &= \langle \rho, \zeta, \chi, \mathbf{r}'_2, \ldots, \mathbf{r}'_N, r, \theta, \phi \,|nvJMS\Lambda\Sigma\Omega\rangle \\ &= \sqrt{\frac{2J+1}{8\pi^2}} \langle \rho, \zeta, \mathbf{r}'_2, \ldots, \mathbf{r}'_N, r \,|nv\rangle \,|S\Sigma\rangle\, D^{J*}_{M\Omega}(\phi, \theta, \chi). \end{aligned} \tag{7.12}$$

The case (a) basis function is one of the more useful mathematical tools of diatomic theory. Using semi-empirical molecular parameters and case (a) basis functions one can compute the case (a) matrix representation of the Hamiltonian, which is then numerically diagonalized. If this is done for the upper and lower states of a band system, the vacuum wavenumbers $\tilde{\nu}$ are the differences between the upper and lower energy eigenvalues, and the Hönl–London line-strength factors $S(J', J)$ are computed from the orthogonal matrices that diagonalized the upper and lower Hamiltonians and the case (a) matrix elements of the transition operator (e.g., the electric dipole operator). A nonvanishing $S(J', J)$ identifies an allowed spectral line.

Like the case (b) basis, the case (a) basis functions are not eigenfunctions of energy, angular momentum, or parity. The complex conjugate of the Wigner D-function has some properties of a $|JM\rangle$ state. For example,

$$\mathbf{J}^2\, D^{J*}_{M\Omega}(\phi, \theta, \chi) = J(J+1)\, D^{J*}_{M\Omega}(\phi, \theta, \chi), \tag{7.13}$$

$$J_z D^{J*}_{M\Omega}(\phi, \theta, \chi) = M\, D^{J*}_{M\Omega}(\phi, \theta, \chi), \tag{7.14}$$

but then the matter becomes complicated when $J_{z'}$ operates on $D^{J*}_{M\Omega}(\phi, \theta, \chi)$,

$$J_{z'} D^{J*}_{M\Omega}(\phi, \theta, \chi) = \Omega\, D^{J*}_{M\Omega}(\phi, \theta, \chi). \tag{7.15}$$

Now there are two magnetic quantum numbers. $|JM\rangle$ and $|J\Omega\rangle$ might possibly represent states of angular momentum, but $|JM\Omega\rangle$ cannot. Being mathematically precise about what is or is not an angular momentum eigenfunction is not wasted effort. For example, if one treats the Wigner D-function as an angular momentum eigenfunction, then the operation of $J'_\pm$ on it produces a train wreck,

$$J'_\pm D^{J*}_{M\Omega}(\phi, \theta, \chi) = \sqrt{J(J+1) - \Omega(\Omega \mp 1)}\, D^{J*}_{M,\,\Omega\mp 1}(\alpha, \beta, \gamma), \tag{7.16}$$

in which raising/lowering operators whose standard behavior is

$$J'_{\pm}\,|J\Omega\rangle = \sqrt{J(J+1) - \Omega(\Omega \pm 1)}\,|J, \Omega \pm 1\rangle \tag{7.17}$$

become when operating on $D^{J*}_{M\Omega}(\phi, \theta, \chi)$ lowering/raising operators. If one accepts that a function carrying two magnetic quantum numbers cannot represent an angular momentum state, then the mystery concerning the above equation (7.16) disappears. Equation (7.16) is simply a mathematical result, not a conundrum in diatomic theory. Equation (7.16) is extensively used in the calculation of case (a) matrix elements. There is also clearly a $N'_{\pm}\, D^{N*}_{M_N\Lambda}(\phi, \theta, \chi)$ version of this result which is used in case (b) calculations. The importance of equation (7.16) requires that its derivation be given in the following section.

Like the case (b) basis, the case (a) basis is complete. In principle, any calculation can be successfully carried out in the case (a) basis. Typically, calculations in the case (a) basis are analytically less messy than in the case (b) basis.

References

[1] Zettili N 2009 *Quantum Mechanics* 2nd edn (West Sussex: Wiley)
[2] Cohen-Tannoudji C, Diu B and Laloë F 1977 *Quantum Mechanics* (New York: Wiley)
[3] Zare R N 1988 *Angular Momentum* (New York: Wiley)

IOP Publishing

Quantum Mechanics of the Diatomic Molecule (Second Edition)

Christian G Parigger and James O Hornkohl

Chapter 8

Basis set for the diatomic molecule

This chapter uses the Hund's case (a) representation, equation (7.12) of the previous chapter, as the mathematical model of the diatomic molecule. A pictorial model of Hund's case (a) is often presented, e.g., figure 97 in [1], or see Bransden and Joachain [2].

The Born–Oppenheimer approximation gives an approximate separation of the diatomic eigenfunction into electronic, vibrational, and rotational parts,

$$\Psi \approx \psi_{el}\psi_{vib}\psi_{rot}. \tag{8.1}$$

In the Wigner–Witmer diatomic eigenfunction, the coordinates ϕ, θ, and χ are exactly separated from the electronic–vibrational coordinates,

$$\begin{aligned}\langle \rho, \zeta, \chi, \mathbf{r}_2, \ldots, \mathbf{r}_N, r, \theta, \phi \,|nvJM\rangle &= \sum_{\Omega=-J}^{J} \langle \rho, \zeta, \mathbf{r}'_2, \ldots, \mathbf{r}'_N, r \,|nv\rangle\, D^{J*}_{M\Omega}(\alpha, \beta, \gamma) \\ &= \sum_{\Omega=-J}^{J} \langle \rho, \zeta, \mathbf{r}'_2, \ldots, \mathbf{r}'_N, r \,|nv\rangle)\, \langle \phi, \theta, \chi \,|JM\Omega\rangle.\end{aligned} \tag{8.2}$$

The mathematical function $\langle \rho, \zeta, \mathbf{r}_2', \ldots, \mathbf{r}_N', r \,|nv\rangle$ is the *electronic–vibrational basis function*. The mathematical function $\langle \phi, \theta, \chi \,|JM\Omega\rangle$,

$$\langle \phi, \theta, \chi \,|JM\Omega\rangle \equiv D^{J*}_{M\Omega}(\alpha, \beta, \gamma), \tag{8.3}$$

is the *angular momentum basis function*. Calculation of the diatomic eigenfunction requires that one sum the product of electronic–vibrational and angular momentum basis functions. Equation (8.2) is obviously more complicated than equation (8.1) but the latter is exact, and also more specific. The precise meaning of the word rotational is sometimes unclear. Is ψ_{rot} the angular momentum eigenfunction of the diatomic molecule? The angle χ is both an electronic and and angular momentum variable. Is χ an argument of ψ_{el} or ψ_{rot}? Similarly, is the total electronic spin $\mathbf{S}$ associated with ψ_{el} or ψ_{rot}?

One knows from first principles of angular momentum theory that for a conservative system the Hamiltonian will be a $(2J + 1) \times (2J + 1)$ matrix.

doi:10.1088/978-0-7503-6204-7ch8

Table 8.1. The Hund's case (a) matrix representation of a hypothetical diatomic Hamiltonian for $J = 2$ and $S = 1$.

$H_{-1,-1}$	$H_{-1,0}$	$H_{-1,1}$	0	0	0	0	0	0
$H_{0,-1}$	$H_{0,0}$	$H_{0,1}$	0	0	0	0	0	0
$H_{1,-1}$	$H_{1,0}$	$H_{1,1}$	0	0	0	0	0	0
0	0	0	$H_{-2,-2}$	$H_{-2,-1}$	$H_{-2,0}$	$H_{-2,0}$	$H_{-2,1}$	$H_{-2,2}$
0	0	0	$H_{-1,-2}$	$H_{-1,-1}$	$H_{-1,0}$	$H_{-1,0}$	$H_{-1,1}$	$H_{-1,2}$
0	0	0	$H_{0,-2}$	$H_{0,-1}$	$H_{0,0}$	$H_{0,0}$	$H_{0,1}$	H_{02}
0	0	0	$H_{0,-2}$	$H_{0,-1}$	$H_{0,0}$	$H_{0,0}$	$H_{0,1}$	$H_{0,2}$
0	0	0	$H_{1,-2}$	$H_{1,-1}$	$H_{1,0}$	$H_{1,0}$	$H_{1,1}$	$H_{1,2}$
0	0	0	$H_{2,-2}$	$H_{2,-1}$	$H_{2,0}$	$H_{2,0}$	$H_{2,1}$	$H_{2,2}$

Rotational invariance in isotropic space leads to conservation of the total angular momentum. The quantum numbers J, M, and Ω include spin. However, angular momentum theory does not tell us how the total is composed of its orbital and spin components. One must postulate an angular momentum coupling model that describes how the total $\mathbf{J}$ is formed from, for example, the total orbital angular momentum $\mathbf{N}$ and the total electronic spin $\mathbf{S}$, and, as described in chapter 7, construct a mathematically complete basis for the $|JM\rangle$ eigenstates of the total angular momentum. If the total electronic spin is to a useful degree of approximation a constant of the motion (i.e., if S is a so-called 'good quantum number'), then the the multiplicity $2S + 1$ expands the dimensions of the Hamiltonian to $(2S + 1)(2J + 1) \times (2S + 1)(2J + 1)$. For the commonly occurring values $J = 20$ and $S = 1$, one is contemplating 123×123 matrices. The smallness of the mass of the electron comes to the rescue, and not in exactly the same ways it does in the Born–Oppenheimer approximation. Table 8.1 illustrates a hypothetical Hamiltonian for $J = 2$ and $S = 1$.

The angle χ describes rotation of one of the electrons about the internuclear vector at the distance ρ. The Hamiltonian will contain the kinetic energy term

$$T_\chi = -\sum_{\Omega=-J}^{J} \frac{\hbar^2}{2\, m_{\mathrm{e}}\, \rho^2} \frac{\partial^2 \Psi_{nvJ\Omega}}{\partial \chi^2} = \sum_{\Omega=-J}^{J} \frac{\hbar^2 \Omega^2}{2\, m_{\mathrm{e}}\, \rho^2}. \tag{8.4}$$

This result holds if rotation of the nuclei about the internuclear axis is ignored.[1]

[1] The term $m_e \rho^2$ in equation (8.4) is the moment of inertia of an electron about the internuclear axis. One might falsely assume that the moment of inertia of a nucleus about the internuclear axis is much larger than that of an electron because nuclear masses are roughly $10^3 - 10^5$ larger that the electronic mass m_e. Actually the small size of a nucleus overwhelms its large mass, and the moment of inertia of a nucleus about the internuclear axis is much smaller than the electron's. The energy of a state of nuclear rotation about the internuclear axis would be very large, so large in fact that attempting to excite it would destroy the molecule. Thus, rotation of the nuclei about the internuclear axis can be ignored in an existing molecule.

In the Born–Oppenheimer approximation, the smallness of electronic mass has the nuclei moving much more rapidly than the nuclei. This consideration from classical mechanics leads to the approximate separation of the electronic eigenfunction ψ_e. In the context of the Wigner–Witmer diatomic eigenfunction, the smallness of electronic mass adds a large number of large kinetic energy terms to the Hamiltonian. These large kinetic energy terms come in pairs, the first associated with the basis function $D^{J*}_{M\Omega}(\alpha, \beta, \gamma)$ and the second with $D^{J*}_{M, -\Omega}(\alpha, \beta, \gamma)$, and from equation (8.4) one sees that the $\langle\phi, \theta, \chi\ |JM\Omega\rangle$ and $\langle\phi, \theta, \chi\ |JM, -\Omega\rangle$ basis states contribute equal amounts of kinetic energy to the Hamiltonian. As a first degree of approximation, we assume that $\langle\phi, \theta, \chi\ |JM\Omega\rangle$ and $\langle\phi, \theta, \chi\ |JM, -\Omega\rangle$ make equal contributions. This turns out be be a close approximation.

References

[1] Herzberg G 1950 Molecular Spectra and Molecular Structure I *Spectra of Diatomic Molecules* 2nd edn (New York: Van Nostrand Reinhold)

[2] Bransden B H and Joachain C J 2003 *Physics of Atom and Molecules* 2nd edn (Harlow: Prentice-Hall)

IOP Publishing

Quantum Mechanics of the Diatomic Molecule (Second Edition)

Christian G Parigger and James O Hornkohl

Chapter 9

Angular momentum states of diatomic molecules

9.1 Introduction

The interpretation of atomic and diatomic spectra were early challenges in the development of quantum mechanics and led to the quantum theory of angular momentum. The isotropy of space, rotational invariance, and conservation of the total angular momentum are such general concepts that angular momentum is now omnipresent in physical theory. Angular momentum is treated in quantum mechanics textbooks (e.g., [1–6], and there are several monographs on the topic, i.e., [7–13]). The compendium of [14] is a valuable reference.

In this chapter, and throughout this book, a very strict adherence to the mathematical definition of angular momentum is maintained. The mathematical nature of the theory of quantum mechanics can be viewed as a curse or as a valuable tool for recognizing spurious results produced by classical concepts and associated classical intuition. For example, the rotation matrix element $D^{J}_{MM'}(\alpha, \beta, \gamma)$ and its complex conjugate $D^{J*}_{MM'}(\alpha, \beta, \gamma)$ are functions of only angular variables, carry only angular momentum quantum numbers, and are eigenfunctions of a differential equation for the angles. Therefore, the rotation matrix element may be viewed as an angular momentum eigenfunction. The rotation matrix element, written in shorthand as $|JM\Omega\rangle$, is obviously not the eigenstate $|JM\rangle$ or $|J\Omega\rangle$, or, in other words, $|JM\Omega\rangle$ is not an eigenstate in a standard description of angular momentum. In the standard description, $\mathbf{J}$ and only one component of $\mathbf{J}$, conventionally chosen to the J_z, describes a quantum mechanical angular momentum state, either $|JM\rangle$ or $|J\Omega\rangle$. In terms of quantum mechanics operators, $\mathbf{J}^2$ and only one component commute, and hence can be measured simultaneously. The mathematical entity $|J\Omega\rangle$ carries two magnetic quantum numbers, and therefore does not satisfy the mathematical properties of angular momentum.

The concise distinctions discussed in the previous paragraph are actually essential for diatomic spectroscopy. The formulas for operation of raising J_+ and lowering J_- operators on standard $|JM\rangle$ states are rotationally invariant, like

doi:10.1088/978-0-7503-6204-7ch9

Schrödinger's equation and several other equations encountered in quantum physics,

$$J_{\pm}|JM\rangle = \sqrt{J(J+1) - M(M \pm 1)}\; |J, M \pm 1\rangle, \tag{9.1a}$$

$$J'_{\pm}|J\Omega\rangle = \sqrt{J(J+1) - \Omega(\Omega \pm 1)}\; |J, \Omega \pm 1\rangle, \tag{9.1b}$$

$$\vdots$$

$$J''_{\pm}|JM''\rangle = \sqrt{J(J+1) - M''(M'' \pm 1)}\; |J, M'' \pm 1\rangle. \tag{9.1c}$$

However, what is the origin of the formula of applying angular momentum raising and lowering operators,

$$J'_{\pm} D^{J*}_{M\Omega}(\phi, \theta, \chi) = \sqrt{J(J+1) - \Omega(\Omega \mp 1)}\; D^{J*}_{M,\, \Omega\mp 1}(\phi, \theta, \chi), \tag{9.2}$$

frequently applied in diatomic spectroscopy? Equations (9.1) hold for angular momentum states. Equation (9.2), which appears to have a functionality as equation (9.1b), can be correctly viewed as mathematical evidence that the rotation matrix element does not represent an angular momentum state. However, $J'_{\pm}$ operating on $|JM\Omega\rangle = D^{J*}_{M,\,\Omega}(\phi, \theta, \chi)$ does not equal $J'_{\pm}$ operating on $|J\Omega\rangle$ because $|JM\Omega\rangle$ cannot be identified with $|J\Omega\rangle$. A sum of $|JM\Omega\rangle$ basis functions, i.e., rotation matrix elements, is required to relate representations of the $|JM\rangle$ and $|J\Omega\rangle$ angular momentum states. In view of diatomic spectroscopy, equation (9.2) is not a mysterious agreed upon convention [12, 15, 16], but is rather a mathematical result.

9.2 The standard $|JM\rangle$ angular momentum representation

In addition to equations (9.1) in which

$$J_{\pm} = J_x \pm i\, J_y, \tag{9.3}$$

the standard representation provides

$$\mathbf{J}^2|JM\rangle = J(J+1)\,|JM\rangle \tag{9.4}$$

and

$$J_z|JM\rangle = M\,|JM\rangle, \tag{9.5}$$

there are $2J + 1$ values of M,

$$M = -J, -J+1, -J+1, \ldots, J-1, J, \tag{9.6}$$

satisfying equations (9.4)and (9.5). An energy eigenstate $|nJM\rangle$ with n representing quantum numbers required in addition to J and M has a degeneracy of $2J + 1$.

The above results are traditionally derived from the angular momentum commutators,

$$L_x L_y - L_y L_x = i\, L_z \tag{9.7a}$$

$$L_y L_z - L_z L_y = i\, L_x, \tag{9.7b}$$

$$L_z L_x - L_x L_z = i\, L_y. \tag{9.7c}$$

which follow from the classical formula for orbital angular momentum **L** in terms of the linear momentum **p** and distance **r**,

$$\mathbf{L} = \mathbf{r}\, \mathbf{x}\, \mathbf{p}, \tag{9.8}$$

and the quantum mechanical linear momentum operators,

$$p_x = \frac{\hbar}{i}\frac{\partial}{\partial x}, \tag{9.9a}$$

$$p_y = \frac{\hbar}{i}\frac{\partial}{\partial y}, \tag{9.9b}$$

$$p_z = \frac{\hbar}{i}\frac{\partial}{\partial z}. \tag{9.9c}$$

Spin cannot be expressed in terms of spatial coordinates but is nevertheless identifiable as angular momentum. With the sum **J** of orbital plus spin momenta replacing **L**, the commutators are often taken to be the quantum mechanical definition of angular momentum,

$$J_x J_y - J_y J_x = i\, J_z, \tag{9.10a}$$

$$J_y J_z - J_z J_y = i\, J_x, \tag{9.10b}$$

$$J_z J_x - J_x J_z = i\, J_y. \tag{9.10c}$$

The operator J_x commutes with its square J_x^2, and **J** therefore commutes with $\mathbf{J}^2$. Although the nonvanishing commutators (9.10) dictate that J_x, J_y, and J_z cannot all be members of the complete set of commuting observables (CSCO), $\mathbf{J}^2$ and one component of **J**, conventionally chosen to be J_z, can be constants of the motion. Again, a mathematical object indicating that two or more components of **J** are constants of motion is not an eigenstate of angular momentum.

9.3 Rotations

The topic of angular momentum is of course closely related to the topic of rotations; in other words, angular momentum can be utilized as generators of rotations. There are two distinct types of rotation, i.e., coordinate rotation and mechanical rotation. Invariance is the concept that one can change the details of the mathematical description of a physical state without altering the defining mathematical description of that state. Conservation is the concept that the numerical values of certain variables, e.g., energy, linear momentum, and angular momentum, are constant during mechanical motion. Invariance with respect to transformations implies conservation laws. The homogeneity of time leads to conservation of energy, the homogeneity of space leads to conservation of linear momentum, and the isotropy of space leads to conservation of angular momentum. The concept of angular momentum states of the diatomic molecule is closely related to rotational invariance of the equations that describe the molecule.

Angular momentum $\mathbf{J}(x, y, z)$ is the generator of coordinate rotations α, β, and γ,

$$\mathcal{R}(\alpha, \beta, \gamma) = e^{-i\alpha J_z} e^{-i\beta J_y} e^{-i\gamma J_z}. \tag{9.11}$$

The Hermitian operators J_x, J_y, and J_z in the exponents generate unitary symmetry operators $\mathcal{R}(\alpha)$, $\mathcal{R}(\beta)$, and $\mathcal{R}(\gamma)$. The exponential function that contains the imaginary unit and the angles are as usual interpreted formally as a sum. As shown in quantum mechanics textbooks, for operators A and B,

$$e^A e^B \neq e^{A+B} \tag{9.12}$$

which means that the right-hand side of equation (9.11) cannot be written as a single exponential. The standard representations of $|JM\rangle$ and $|J\Omega\rangle$ can be related,

$$\langle \mathbf{r}_1, \mathbf{r}_2, \ldots, \mathbf{r}_N \,|nJM\rangle = \sum_{\Omega=-J}^{J} \langle \mathbf{r}_1, \mathbf{r}_2, \ldots, \mathbf{r}_N \,|nJ\Omega\rangle \,\langle J\Omega\,|JM\rangle, \tag{9.13a}$$

with n denoting other quantum numbers and kept in the equations as needed. Unity in the form $\mathcal{R}(\alpha, \beta, \gamma)\,\mathcal{R}(\alpha, \beta, \gamma)^{-1} = \mathcal{R}(\alpha, \beta, \gamma)\,\mathcal{R}^{\dagger}(\alpha, \beta, \gamma)$ can be applied to both sides,

$$\begin{aligned}
&\langle \mathbf{r}_1, \mathbf{r}_2, \ldots, \mathbf{r}_N \,|nJM\rangle \\
&= \sum_{\Omega=-J}^{J} \langle \mathbf{r}_1, \mathbf{r}_2, \ldots, \mathbf{r}_N \,|\mathcal{R}(\alpha, \beta, \gamma)\,|nJ\Omega\rangle \,\langle J\Omega\,|\mathcal{R}^{\dagger}(\alpha, \beta, \gamma)|JM\rangle \\
&= \sum_{\Omega=-J}^{J} \langle \mathbf{r}'_1, \mathbf{r}'_2, \ldots, \mathbf{r}'_N \,|nJ\Omega\rangle \,\langle JM\,|\mathcal{R}(\alpha, \beta, \gamma)|J\Omega > * \\
&= \sum_{\Omega=-J}^{J} \langle \mathbf{r}'_1, \mathbf{r}'_2, \ldots, \mathbf{r}'_N \,|nJ\Omega\rangle \, D^{J*}_{M\Omega}(\phi, \theta, \chi).
\end{aligned} \tag{9.13b}$$

The previously discussed equations (9.10) are traditional definitions of angular momentum [17]). However, the comparison of the generators of rotations in spherical coordinates,

$$J_\phi = \frac{\hbar}{i}\frac{\partial}{\partial\phi} \tag{9.14a}$$

$$J_\theta = \frac{\hbar}{i}\frac{\partial}{\partial\theta} \tag{9.14b}$$

$$J_\chi = \frac{\hbar}{i}\frac{\partial}{\partial \chi} \tag{9.14c}$$

and of equations (9.9) that describe linear motion indicate that equations (9.14) indeed describe angular motion. It may be preferred to discuss angular momentum utilizing its generators because equations (9.14) appear to associate angular transformation rather than the abstract commutators (9.10).

As one reviews equations (9.14), care must be exercised in the interpretations. Whereas the three commutator equations (9.10) are written in a single coordinate system, each of the three equations (9.14) is written in a different coordinate system. The rotation ϕ is about the z-axis in an xyz system of coordinates. The second rotation θ is about the first intermediate y_1-axis in an Euler rotation of coordinates (see below). The third rotation is about the final z'-axis in the full Euler rotation of coordinates. The question then arises of whether one should work with angular momentum $\mathbf{J}(x, y, z)$ in a single coordinate system or with $\mathbf{J}(\phi, \theta, \chi)$ that equally reflects the nature of angular momentum?

9.4 Generators of coordinate transformations

The total energy operator H (i.e., the Hamiltonian), the total linear momentum operator $\mathbf{P}$, and the total angular momentum operator $\mathbf{J}$ are the generators of translations in time, translations in space, and rotations in space (see chapter 5). The unitary time translation symmetry operator $\mathcal{U}(t_0)$ gives the time coordinate a new origin, the unitary spatial translation symmetry operator $\mathcal{T}(\mathbf{R}_0)$ gives Cartesian spatial coordinates a new origin, and the unitary rotation symmetry operator $\mathcal{R}(\alpha, \beta, \gamma)$ gives angular coordinates a new origin. In a conservative system, one can always equate the time origin to some physical event. The inverse time translation operator becomes the evolution operator describing the temporal evolution of the system, and the time variable is separated. Invariance during time translation becomes the conservation of energy during temporal changes in the system. One can also always reference the new Cartesian coordinate origin to the electrons and nuclei composing the system, and the coordinates of the center of total mass are separated. The inverse of the spatial translation operator becomes the operator describing the total linear motion of the system. Invariance during spatial translation becomes the conservation of linear momentum during linear physical motion.

This scheme of using a unitary symmetry operator to separate variables and describe mechanical motion is not applicable for the rotation operator $\mathcal{R}(\alpha, \beta, \gamma)$. The total angular momentum $\mathbf{J}$ can always be defined in terms of the mechanical rotations ϕ, θ, and χ of equations (9.14), and that equations (9.1), (9.4), (9.5), and (9.10) all can be mathematically derived from $\mathbf{J}(\phi, \theta, \chi)$, but almost without exception no real physical systems executes the internal mechanical rotations ϕ, θ, and χ. The total angular momentum is always a constant of the motion for a closed system. Almost never can the total angular momentum momentum for such a system be defined in terms of three exactly separable variables. As shown in the previous chapter, the Wigner–Witmer diatomic eigenfunction is merely the very general equation for coordinate (passive) rotations written in terms of the internal variables of the diatomic molecule.

The geometry of the diatomic molecule allows one to equate the Euler angles of coordinate rotation to the Euler angles of mechanical rotation, thereby removing all angular dependence from the electronic–vibrational eigenfunction $\langle \rho, \zeta, \mathbf{r}'_2, \ldots, \mathbf{r}'_N, r \,|nv\rangle$, and the angles of mechanical rotation are exactly separated. The general equation (4.8) holds only when J, M, and Ω refer to the total angular momentum. Thus, the same quantum numbers in the Wigner–Witmer eigenfunction (4.12) also include all orbital and spin momenta.

References

[1] Schiff L I 1968 *Quantum Mechanics* 3rd edn (London: McGraw-Hill)

[2] Messiah A 1964 *Quantum Mechanics* (North Holland: Amsterdam)

[3] Landau L D and Lifshitz E M 1977 *Quantum Mechanics* 3rd edn (Amsterdam: Butterworth-Heinemann)

[4] Gottfried K 1998 and Ting-Mow Yan *Quantum Mechanics: Fundamentals* 2nd edn (New York: Springer)

[5] Zettili N 2009 *Quantum Mechanics* 2nd edn (West Sussex: Wiley)

[6] Weinberg S 2015 *Lectures on Quantum Mechanics* 2nd edn (Cambridge: Cambridge University Press)

[7] Rose M E 1995 *Elementary Theory of Angular Momentum* (Mineola, NY: Dover)

[8] Edmonds A R 1974 *Angular Momentum in Quantum Mechanics* 2nd edn (Princeton, NJ: Princeton University Press)

[9] Brink D M and Satchler G R 1993 *Angular Momentum* 3rd edn (Oxford: Oxford University Press)

[10] Biedenharn L C and Louck J D 2009 *Angular Momentum in Quantum Physics* (Cambridge: Cambridge University Press)

[11] Thompson W J 1994 *Angular Momentum* (New York: Wiley)

[12] Zare R N 1988 *Angular Momentum* (New York: Wiley)

[13] Chaichain M and Hagedorn R 1998 *Symmetries in Quantum Mechanics* (New York: Taylor and Francis)

[14] Varshalovich D A, Moskalev A N and Khersonskii V K 1988 *Quantum Theory of Angular Momentum* (Singapore: World Scientific)

[15] Brown J M and Carrington A 2003 *Rotational Spectroscopy of Diatomic Molecules* (Cambridge: Cambridge University Press)

[16] Lefebvre-Brion H and Field R W 2004 *The Spectra and Dynamics of Diatomic Molecules* (New York: Elsevier/Academic)

[17] Born M, Heisenberg W and Jordan P 1926 *Z. Phys.* **35** 557
Sources of Quantum Mechanics 1967 ed B L Van Der Waerden p 321 (New York: Dover)

IOP Publishing

Quantum Mechanics of the Diatomic Molecule (Second Edition)

Christian G Parigger and James O Hornkohl

Chapter 10

Diatomic parity

How does the parity eigenvalue $p = \pm 1$ vary with the total angular momentum quantum number J? This is a general quantum mechanics question, not a specifically diatomic question. The parity eigenvalue splits into two parts, an angular momentum dependent part and a constant called the intrinsic parity. The following gives the first identification of the intrinsic parity of the diatomic molecule. A Hund basis state for which $\Lambda = 0$ is called a Σ state. The parity operator is the product of an improper rotation that inverts the sign of a single Cartesian component of a coordinate vector and proper rotation that inverts the signs of two Cartesian components. Current diatomic theory places the eigenvalue p_Σ of the one-component parity operator as a superscript on the symbol Σ denoting the $\Lambda = 0$ basis, but currently the one-component eigenvalue is said to have relevance to $\Lambda \neq 0$ basis states. It is seen below that p_Σ is the intrinsic parity of the diatomic molecule, a global value applying to all basis states of the molecule.

The Born–Oppenheimer approximation does not allow one to definitively determine the diatomic parity eigenvalue. The exact separation of the total angular momentum basis in the Wigner–Witmer diatomic eigenfunction is required for exact determination of how the diatomic parity eigenvalue depends on the total angular momentum quantum number J.

The application of diatomic parity to the general diatomic eigenfunction is summarized in recent papers [1, 2], and in appendices E and F. Intrinsic parity is further elaborated in appendix G.

10.1 Parity details

Before showing how the parity of diatomic states follows in a straightforward way from the Wigner–Witmer diatomic eigenfunction, some methods not to be used in the treatment diatomic parity will be itemized.

10.1.1 Parity is rotationally invariant

The parity operator $\mathcal{P}$ changes the signs of Cartesian coordinates,

$$\mathcal{P}\Psi(x, y, z) = \Psi(-x, -y, -z) \tag{10.1}$$

Despite the simple matrix relationship

$$\begin{bmatrix} -1 & 0 & 0 \\ 0 & -1 & 0 \\ 0 & 0 & -1 \end{bmatrix} \begin{bmatrix} x \\ y \\ z \end{bmatrix} = \begin{bmatrix} -x \\ -y \\ -z \end{bmatrix}, \tag{10.2}$$

and the relationship between rotated and fixed coordinates in terms of the direction cosines,

$$\begin{bmatrix} x' \\ y' \\ z' \end{bmatrix} = \begin{bmatrix} c_{11} & c_{12} & c_{13} \\ c_{21} & c_{22} & c_{23} \\ c_{31} & c_{32} & c_{33} \end{bmatrix} \begin{bmatrix} x \\ y \\ z \end{bmatrix}. \tag{10.3}$$

the claim is made that, '*It is evidently necessary to distinguish clearly between the "molecule fixed" and the "laboratory fixed" inversion operation I, since these two operations are not equivalent*' [3]. Equations (10.2) and (10.3) are linear transformations. Changing the signs of x, y, and z will obviously change the signs of x', y', and z', and vice versa. Parity is a member of the complete set of commuting observables. The parity operator $\mathcal{P}$, like the other operators of the complete set of commuting observables, is rotationally invariant. Nevertheless, the idea that inverting signs of fixed coordinates will not invert signs of rotated coordinates has become a standard part of diatomic theory. The reason for this false claim is fairly easy to discern. The discrete inversion of Cartesian signs does not reveal how the parity eigenvalue p varies with the the total angular momentum quantum number J. As mentioned elsewhere in this book, angular momentum is angular. How the parity eigenvalue varies with J is not determined by a discrete transformation of Cartesian coordinates but by a discrete transformation of angular coordinates, as shown below.

10.1.2 Spin is immune to the parity operator

The Hund's basis contains the spin ket $|S\Sigma\rangle$ referenced to the rotated coordinate system in which the z'-axis follows the internuclear vector $\mathbf{r}(r, \theta, \phi)$. The spin ket $|S\Sigma\rangle$ is locked to the internuclear axis. Under coordinate rotation the spin kets $|SM_S\rangle$ and $|S\Sigma\rangle$ are related by

$$|SM_S\rangle = \sum_{\Sigma=-S}^{S} |S\Sigma\rangle\, D^{J*}_{MM'}(\phi, \theta, \chi) \tag{10.4}$$

Spin $\mathbf{S}$ is not expressible in terms of spatial coordinates. The parity operator $\mathcal{P}$ has no defining characteristic other than to alter spatial coordinates. The intrinsic parity, that part of the parity eigenvalue not dependent on J, exists independently of spatial coordinates, but the parity operator does not. Spin is immune to the parity operator. Equation [3–7]

$$\mathcal{P}\,|S\Sigma\rangle = (-)^{S-\Sigma}|S, -\Sigma\rangle \tag{10.5}$$

lacks meaning because $|S\Sigma\rangle$ is totally immune to $\mathcal{P}$.

10.1.3 Parity operates on Cartesian coordinates, not angles

The mistake shown in equation (10.5) is the result of thinking that the parity operator can act on Euler angles. The discrete parity transformation acts on Cartesian coordinates. A discrete transformation of the Euler angles can invert the signs of two of the Cartesian coordinates. Thus, the parity operator can be constructed from a discrete transformation of the Euler angles and a one-component parity operator.

10.1.4 Intrinsic parity and Λ doublets

The symbol Σ denotes a diatomic state for which the internuclear component of the total orbital angular momentum **N** is zero (i.e. when the quantum number Λ is zero). A trailing superscript sign is attached, Σ^+ and Σ^-, to denote the eigenvalue p_Σ of a one-dimensional parity operator that inverts the signs of one Cartesian component of the rotated coordinate vectors. The operator $\sigma_v(y'z')$ is used below to invert the signs of the x' components. The eigenvalue p_Σ is the intrinsic parity of the diatomic molecule, which is a global value. Every state of a given molecule has the intrinsic parity p_Σ, i.e., Π states are either Π^+ or Π^-, Δ states are either Δ^+ or Δ^-, and so forth. Every diatomic state for which the internuclear component of the total orbital angular momentum is nonzero is 'Λ doubled'. What this means is that when $\Lambda \neq 0$ each diatomic state comes in two versions, the positive parity version and the negative parity version. This follows directly from the Wigner–Witmer eigenfunction. In current diatomic theory, the value of p_Σ, a value fixed for every state in a given molecule, is considered not applicable when $\Lambda \neq 0$.

10.1.5 Summary of parity details

How the parity eigenvalue $p = \pm 1$ varies with J will be derived below. In this derivation, inversion of the signs of fixed coordinates will be accomplished by inverting the signs of the rotated coordinates, the parity operator will not operate on spin, the parity operator will not operate on the Euler angles, inversion of the signs of two rotated Cartesian coordinates will be accomplished with a discrete transformation of the Euler angles, and intrinsic parity and Λ doublets are treated as separate topics.

10.2 Parity designation

In current practice a diatomic line list typically includes the vacuum wavenumber $\tilde{\nu}$, the upper J' and lower J quantum numbers, and the upper and lower *e*/*f* parity designations. The upper and lower parities are input data to currently used line position fitting programs. In this book, diatomic parity is an optional program output, a computed quantity, not required input data. The same eigenvectors that diagonalize the Hamiltonian matrix will also diagonalize the parity matrix. Computation of the latter requires the value p_{Σ}, the intrinsic parity of the diatomic molecule, and if the molecule is homonuclear it requires the value of gerade/ungerade p_{gu} symmetry of the molecule.

The conversion of parity from required program input data to an optional program output is a major departure from current practice.

10.3 The parity operator

A mathematical function like $\cos(x)$ is called even because $\cos(-x) = \cos(x)$ and another like the $\sin(x)$ is called odd because $\sin(-x) = -\sin(x)$. A function like $f(x) = \cos(x) + \sin(x)$ is neither even nor odd, but its even and odd parts can be found from

$$F(x)_{\text{even}} = \frac{1}{2}[f(x) + f(-x)], \tag{10.6a}$$

$$F(x)_{\text{odd}} = \frac{1}{2}[f(x) - f(-x)]. \tag{10.6b}$$

The parity of a quantum mechanical state is the evenness or oddness of its eigenfunction. Equation (4.16) for the Wigner–Eckart diatomic eigenfunction uses fixed coordinates on the left-hand side and rotated coordinates on the right-hand side. The following is concerned with construction of a parity operator that can be applied to either side of the equation. We will find

$$\mathcal{P}_{\text{left}} = \sigma_v(yz)C_2(x) \tag{10.7a}$$

$$\mathcal{P}_{\text{right}} = \sigma_v(y'z')C_2(x'). \tag{10.7b}$$

Note that the above equations are rotationally invariant (i.e., the same equation).

The parity operator $\mathcal{P}$ inverts the signs of each component of all Cartesian coordinates in the Hamiltonian and its eigenfunction. For example, $\mathcal{P}$ can be defined by its operation on the generic eigenfunction $\langle \mathbf{r}_1, \mathbf{r}_2, \ldots, \mathbf{r}_{\mathcal{N}} | nJM \rangle$,

$$\langle \mathbf{r}_1, \mathbf{r}_2, \ldots, \mathbf{r}_{\mathcal{N}} | \mathcal{P} | nJM \rangle \equiv \langle -\mathbf{r}_1, -\mathbf{r}_2, \ldots, -\mathbf{r}_{\mathcal{N}} | nJM \rangle \tag{10.8}$$

in which each spatial coordinate vector $\mathbf{r}_i(x_i, y_i, z_i)$ is represented by its Cartesian components. The parity operation changes the handedness of the coordinate system, but no continuous mathematical procedure can change the handedness of a coordinate system. Mechanical motion cannot accomplish sign inversion for all coordinates. Perhaps out of a concern that the term *discontinuous operator* is too harsh for inclusion in a legitimate physical theory, most call $\mathcal{P}$ a *discrete operator*. A discrete operator has the unusual feature of being both unity,

$$\mathcal{P}^{-1} = \mathcal{P}^{\dagger} \tag{10.9}$$

$$\mathcal{P}^{\dagger}\mathcal{P} = 1 \tag{10.10}$$

and Hermitian with an eigenvalue of ±1 [8–11],

$$\langle \mathbf{r}_1, \mathbf{r}_2, \dots, \mathbf{r}_{\mathcal{N}} | \mathcal{P} | nJM \rangle = p \langle \mathbf{r}_1, \mathbf{r}_2, \dots, \mathbf{r}_{\mathcal{N}} | nJM \rangle \tag{10.11a}$$

$$= \pm \langle \mathbf{r}_1, \mathbf{r}_2, \dots, \mathbf{r}_{\mathcal{N}} | nJM \rangle. \tag{10.11b}$$

The determinant of the $\mathcal{P}$ matrix is -1, which classifies $\mathcal{P}$ as an improper rotation, i.e., a rotation that mechanical rotation and coordinate rotation cannot accomplish. We will see below that coordinate rotation can invert the signs of two of the three components of $\mathbf{r}(x, y, z)$. In addition to being unitary and Hermitian, the parity operator is its own inverse,

$$\mathcal{P}^{-1} = \mathcal{P}. \tag{10.12}$$

The 3 × 3 matrix representation of $\mathcal{P}$ in equation (10.2) is easily seen to be Hermitian, unitary, and its own inverse.

There are one-dimensional parity operators, for example

$$\sigma_v(yz)\begin{bmatrix} x \\ y \\ z \end{bmatrix} = \begin{bmatrix} -1 & 0 & 0 \\ 0 & 1 & 0 \\ 0 & 0 & 1 \end{bmatrix}\begin{bmatrix} x \\ y \\ z \end{bmatrix} = \begin{bmatrix} -x \\ y \\ z \end{bmatrix}, \tag{10.13}$$

which is said to be 'reflection of the x-axis through the yz-plane.' Obfuscation which hides the simple mathematical facts that $\sigma_v(yz)$ inverts the sign on x but not on y and z. The determinant of the $\sigma_v(yz)$ matrix is clearly -1 marking it as a parity operator or improper rotation. There are also similar two-dimensional symmetry operations, for example

$$C_2(x)\begin{bmatrix} x \\ y \\ z \end{bmatrix} = \begin{bmatrix} 1 & 0 & 0 \\ 0 & -1 & 0 \\ 0 & 0 & -1 \end{bmatrix}\begin{bmatrix} x \\ y \\ z \end{bmatrix} = \begin{bmatrix} x \\ -y \\ -z \end{bmatrix}, \tag{10.14}$$

but here the determinant of the $C_2(x)$ matrix is $+1$, a proper rotation and not a parity operation. The product $\sigma_v(yz)$ and $C_2(x)$ matrices yields the $\mathcal{P}$ matrix of equation (10.2), and the operator $\mathcal{P}$ can be written as the product of operators,

$$\mathcal{P} = \sigma_v(yz)\, C_2(x). \tag{10.15}$$

As shown by, for example [12], parity is multiplicative. Angular momentum operators and eigenvalues are added, parity operators and eigenvalues are multiplied. The product $\sigma_v(yz)C_2(x)$ is appropriately applied to an eigenfunction depending on unprimed coordinates while $\sigma_v(y'z')\, C_2(x')$ is applied to an eigenfunction depending on rotated coordinates.

One can verify by direct substitution that the discrete transformations of the Euler angles

$$\alpha \rightarrow \pi + \alpha \tag{10.16a}$$

$$\beta \rightarrow \pi - \beta \tag{10.16b}$$

$$\gamma \rightarrow -\gamma \tag{10.16c}$$

change the signs of columns 2 and 3 in the 3×3 matrices in equations (10.3). With these discrete transformations of the Euler angles, the parity operator becomes

$$\mathcal{P} = \sigma_v(yz)\, C_2(\pi + \alpha,\, \pi - \beta,\, -\gamma). \tag{10.17}$$

An exact formula for diatomic parity eigenvalues will be obtained below by applying the product $\sigma_v(yz)\, C_2(x)$ to the Wigner–Witmer eigenfunction.

10.4 Parity and angular momentum

Each wave in the spherical wave expansion of the electromagnetic wave carries specific values of parity and angular momentum [9, 13, 14]. Calculation of the transition moments of each multipole in the multipole expansion of the electromagnetic field yields selection rules for angular momentum and parity. For electromagnetic transitions, each angular momentum selection rule is paralleled by a parity selection rule. There is another connection between parity and angular momentum. The parity eigenvalue always separates into two parts, a constant part and an angular momentum dependent part. Before applying the parity operator to the Wigner–Witmer eigenfunction, let us get a better understanding of why the parity eigenvalue splits into two parts by first applying $\mathcal{P}$ to the very general equation

$$\langle \mathbf{r}_1, \mathbf{r}_2, \ldots, \mathbf{r}_{\mathcal{N}} | \mathcal{R}(\alpha, \beta, \gamma) | nJM \rangle = \sum_{M'=-J}^{J} \langle \mathbf{r}'_1, \mathbf{r}'_2, \ldots, \mathbf{r}'_{\mathcal{N}} | nJM' \rangle\, D^{J*}_{MM'}(\alpha, \beta, \gamma). \tag{10.18}$$

This equation essentially defines the Wigner D-function for coordinate (passive) rotations (see, e.g., equations (4.1.1)–(4.1.3) of [15]). The above equation describes the influence of coordinate rotation, not mechanical rotation, on an eigenfunction and this explains the appearance of the complex conjugate of $D^{J}_{MM'}(\alpha, \beta, \gamma)$

When one forms the product of the parity operator $\mathcal{P}$ and rotation operator $\mathcal{R}(\alpha, \beta, \gamma)$, one has built an operator whose effect will depend on both parity and angular momentum. The angular momentum dependence comes because the components of angular momentum $\mathbf{J}$ are the generators of the rotations α, β, and γ [see, e.g., [11, 12, 16–20]].

Application of $\mathcal{P}$ to the left-hand side of the general equation (10.18) and application $\sigma_v(y'z')\, C_2(\pi + \alpha,\, \pi - \beta,\, -\gamma)$ to the right-hand side yields

$$
\begin{aligned}
&\langle \mathbf{r}_1, \mathbf{r}_2, \dots, \mathbf{r}_{\mathcal{N}} | \mathcal{P} | nJM \rangle \\
&= \sum_{M'=-J}^{J} \langle \mathbf{r}'_1, \mathbf{r}'_2, \dots, \mathbf{r}'_{\mathcal{N}} | \sigma_v(y', z') | JM' \rangle \\
&\times D^{J*}_{MM'}(\pi + \alpha, \pi - \beta, -\gamma) \\
&= p_\sigma (-)^{J+2M} \sum_{M'=-J}^{J} \langle -x'_1, -y'_1, -z'_1, -x'_2, -y'_2, -z'_2, \\
&\dots, -x'_{\mathcal{N}}, -y'_{\mathcal{N}}, -z'_{\mathcal{N}} | nJM \rangle \, D^{J*}_{M,-M'}(\alpha, \beta, \gamma) \\
&= p \sum_{M'=-J}^{J} \langle \mathbf{r}'_1, \mathbf{r}'_2, \dots, \mathbf{r}'_{\mathcal{N}} | JM' \rangle \, D^{J*}_{M,-M'}(\alpha, \beta, \gamma)
\end{aligned}
\tag{10.19}
$$

or

$$p = p_\sigma (-)^{J+2M}. \tag{10.20}$$

This equation gives the general result that the parity eigenvalue splits into two parts, the angular momentum dependent part $(-)^{J+2M}$ and a constant part called the intrinsic parity, but there is a flaw in the angular momentum dependent part. A general result from angular momentum theory is that the $2J + 1$ states $|J, -J\rangle$, $|J, -J + 1\rangle$, … , $|J, J - 1\rangle$, $|JJ\rangle$ all have the same parity. Parity does not depend on the magnetic quantum number M. For integer angular momenta, the factor $(-)^{2M}$ in equation (10.20) simply disappears, but for half-integer momenta the M dependence introduces a minus sign,

$$p = -p_\sigma (-)^J \qquad J \text{ half-integer}. \tag{10.21}$$

If one agrees to the convention to always subtract 1/2 from half-integer values of J, the above becomes

$$p = -p_\sigma (-)^{J-1/2} (-)^{1/2} \tag{10.22a}$$

$$= -i \, p_\sigma (-)^{J-1/2} \qquad J \text{ half-integer}, \tag{10.22b}$$

and one sees that the intrinsic parity p_σ is purely imaginary when J is half-integer. If one further extends the convention to ignore the leading i on the right-hand side of the above equation, one obtains

$$p = p_\sigma (-)^J \qquad J \text{ integer} \tag{10.23a}$$

$$= -p_\sigma (-)^{J-1/2} \qquad J \text{ half-integer} \tag{10.23b}$$

in which all terms can be regarded as real.

10.5 Diatomic parity

Operation of $\sigma_v(y'z') \, C_2(\pi + \alpha, \pi - \beta, -\gamma)$ on the right-hand side of equation (4.16) gives

$$
\begin{aligned}
&\sum_{\Omega=-J}^{J} \langle \rho,\, \zeta,\, \mathbf{r}'_2,\, \dots,\, \mathbf{r}'_N,\, r\, |\sigma_v(y'z')| nv\rangle\, D^{J*}_{M\Omega}(\pi+\phi,\, \pi-\theta,\, -\chi) \\
&= \sum_{\Omega=-J}^{J} \langle \rho,\, -\zeta,\, -x'_2,\, -y'_2,\, -z'_2,\, \dots,\, -x'_N,\, -y'_N,\, -z'_N,\, r\, |nv\rangle\, D^{J*}_{M,\,-\Omega}(\phi,\, \theta,\, \chi) \\
&= p_{\Sigma}\, (-)^{J+2\Omega} \sum_{\Omega=-J}^{J} \langle \rho,\, \zeta,\, \mathbf{r}'_2,\, \dots,\, \mathbf{r}'_N,\, r\, |nv\rangle\, D^{J*}_{M,\,-\Omega}(\phi,\, \theta,\, \chi) \\
&= p \sum_{\Omega=-J}^{J} \langle \rho,\, \zeta,\, \mathbf{r}'_2,\, \dots,\, \mathbf{r}'_N,\, r\, |nv\rangle\, D^{J*}_{M\Omega}(\phi,\, \theta,\, \chi).
\end{aligned}
\tag{10.24}
$$

In the last line, the minus sign on Ω in $D^{J*}_{M\Omega}(\phi,\, \theta,\, \chi)$ was dropped because changing this sign merely reverses the order in which the sum over Ω is formed. Comparison of the last two lines gives

$$p = p_{\Sigma}\, (-)^{J+2\Omega} \tag{10.25}$$

in which p_{Σ} is the intrinsic parity of the diatomic molecule. The conventions that lead from equation (10.18) to equation (10.23) give

$$p = p_{\Sigma}(-)^{J} \qquad J \text{ integer} \tag{10.26a}$$

$$= -p_{\Sigma}(-)^{J-1/2} \qquad J \text{ half–integer} \tag{10.26b}$$

10.6 Λ doublets

The separation of ϕ, θ, and χ becomes apparent when $D^{J*}_{M\Omega}(\phi,\, \theta,\, \chi)$ is broken into it parts,

$$D^{J*}_{M\Omega}(\phi,\, \theta,\, \chi) = e^{iM\phi}\, d^{J}_{M\Omega}(\beta)\, e^{i\Omega\chi} \tag{10.27}$$

We saw above that the diatomic parity operator changes the sign of χ. From equations (10.6), one sees that the sum over Ω from $-J$ to J in the Wigner–Witmer eigenfunction (4.16) guarantees that each diatomic state of a given parity has a parity twin.

References

[1] Hornkohl J O and Parigger C G 2017 *Int. J. Mol. Theor. Phys.* **1** 00103

[2] Hornkohl J O and Parigger C G 2017 *Int. J. Mol. Theor. Phys.* **1** 00102

[3] Hougen J T 1962 *J. Chem. Phys.* **36** 519

[4] Larsson M 1981 *Phys. Scr.* **23** 835

[5] Zare R N 1988 *Angular Momentum* (New York: Wiley)

[6] Lefebvre-Brion H and Field R W 2004 *The Spectra and Dynamics of Diatomic Molecules* (New York: Elsevier/Academic)

[7] Brown J M and Carrington A 2003 *Rotational Spectroscopy of Diatomic Molecules* (Cambridge: Cambridge University Press)

[8] Kemble E C 2005 *The Fundamental Principles of Quantum Mechanics* (Mineola, NY: Dover)

[9] Messiah A 1964 *Quantum Mechanics* (Amsterdam: North Holland)
[10] Cohen-Tannoudji C, Diu B and Laloë F 1977 *Quantum Mechanics* (New York: Wiley)
[11] Zettili N 2009 *Quantum Mechanics* 2nd edn (West Sussex: Wiley)
[12] Gottfried K 1998 and Ting-Mow Yan *Quantum Mechanics: Fundamentals* 2nd edn (New York: Springer)
[13] Rose M E 1995 *Elementary Theory of Angular Momentum* (Mineola, NY: Dover)
[14] Edmonds A R 1974 *Angular Momentum in Quantum Mechanics* 2nd edn (Princeton, NJ: Princeton University Press)
[15] Varshalovich D A, Moskalev A N and Khersonskii V K 1988 *Quantum Theory of Angular Momentum* (Singapore: World Scientific)
[16] Landau L D and Lifshitz E M 1977 *Quantum Mechanics* 3rd edn (Amsterdam: Butterworth-Heinemann)
[17] Tung W-K 1985 *Group theory in physics* (Singapore: World Scientific)
[18] Merzbacher E 1998 *Quantum Mechanics* 3rd edn (New York: Wiley)
[19] Chaichain M and Hagedorn R 1998 *Symmetries in Quantum Mechanics* (New York: Taylor and Francis)
[20] Weinberg S 2015 *Lectures on Quantum Mechanics* 2nd edn (Cambridge: Cambridge University Press)

IOP Publishing

Quantum Mechanics of the Diatomic Molecule (Second Edition)

Christian G Parigger and James O Hornkohl

Chapter 11

The Condon and Shortley line strength

The treatment of the diatomic molecule leads to line strengths that are utilized for the computation of molecular spectra. The *strength* of a line [1], or the line strength, is introduced in this chapter.

The geometric configuration of the charged electrons and nuclei in the molecule produces electric and magnetic moments. In a molecule governed by classical mechanics and electrodynamics, molecular motion would produce multipole radiation (and the rapid collapse of the molecule). In quantum mechanics, a multipole of charge radiates only during quantum jumps to a lower energy level and absorbs radiation only during a quantum jump to a higher level. The total Hamiltonian for the system of a molecule plus electromagnetic field is the sum of a quantized molecular Hamiltonian and a quantized electromagnetic field.

Emission and absorption of light is produced by creation and annihilation operators acting on a quantized electromagnetic field. The quantum mechanical model being used here is that of a molecule residing in the always present quantized radiation field. This model of molecule plus field is conservative. Energy and momentum gained by the molecule are lost by the field, and vice versa. The molecule and radiation field interact only during quantum jumps, and each can be treated as an independent conservative system before and after quantum jumps.

A multipole expansion can be made of the free electromagnetic field. Transition matrix elements (transition moments) of the individual terms $T_k^{(q)}$ in the multipole expansion of the electromagnetic field control the probability of radiative transitions, and the source of selections rules.

Condon and Shortley write in the preface to their book [1] '*We have defined in section 7*[4] *a quantity, called strength of a line, which we find to give a more convenient theoretical specification of the radiation intensity than either of the Einstein transition probabilities. We hope that this new usage will find favour among spectroscopists.*' In the following, the *strength* definition is stated as provided by Condon and Shortley [1].

doi:10.1088/978-0-7503-6204-7ch11

The line strength [1] is introduced by considering first the actual emitted intensity of the component from state a to state b,

$$I(a, b) = N(a)h\nu\mathbf{A}(a, b), \tag{11.1}$$

where $N(a)$ is the number of atoms in state a and $\mathbf{A}(a, b)$ is the spontaneous emission probability for that transition. This equation holds only when the radiation density present is so small that the induced emission is negligible.

The total *intensity* of a line is the sum of the intensities of its components, hence in natural excitation (sufficiently isotropic excitation) for the line from level A to B,

$$I(A, B) = N(a)h\nu\frac{64\pi^4\sigma^3}{3h}\sum_{a,b}|(a|\mathbf{P}|b)|^2, \qquad \sigma = \frac{\nu}{c} = \tilde{\nu}, \tag{11.2}$$

in the case of dipole radiation. The sum of the squared matrix components occurring here we shall define as the *strength* of the line, $\mathbf{S}(A, B)$, i.e., sum over a and b. The intensity of a particular line is therefore proportional to the number of atoms in any one of the initial states, to σ^4 and to $\mathbf{S}(A, B)$.

The intensity is written as proportional to the total number of atoms in the initial level, $N(A)$,

$$I(A, B) = N(A)h\nu\mathbf{A}(A, B). \tag{11.3}$$

As $N(A) = (2j_A + 1)N(a)$, this gives

$$\mathbf{A}(A, B) = \frac{1}{2j_A + 1}\frac{64\pi^4\sigma^3}{3h}\mathbf{S}(A, B). \tag{11.4}$$

Here, $\mathbf{A}$ is the spontaneous-transition probability of Einstein. Of course, in SI-units and conventional writing of wave-numbers, $\tilde{\nu}$, equation (11.4) reads

$$\mathbf{A}(A, B) = \frac{1}{2j_A + 1}\frac{16\pi^3\tilde{\nu}^3}{3\varepsilon_0 h}\mathbf{S}(A, B). \tag{11.5}$$

Reference

[1] Condon E U and Shortley G 1953 *The Theory of Atomic Spectra* (Cambridge: Cambridge University Press)

IOP Publishing

Quantum Mechanics of the Diatomic Molecule (Second Edition)

Christian G Parigger and James O Hornkohl

Chapter 12

Hönl–London line-strength factors in Hund's Cases (a) and (b)

12.1 Case (a) basis functions

The Hund's case (a) basis function is obtained when one drops the summation in the Wigner–Witmer eigenfunction (4.16), renormalizes the result, and inserts the spin ket $|S\Sigma\rangle$,

$$
\begin{aligned}
|a\rangle &= \langle \mathbf{r}'_1, \mathbf{r}'_2, \ldots, \mathbf{r}'_{N-1}, r, \rho, \zeta, \phi, \theta, \chi\, |nvJM\Omega\Lambda S\Sigma\rangle \\
&= \sqrt{\frac{2J+1}{8\pi^2}}\, \langle \mathbf{r}'_1, \mathbf{r}'_2, \ldots, \mathbf{r}'_{N-1}, r, \rho, \zeta\, |nv\rangle\, |S\Sigma\rangle\, D^{J*}_{M\Omega}(\phi, \theta, \chi).
\end{aligned}
\tag{12.1}
$$

If nuclear spin is ignored and one writes the total angular momentum **J** in terms of the total orbital angular momentum **N**,

$$\mathbf{N} = \mathbf{L} + \mathbf{R} \tag{12.2}$$

where **L** is the total orbital angular momentum of the electrons, **R** is the orbital angular momentum of the nuclei, and **S** is the total spin of the electrons,

$$\mathbf{J} = \mathbf{N} + \mathbf{S} \tag{12.3}$$

and then makes the inverse Clebsch–Gordan series expansion of $D^{J*}_{M\Omega}(\phi, \theta, \chi)$,

$$
\begin{aligned}
D^{J*}_{M\Omega}(\phi, \theta, \chi) = &\sum_{M_N=-N}^{N} \sum_{M_S=-S}^{S} \sum_{\Sigma=-S}^{S} \sum_{\Lambda=-N}^{N} \\
&\times \langle NM_N;\, SM_S|JM\rangle\, \langle N\Lambda;\, S\Sigma|J\Omega\rangle\, D^{N*}_{M_N\Lambda}(\phi, \theta, \chi)\, D^{S*}_{M_S\Sigma}(\phi, \theta, \chi)
\end{aligned}
\tag{12.4}
$$

then one sees that the same set of Euler angles appear in $D^{N*}_{M_N\Lambda}(\phi, \theta, \chi)$ and $D^{S*}_{M_S\Sigma}(\phi, \theta, \chi)$, indicating that the z' component of **S** (quantum number Σ) and z'

doi:10.1088/978-0-7503-6204-7ch12

component of **L** (quantum number Λ) are perfectly coupled, and the the quantum number Ω is the sum of Λ and Σ,

$$\Omega = \Lambda + \Sigma. \tag{12.5}$$

12.2 Case (b) basis functions

In Hund's case (b), the electronic spin states are independent of the electronic orbital angular momentum states. The $|JM\rangle$ states cannot be built using the inverse Clebsch–Gordan series because the spin states do not rotate in sync with the orbital rotations. The case (b) eigenfunction can be built by coupling $|NM_N\rangle$ states with $|SM_S\rangle$ states with the Clebsch–Gordan coefficient $\langle NM_N;\, SM_S\, |JM\rangle$. The $|NM_N\rangle$ states are obtained by replacing the quantum numbers J, M, and Ω in the Wigner–Witmer eigenfunction with N, M_N, and Λ, dropping the summation, and renormalizing the result. When this spin free result is coupled to the electronic spin states, one obtains $|SM_S\rangle$

$$|b\rangle = \langle \mathbf{r}'_1, \mathbf{r}'_2, \ldots, \mathbf{r}'_{N-1}, r, \rho, \zeta, \phi, \theta, \chi\, |nvJMN\Lambda S\rangle = \sqrt{\frac{2N+1}{8\pi^2}} \sum_{M_N=-N}^{N} \sum_{M_S=-S}^{S} \tag{12.6}$$
$$\times \langle \mathbf{r}'_1, \mathbf{r}'_2, \ldots, \mathbf{r}'_{N-1}, r, \rho, \zeta\, |n\rangle\, \langle NM_N;\, SM_S|JM\rangle\, |SM_S\rangle\, D^{N*}_{M_N\Lambda}(\phi, \theta, \chi).$$

12.3 Mathematical properties of case (a) and case (b) basis functions

The case (a) and case (b) basis functions are complete, orthogonal eigenfunctions. However, they do not possess the mathematical properties of quantum mechanical angular momentum states. For example, the familiar behavior of angular momentum states under operation by the raising and lowering operators,

$$J_\pm |JM\rangle = C_\pm(JM)\, |J, M \pm 1\rangle \tag{12.7}$$

$$J_\pm |J\Omega\rangle = C_\pm(J\Omega)\, |J, \Omega \pm 1\rangle \tag{12.8}$$

where

$$C_\pm(JM)\, \sqrt{J(J+1) - M(M \pm 1)} \tag{12.9}$$

is not duplicated when the raising and lowering operators are applied to rotation matrix elements.

$$J_\pm D^J_{M\Omega}(\phi, \theta, \chi) = -\sqrt{J(J+1) - M(M \mp 1)}\, D^J_{M\mp1,\, \Omega}(\phi, \theta, \chi) \tag{12.10}$$

$$J_\pm D^{J*}_{M\Omega}(\phi, \theta, \chi) = \sqrt{J(J+1) - M(M \pm 1)}\, D^{J*}_{M\pm1,\, \Omega}(\phi, \theta, \chi) \tag{12.11}$$

$$J'_\pm D^J_{M\Omega}(\phi, \theta, \chi) = \sqrt{J(J+1) - \Omega(\Omega \pm 1)}\, D^J_{M,\, \Omega\pm1}(\phi, \theta, \chi) \tag{12.12}$$

$$J'_\pm D^{J*}_{M\Omega}(\phi, \theta, \chi) = -\sqrt{J(J+1) - \Omega(\Omega \mp 1)}\, D^{J*}_{M,\, \Omega\mp1}(\phi, \theta, \chi) \tag{12.13}$$

One sees from the above that raising operators sometimes lower the magnetic quantum number, and sometimes introduce an unexpected sign change. Whereas in an angular momentum state the total and one component of the total are constants of the motion, in a rotation matrix element basis function the total and *two* of its components are constants

$$J_z\, D^{J*}_{M\Omega}(\phi,\, \theta,\, \chi) = J_z\, e^{iM\phi} d^{J}_{M\Omega}(\theta)\, e^{i\Omega\chi} \tag{12.14}$$

$$= -i\frac{\partial}{\partial\phi}\, D^{J*}_{M\Omega}(\phi,\, \theta,\, \chi) \tag{12.15}$$

$$= M\, D^{J*}_{M\Omega}(\phi,\, \theta,\, \chi) \tag{12.16}$$

$$J_{z'}\, D^{J*}_{M\Omega}(\phi,\, \theta,\, \chi) = J_{z'}\, e^{iM\phi} d^{J}_{M\Omega}(\theta)\, e^{i\Omega\chi} \tag{12.17}$$

$$= -i\frac{\partial}{\partial\chi}\, D^{J*}_{M\Omega}(\phi,\, \theta,\, \chi) \tag{12.18}$$

$$= \Omega\, D^{J*}_{M\Omega}(\phi,\, \theta,\, \chi) \tag{12.19}$$

A self-evident but mathematically rigorous explanation of why $D^{J*}_{M\Omega}(\phi,\, \theta,\, \chi)$ does not represent a state of angular momentum is that it carries two magnetic quantum numbers. As the general equation equation (4.8) and Wigner–Witmer equation (4.16) show, one must sum the rotation matrix elements over one of the extra magnetic quantum numbers to obtain a true angular momentum eigenstate.

12.4 Diatomic parity operator

The diatomic parity operator is the product to two parity operators because Euler rotations are proper but the parity operator (inversion of the sign of spatial coordinates) is an improper rotation (i.e., one that changes the handedness of the coordinate system and therefore cannot be the result of physical rotation). Three different transformations of the Euler angles can change the sign of two of the three components of a molecule-fixed spatial coordinate vector. The transformations $\phi \to \pi + \phi$, $\theta \to \pi - \theta$ and $\chi \to -\chi$ change the signs of the y' and x' components of all molecule-fixed spatial vectors. Thus, if one defines the operator P_Σ to invert the signs of the x' components, the parity diatomic parity operator can be written as

$$P = P_\Sigma\, P(\phi,\, \theta,\, \chi). \tag{12.20}$$

Scaler invariants (distances), such as r and ρ, and spin, which cannot be expressed in terms of spatial coordinates, are immune to the parity operator. Application of the parity operator to the case (a) basis yields

$$
\begin{aligned}
P_\Sigma\, P(\phi, \theta, \chi) & \sqrt{\frac{2J+1}{8\pi^2}}\, \langle \mathbf{r}_1', \mathbf{r}_2', \ldots, \mathbf{r}_{N-1}', r, \rho, \zeta\, |nv\rangle\, D^{J*}_{M\Omega}(\phi, \theta, \chi) \\
&= \sqrt{\frac{2J+1}{8\pi^2}}\, \langle -\mathbf{r}_1', -\mathbf{r}_2', \ldots, -\mathbf{r}_{N-1}', r, \rho, \zeta\, |n\rangle\, D^{J*}_{M,\,-\Omega}(\phi, \theta, \chi) \\
&= p_\Sigma(-)^{J+2M} \sqrt{\frac{2J+1}{8\pi^2}}\, \langle \mathbf{r}_1', \mathbf{r}_2', \ldots, \mathbf{r}_{N-1}', r, \rho, \zeta\, |nv\rangle\, D^{J*}_{M,\,-\Omega}(\phi, \theta, \chi).
\end{aligned}
\tag{12.21}
$$

When applied to a parity state, the parity operator can at most change the sign of the eigenfunction, but as seen above the parity operator converts $D^{J*}_{M\Omega}(\phi, \theta, \chi)$ into the orthogonal function $D^{J*}_{M,\,-\Omega}(\phi, \theta, \chi)$. A rotation matrix element lacks the mathematical properties of a parity eigenfunction. Thus, the case (a) and case (b) basis functions are not parity eigenfunctions.

Our molecular model here includes only the electromagnetic field. In particular, the weak force is excluded. Parity is a member of the collection of commuting observables. In addition, the multipole expansion of the electromagnetic field is an eigenfunction of the total angular momentum, one of its components, and parity [1]. In summary, angular momentum and parity are in perfect accord. It comes as no surprise that if $D^{J*}_{M\Omega}(\phi, \theta, \chi)$ is not a parity state, then it also not an angular momentum state, and vice versa.

Matrix elements of the parity operator in Hund's cases (a) and (b) are given by

$$
p_{ij}^{(a)} = p_\Sigma\, (-)^J\, \delta(J_i J_j)\, \delta(\Omega_i, -\Omega_j)\, \delta(\Lambda_i, -\Lambda_j)\, \delta(n_i\, n_j) \tag{12.22a}
$$

$$
p_{ij}^{(b)} = p_\Sigma\, (-)^{N_i}\, \delta(N_i\, N_j)\, \delta(\Lambda_i, -\Lambda_j)\, \delta(n_i\, n_j) \tag{12.22b}
$$

12.5 Hönl–London line-strength factors

The Hönl–London line-strength factor, sometimes called the rotational line-strength factor, is the angular momentum part of the Condon and Shortley line strength for a diatomic molecule. The probability of an exchange of energy between a molecule and the electromagnetic field (i.e., the probability of a radiative transition) is controlled by the matrix elements of the operators for the interactions between the molecule and radiation field. Typically, the matrix elements of one tensor operator $\mathbf{T}^{(q)}$ are much larger than the matrix elements for all other operators, and the line strength is defined as the absolute square of the matrix element $\langle f\,|\mathbf{V}|\,i\rangle$ summed over all transitions leading to the same observation (e.g., same spectral line.).

It was noted above that equations (4.8) and (4.16) merely answers the question of what happens to the eigenfunction when a coordinate changes the direction of the z-axis. One can ask the same question of operators. An operator that transforms in the same way as angular momentum eigenfunctions transform is called a spherical tensor operator. For example, the spherical tensor components of the the vector operator $\mathbf{V}(xyz)$ are defined as

$$T_1^{(1)} = -\sqrt{\frac{1}{2}}\,(V_x + iV_y) \tag{12.23}$$

$$T_0^{(1)} = V_z \tag{12.24}$$

$$T_{-1}^{(1)} = \sqrt{\frac{1}{2}}\,(V_x - iV_y) \tag{12.25}$$

and its components in the laboratory coordinate system are related to those in the rotated coordinate system by

$$T_k^{(1)}(xyz) = \sum_{\kappa=-1}^{1} T_\kappa^{(1)}(x'y'x')\, D_{k\kappa}^{1*}(\phi,\, \theta,\, \chi). \tag{12.26}$$

One sees that this is just equation (4.16) written for the vector operator $\mathbf{V}(xyz)$ instead of for the diatomic eigenfunction.

12.6 Triple integral of three rotation matrix elements

$$\int_0^{2\pi}\int_0^{\pi}\int_0^{2\pi} D_{m_3\omega_3}^{j_3*}(\phi,\, \theta,\, \chi)\, D_{m_2\omega_2}^{j_2}(\phi,\, \theta,\, \chi)\, D_{m_1\omega_1}^{j_1}(\phi,\, \theta,\, \chi)\, \sin(\theta)\, d\phi\, d\theta\, d\chi$$
$$= \frac{8\pi^2}{2j_3 + 1}\langle j_1\ m_1;\ j_2\ m_2 | j_3\ m_3\rangle\ \langle j_1\ \omega_1;\ j_2\ \omega_2 | j_3\ \omega_3\rangle \tag{12.27}$$

The Clebsch–Gordan coefficients are real. Therefore, the above integral is real. From the formula relating the rotation matrix element to its complex conjugate,

$$D_{j\omega}^{j}(\alpha,\, \beta,\, \gamma) = (-)^{\omega-m} D_{-m,\,-\Omega}^{J*}(\alpha,\, \beta,\, \gamma), \tag{12.28}$$

it is seen that complex conjugation of one or more of the rotation matrix elements might change the sign of the integral.

12.7 Calculation of the Hönl–London line-strength factors for cases (a) and (b)

Given here are mathematically simple (and somewhat simple minded) calculations of Hönl–London line-strength factors for cases (a) and (b). The calculation is described as somewhat simple minded because one knows that any quantum state must possess the mathematical properties of angular momentum and if the system is controlled by electromagnetic forces it must also possess parity. The basis functions equations(12.1) and (12.6) possess neither.

Placing the case (a) basis functions into the definition of the line strength (11) gives

$$S(n'J', nJ) = \sum_{M=-J}^{J} \sum_{M'=-J'}^{J'} \sum_{k=-q}^{q} \sum_{\kappa=-q}^{q}$$
$$\left| \frac{1}{8\pi^2}\sqrt{(2J+1)(2J'+1)}\ \langle nv|T_{k\kappa}^{(q)}(\mathbf{r}'_1, \mathbf{r}'_2, \dots, \mathbf{r}'_{N-1}, r, \rho, \zeta)|n'v'\rangle \right. \tag{12.29}$$
$$\left. \int_0^{2\pi}\int_0^{2\pi}\int_0^{2\pi} D_{M\Omega}^{J}(\phi, \theta, \chi)\, D_{k\kappa}^{q*}(\phi, \theta, \chi)\, D_{M'\Omega'}^{J'*}(\phi, \theta, \chi)\sin(\theta)\, d\phi\, d\theta\, d\chi \right|^2.$$

Whereas in the standard integral (12.27) the complex conjugate of only one Wigner D-function appears, in the above two D-functions appear as their complex conjugates. As noted above, this can only change the sign of the integral but because the integral will be squared we can ignore the possible sign change. The integral signs become vitally important below where sums of the integrals are squared. Evaluation of the triple integral above gives

$$\begin{aligned}
S(n'v'J', nvJ) &= \frac{2J'+1}{2J+1}\sum_{M=-J}^{J}\sum_{M'=-J'}^{J'}\sum_{k=-q}^{q}\sum_{\kappa=-q}^{q} \\
&\times \left| \langle nv\,|T_{k\kappa}^{(q)}\,|(\mathbf{r}'_1\mathbf{r}'_2 \dots \mathbf{r}'_{N-1}\, r\, \rho\, \zeta)|n'v'\rangle\, \langle J'\, M';\, q\, k\,|J\, M\rangle\, \langle J'\, \Omega';\, q\, \kappa\,|J\, \Omega\rangle \right|^2 \\
&= \frac{2J'+1}{2J+1}\sum_{\kappa=-q}^{q} \left| \langle nv|T_{k\kappa}^{(q)}(\mathbf{r}'_1, \mathbf{r}'_2, \dots, \mathbf{r}'_{N-1}, r, \rho, \zeta)|n'v'\rangle \right|^2 \langle J'\, \Omega';\, q\, \kappa\,|J\, \Omega\rangle^2 \sum_{M=-J}^{J} 1 \\
&= (2J'+1) \left| \langle nv\,|T_{k,\,\Omega'-\Omega}^{(q)}|n'v'\rangle(\mathbf{r}'_1\mathbf{r}'_2 \dots \mathbf{r}'_{N-1}\, r\, \rho\, \zeta)|n'\rangle \right|^2 \langle J'\, \Omega'\, q,\, \Omega - \Omega'|J\, \Omega\rangle^2 \\
&= S(n'v', nv)\, S(J', J)
\end{aligned} \tag{12.30}$$

where

$$S(n'v', nv) = \left| \sum_{k=-q}^{q} \langle n|T_{k,\,\Omega'-\Omega}^{(q)}(\mathbf{r}'_1, \mathbf{r}'_2, \dots, \mathbf{r}'_{N-1}, r, \rho, \zeta)\, |n'v'\rangle \right|^2 \tag{12.31}$$

$$S(J', J) = (2J'+1)\, \langle J'\, \Omega';\, q,\, \Omega - \Omega'\,|J\, \Omega\rangle^2 \tag{12.32}$$

In the above, we used

$$\begin{aligned}
&\sum_{m_3=-j_3}^{j_3}\sum_{m_1=-j_1}^{j_1}\sum_{m_2=-j_2}^{j_2} \langle j_1\, m_1;\, j_2\, m_2\, |j_3\, m_3\rangle\, \langle j_1\, m_1;\, j_2\, m_2\, |j'_3\, m'_3\rangle \\
&= \sum_{m_3=-j_3}^{j_3} \delta_{j_3 j'_3}\delta_{m_3 m'_3} = 2j_3 + 1.
\end{aligned} \tag{12.33}$$

in

$$\sum_{M=-J}^{J}\sum_{M'=-J'}^{J'}\sum_{k=-q}^{q} \langle J'\, M';\, q\, k\,|J\, M\rangle\, \langle J'\, M';\, q\, k\,|J\, M\rangle = \sum_{M=-J}^{J} \delta_{JJ}\delta_{MM} = 2J+1. \tag{12.34}$$

We could have used the symmetry of the Clebsch–Gordan coefficients in which $j_2\, m_2$ and $j_3\, m_3$ are exchanged to write

$$
\begin{aligned}
&\sum_{m_3=-j_3}^{j_3} \sum_{m_1=-j_1}^{j_1} \sum_{m_2=-j_2}^{j_2} \langle j_1\, m_1;\, j_2\, m_2\, |j_3\, m_3\rangle\, \langle j_1\, m_1;\, j_2\, m_2\, |j_3\, m_3\rangle \\
&= \left[(-)^{j_1 - m_1} \sqrt{\frac{2j_3+1}{2j_2+1}} \right]^2 \sum_{m_2=-j_2}^{j_2} \sum_{m_1=-j_1}^{j_1} \sum_{m_3=-j_3}^{j_3} \langle j_1\, m_1;\, j_3,\, -m_3\, |j_2,\, -m_2\rangle^2 \\
&= \frac{2j_3+1}{2j_2+1} \sum_{m_2=-j_2}^{j_2} \delta_{j_2 j_2}\, \delta_{m_2 m_2} \\
&= 2j_3 + 1
\end{aligned} \tag{12.35}
$$

The line strength is symmetrical with respect to the upper and lower levels,

$$S(n'v_n' J',\, nv_n J) = S(nv_n J,\, n'v_{n'}' J') \tag{12.36}$$

and the Hönl–London line-strength factor should share this symmetry. From two symmetries of the Clebsch–Gordan coefficient, the first which swaps $\langle j_1 m_1|$ and $|j_3 m_3\rangle$,

$$\langle j_1\, m_1;\, j_2\, m_2\, |j_3\, m_3\rangle = \sqrt{\frac{2j_3+1}{2j_1+1}}\, (-)^{j_1 - m_1 + j_3 - m_3} \langle j_3\, m_3;\, j_2\, m_2\, |j_1\, m_1\rangle, \tag{12.37}$$

and the second which changes the signs of all three magnetic quantum numbers,

$$\langle j_1\, m_1;\, j_2\, m_2\, |j_3\, m_3\rangle = (-)^{j_1 - j_2 - j_3} \langle j_1,\, -m_1;\, j_2,\, -m_2\, |j_3,\, -m_3\rangle, \tag{12.38}$$

and because the case (a) Hönl–London line-strength factor, equation (12.32), is expressed as the square of the Clebsch–Gordan coefficient, the Hönl–London line-strength factors for case (a) to case (a) transitions are symmetrical as required,

$$(2J'+1)\, \langle J'\, \Omega';\, q,\, \Omega - \Omega'\, |J\, \Omega\rangle^2 = (2J+1) \langle J\, \Omega;\, q,\, \Omega' - \Omega\, |J' \Omega'\rangle^2. \tag{12.39}$$

12.8 Hund's case (b) Hönl–London line-strength factors

The probability of the transition $nvJM \leftrightarrow n'v'J'M'$ is controlled by the matrix elements $\langle nvJM\, | T^{(q)}{}_k\, | n'v'J'M'\rangle$, sometimes called transition moments, where $T_k^{(q)}$ is the kth component of the tensor of degree q responsible for the transition. If the system in question depends upon the internal variables ρ, ζ, χ, $\mathbf{r}_2$, … , $\mathbf{r}_N$, r, θ, ϕ, then the operators responsible for the electromagnetic transitions (e.g., electric dipole, magnetic dipole, electric quadrupole, *etc.*) will depend upon the same coordinates as the eigenfunction Φ_{nvJM}, and when expressed as an irreducible tensor, will have the same behavior under coordinate rotation,

$$T_k^{(q)}(\rho,\, \zeta,\, \chi,\, \mathbf{r}_2,\, \ldots,\, \mathbf{r}_N,\, r,\, \theta,\, \phi) = \sum_{\kappa=-q}^{q} T^{(q)}(\rho,\, \zeta,\, \mathbf{r}_2,\, \ldots,\, \mathbf{r}_N,\, r)\, D_{k\kappa}^{q*}(\phi,\, \theta,\, \chi) \tag{12.40}$$

as the eigenfunction Φ_{nvJM}. For Hund's case (b) basis functions, the transition moments become

$$\begin{aligned}
\langle b\,|T_k^{(q)}|b'\rangle &= \sum_{k=-q}^{q}\sum_{\kappa=-q}^{q}\langle nv\,|T_\kappa^{(q)}(\rho,\,\zeta,\,\mathbf{r}'_2,\,\mathbf{r}'_N,\,r)|n'v'\rangle \\
&\times \frac{\sqrt{(2N+1)(2N'+1)}}{8\pi^2}\sum_{M_N-N}^{N}\sum_{M_S=-S}^{S}\sum_{M'_N=-N'}^{N'}\sum_{M'_S=-S'}^{S'}\langle SM_S\,|S'M'_S\rangle \\
&\times \langle NM_N;\,SM_S\,|JM\rangle\,\langle N'M'_N;\,SM_S\,|J'M'\rangle \\
&\times \int_0^{2\pi}\int_0^{\pi}\int_0^{2\pi} D_{M_N\Lambda}^{N*}(\phi,\,\theta,\,\chi)\,D_{k\kappa}^{q}(\phi,\,\theta,\,\chi)\,D_{M'_N\Lambda'}^{N'}(\phi,\,\theta,\,\chi)\sin(\theta)\,d\phi\,d\theta\,d\chi \\
&= \sqrt{2N+1}\sum_{\kappa=-q}^{q}\langle nv\,|T_\kappa^{(q)}(\rho,\,\zeta,\,\mathbf{r}'_2,\,\mathbf{r}'_N,\,r)|n'v'\rangle\sum_{M_N=N}^{N}\sum_{M_S=-S}^{S}\sum_{M'_N=-N'}^{N'} \\
&\times \langle NM_N;\,SM_S\,|JM\rangle\,\langle N'M'_N;\,SM_S\,|J'M'\rangle\langle NM_N;\,q\,k\,|N'M'_N\rangle\,\langle N\Lambda;\,q\,\kappa\,|N'\Lambda'\rangle \\
&= \sqrt{2N+1}\sum_{\kappa=-q}^{q}\sum_{M_N=N}^{N}\sum_{M_S=-S}^{S}\sum_{M'_N=-N'}^{N'}\langle nv\,|T_\kappa^{(q)}(\rho,\,\zeta,\,\mathbf{r}'_2,\,\mathbf{r}'_N,\,r)|n'v'\rangle \\
&\langle NM_N;\,SM_S\,|JM\rangle\,\langle N'M'_N;\,SM_S\,|J'M'\rangle\,\langle NM_N;\,q\,k\,|N'M'_N;\,\rangle\,\langle N\Lambda;\,q\,\kappa\,|N'\Lambda'\rangle \\
&= S(n'v',\,nv)\,S(J',\,J)
\end{aligned} \tag{12.41}$$

$$S(J',\,J) = (2J+1)(2J'+1)(2N+1)\langle N\Lambda\,q\,\kappa|N'\Lambda'\rangle^2\begin{Bmatrix} q & J' & J \\ S & N & N'\end{Bmatrix}^2\delta(S',\,S) \tag{12.42}$$

$$\begin{aligned}
&\sum_{\kappa=-q}^{q}\sum_{M_N=N}^{N}\sum_{M_S=-S}^{S}\sum_{M'_N=-N'}^{N'}\langle NM_N\,SM_S\,|JM\rangle\,\langle N'M'_N\,SM_S\,|J'M'\rangle \\
&\times \langle NM_N\,q\,k\,|N'M'_N\rangle\,\langle N\Lambda\,q\,\kappa\,|N'\Lambda'\rangle \\
&= \sqrt{(2J'+1)(2J+1)}\begin{Bmatrix} q & J' & J \\ S & N & N'\end{Bmatrix}
\end{aligned} \tag{12.43}$$

$$\begin{Bmatrix} a & b & c \\ d & e & f\end{Bmatrix} = \begin{Bmatrix} d & e & c \\ a & b & f\end{Bmatrix} \tag{12.44}$$

$$W(abcd;\,ef) = (-)^{b+e-c-f}\,W(afcd;\,eb) \tag{12.45}$$

$$\begin{Bmatrix} q & J' & J \\ S & N & N'\end{Bmatrix} = (-)^{J'+J-N-N'}\begin{Bmatrix} q & N' & J \\ S & N & J'\end{Bmatrix} \tag{12.46}$$

$$S(J',\,J) = (2J+1)(2J'+1)(2N+1)\langle N\Lambda\,q\,\kappa|N'\Lambda'\rangle^2\begin{Bmatrix} q & N' & J \\ S & N & J'\end{Bmatrix}^2\delta(S',\,S) \tag{12.47}$$

$$\begin{Bmatrix} j_1 = q & j_2 = N' & j_3 = J \\ j_4 = S & j_5 = N & j_6 = J'\end{Bmatrix} \tag{12.48}$$

$$\left\{\frac{j_1 = q \ \ j_2 = N' \ J_{12} = J}{j_3 = S \ \ j = N \ \ j_{23} = J'}\right\} \tag{12.49}$$

$$\left\{\frac{a = q \ b = N' \ e = J}{d = S \ c = N \ f = J'}\right\} \tag{12.50}$$

12.9 The electronic–vibrational strength

The Hönl–London line-strength factors are unitless. The electronic–vibrational strength $S(n'v', nv)$ carries the units of the Condon and Shortley line strength. The HITRAN line strength is not the Condon and Shortley line strength but can be expressed in terms of it.

The Born–Oppenheimer approximation separates the electronic–vibrational eigenfunction into the product of electronic and vibrational eigenfunctions,

$$\langle \mathbf{r}_1', \mathbf{r}_2', \ldots, \mathbf{r}_{N-1}', r, \rho, \zeta \,|nv\rangle \approx \langle \mathbf{r}_1', \mathbf{r}_2', \ldots, \mathbf{r}_{N-1}', \rho, \zeta; r|n\rangle \,\langle r|v_n\rangle, \tag{12.51}$$

In this approximation one can introduce the electronic transition moment,

$$\begin{aligned}\mathcal{R}_{n'n}(r) &= \langle n'|T^{(q)}_{k,\,\Omega-\Omega'}(\mathbf{r}_1'\mathbf{r}_2' \ldots \mathbf{r}_{N-1}' \,\rho\, \zeta;\, r)|n\rangle \\ &= b_0 + b_1 r + b_2 r^2 + \cdots,\end{aligned} \tag{12.52}$$

its vibrational matrix elements,

$$\langle v_{n'}|\mathcal{R}_{n'n}(r)|v_n\rangle = b_0\langle v_{n'}|v_n\rangle + b_1\langle v_{n'}|r|v_n\rangle + b_2\langle v_{n'}|r^2|v_n\rangle + \cdots. \tag{12.53}$$

and these in turn lead to the Franck–Condon factors and r-centroids,

$$|\langle v_{n'}|\mathcal{R}_{n'n}(r)|v_n\rangle|^2 = q(v_{n'}, v_n)[b_0 + b_1\, \bar{r}^{(1)}(v_{n'}, v_n) + b_2\, \bar{r}^{(2)}(v_{n'}, v_n) + \cdots]^2, \tag{12.54}$$

$$q(v_{n'}, v_n) = \langle v_{n'}\,|v_n\rangle^2, \tag{12.55}$$

$$\bar{r}^{(k)}(v_{n'}, v_n) = \frac{\langle v_{n'} v\,|r^k|v_n\rangle}{\langle v_{n'}\,|v_n\rangle}. \tag{12.56}$$

In summary, the Born–Oppenheimer approximation gives

$$S(n'v_{n'}J', nv_nJ) = q(v_{n'}, v_n)[b_0 + b_1\, \bar{r}^{(1)}(v_{n'}, v_n) + b_2\, \bar{r}^{(2)}(v_{n'}, v_n) + \cdots]\, S(J', J'). \tag{12.57}$$

for the diatomic electronic–vibrational line strength. Almost invariably, the accuracy of line position measurement exceeds the accuracy of even relative intensity measurement by several orders of magnitude. The calculation of accurate line positions using the Born–Oppenheimer approximation requires that the vibrational eigenfunction $\langle r\,|v\rangle$ be expressed as a large sum of Born–Oppenheimer vibrational eigenfunctions $\langle r\,|v_n\rangle$,

$$\langle r \,|v\rangle = \sum_{v_n} \langle r \,|v_n\rangle \,\langle v_n \,|v\rangle. \tag{12.58}$$

The sum over v_n produces large Hamiltonian matrices whose dimensions are reduced to manageable size by Van Vleck transformations or other comparable computations, and, when successful, the eigenvalues of the smaller matrices can accurately reproduce experimentally determined terms Typically, only the simple Born–Oppenheimer approximation result (12.57) is required for diatomic intensity predictions.

If one's definition of an ugly number is a number too small in magnitude to be represented in standard single precision floating point digital format, then the line strength can be fairly described as an ugly number. It is normally expressed in units of $(a_0\, e)^2 = 7.188\,2479 \times 10^{-59}$ m^2 Coul^2 where a_0 is the Bohr radius and e is the charge of the electron.

Reference

[1] Rose M E 1995 *Elementary Theory of Angular Momentum* (Mineola, NY: Dover)

IOP Publishing

Quantum Mechanics of the Diatomic Molecule (Second Edition)

Christian G Parigger and James O Hornkohl

Chapter 13

Using the Morse potential in diatomic spectroscopy

13.1 Introduction

A diatomic basis state has a potential energy function $V(r)$. The Morse potential is a simple mathematical formula that one can use to approximate $V(r)$. If the spectral line vacuum wave numbers have been experimentally measured to high J values for all known bands of the band system in question, then the numerical Rydberg-Klein-Rees (RKR) potential energy curves are much better approximations of $V(r)$s for the upper and lower basis functions than the Morse potential. Using RKR $V(r)$s, one can compute the vibrational eigenfunctions, Franck–Condon factors, r-centroids, and tables of G_v and B_v versus v for comparison with the experimental tables of G_v and B_v used to compute the RKR potentials. However, the experimental record for the band system is often incomplete. The Morse potential is particularly useful when one needs the quantities that are computable from vibrational eigenfunctions but the experimental record is too incomplete for computation of RKR potentials.

Only four variables, i.e., ω_e, $\omega_e x_e$, and B_e, and the reduced nuclear mass μ are required for computation of the Morse potential,

$$V^{(M)}(r) = D_e[1 - e^{-\beta(r-r_e)}]^2. \tag{13.1}$$

where r is the internuclear distance and r_e is the value of r at the minimum of $V^{(M)}(r)$. The dissociation energy D_e, measured from the minimum of the $V^{(M)}(r)$ potential energy well, is given by

$$D_e = \frac{\omega_e^2}{4\,\omega_e x_e}. \tag{13.2}$$

The parameter β (unrelated to the Euler angle β) is given in terms of the anharmonicity constant $\omega_e x_e$ and the reduced nuclear mass μ is given in atomic mass units (amu),

doi:10.1088/978-0-7503-6204-7ch13

$$\beta = \sqrt{\frac{4\,\pi\,c\,(\mu/N_A)}{\hbar}\omega_e x_e} \tag{13.3}$$

in which N_A is Avogadro's number. The classical equilibrium internuclear distance r_e is given in terms of the constant B_e,

$$r_e = \sqrt{\frac{\hbar}{4\,\pi\,c\,(\mu/N_A)}\frac{1}{B_e}}. \tag{13.4}$$

The energy eigenvalues of the Morse oscillator are

$$E_v = \omega_e(v + 1/2) - \omega_e x_e(v + 1/2)^2. \tag{13.5}$$

The constant B_e is that from the semiempirical formula for the rotational constant,

$$B_v = B_e - \alpha_e(v + 1/2) + \gamma_e(v + 1/2)^2 + \cdots. \tag{13.6}$$

If one chooses the values of ω_e and $\omega_e x_e$ to be those in the semiempirical formula for vibrational term G_v,

$$G_v = \omega_e(v + 1/2) - \omega_e x_e(v + 1/2)^2 + \omega_e y_e(v + 1/2)^3 + \omega_e z_e(v ++ 1/2)^4, \tag{13.7}$$

then one intuitively expects that the lower region of the the Morse potential $V^{(M)}(r)$ will closely approximate the same lower region of the RKR potential. Computation of the RKR potential requires full experimentally determined tables of G_v and B_v versus v. If experiment has provided G_v and B_v for only the lowest values of the vibrational quantum number v, one can nevertheless compute a Morse potential which usefully approximates the lower region of the RKR potential. Because the eigenfunctions of the Morse oscillator are analytically known (see the following section), one can compute the Franck–Condon factors and r-centroids and, at least for the few lower vibrational states, proceed as if the RKR potential were known. Far from complete tables of G_v and B_v are a common occurrence, and thus the Morse potential is a useful addition to the toolbox of applied spectroscopy.

13.2 Morse eigenfunctions

The variable x

$$x = a\,e^{-\beta(r-r_e)} \tag{13.8}$$

in which

$$a = \frac{\omega_e}{\omega_e x_e} \tag{13.9}$$

is a more convenient variable than the internuclear distance r for expressing the Morse potential

$$V^{(M)}(x) = D_e\left(1 - \frac{x}{a}\right)^2 \tag{13.10}$$

and its energy eigenfunction,

$$\psi_v^{(M)}(x) = \langle x|v\rangle_M = \mathcal{N}_v\, e^{-x/2}\, x^{\alpha_v/2}\, L_v^{(\alpha_v)}(x). \tag{13.11}$$

The normalization factor is

$$\mathcal{N}_v = \sqrt{\frac{\alpha_v\, \beta\, v!}{\Gamma(a - v)}} \tag{13.12}$$

in which the parameter α_v is given by

$$\alpha_v = a - 2\,v - 1 \tag{13.13}$$

and Γ is the gamma function.

The Laguerre polynomial $L_n^{(\alpha)}(x)$ is defined only for $\alpha > -1$. The requirement

$$\alpha_v > -1 \tag{13.14}$$

imposes the limit

$$v_{\max} < \frac{a}{2} \tag{13.15}$$

on the maximum value of the vibrational quantum number v for the Morse oscillator, i.e., the Morse oscillator is bound for v less than $a/2$, but will dissociate if it is excited beyond the $v_{\max}$ bound state. For a diatomic potential energy function this is a significant improvement over the harmonic oscillator potential, which never dissociates.

The Laguerre polynomials $L_v^{(\alpha_v)}(x)$ of the Morse eigenfunction are distinct from the usual Laguerre polynomials $L_n^{(\alpha)}(x)$ in that the parameter α in $L_n^{(\alpha)}(x)$ does not vary with the degree n of the polynomial, whereas α_v in $L_v^{(\alpha_v)}(x)$ does vary with the degree v. The mathematical manifestation of the variation of α_v with v is that $L_v^{(\alpha_v)}(x)$ does not satisfy a three-term recursion formula like that which $L_n^{(\alpha)}(x)$ with fixed α satisfies. This slightly complicates computation of the Morse eigenfunction and limits the practicality of one using a sum of the Morse eigenfunctions as a basis for expressing the eigenfunction of an arbitrary potential $V(r)$. The Lanczos basis is often expressed in terms of the spectroscopic quantities ω_e, $\omega_e x_e$ and r_e, but it uses $L_n^{(\alpha)}(x)$ with a fixed α.

A diatomic potential energy function should satisfy the limits

$$\lim\nolimits_{r\to 0} V(r) \to \infty, \tag{13.16a}$$

$$\lim_{r\to\infty} V(r) = D_e, \tag{13.16b}$$

but the Morse potential function remains finite for fictitious negative values of the internuclear distance[1],

$$\lim_{r\to-\infty} V^{(M)}(r) \to \infty, \tag{13.17a}$$

$$\lim_{r\to\infty} V^{(M)}(r) = D_e. \tag{13.17b}$$

This means that the equations for the Morse energy eigenvalues (13.5) and eigenfunctions (13.11) are approximations. However, for any diatomic molecule the value of the Morse potential ar $r = 0$,

$$V^{(M)}(r = 0) = D_e\, e^{\beta\, r_e} \tag{13.18a}$$

$$= D_e\, e^{\sqrt{\omega_e x_e / Be}} \tag{13.18b}$$

is so large that equations (13.5) and (13.11) are effectively exact for all practical purposes in diatomic spectroscopy.

13.2.1 Computation of Morse eigenfunctions

Equation (13.11) is clearly simple, but its numerical evaluation is not entirely trivial. Computation of $L_v^{(\alpha_v)}(x)$ using the three-term recurrence formula

$$(n+1)L_{n+1}^{(\alpha_v)}(x) = (2n + \alpha_v - x)L_n^{(\alpha_v)}(x) - (n + \alpha_v)L_{n-1}^{(\alpha_v)}(x) \tag{13.19}$$

means that one computes the unneeded $L_0^{(\alpha_v)}(x)$, $L_1^{(\alpha_v)}(x)$, …, $L_{v-1}^{(\alpha_v)}(x)$ just to obtain the desired $L_v^{(\alpha_v)}(x)$. Given the computational power of even a modest digital computer, this wasted computation is likely of no practical consequence.

Of more significant nuisance are the values of $L_n^{(\alpha)}(x)$ large enough to cause numerical overflow. Shen *et al* [2] define 'generalized Laguerre functions (GLFs)',

$$\hat{\mathcal{L}}_n^{(\alpha)}(x) \equiv e^{-x/2} L_n^{(\alpha)}(x) \tag{13.20}$$

which follow the same recurrence formula as the standard $L_n^{(\alpha)}(x)$,

$$(n+1)\hat{L}_{n+1}^{(\alpha_v)}(x) = (2n + \alpha_v - x)\hat{L}_n^{(\alpha_v)}(x) - (n + \alpha_v)\hat{L}_{n-1}^{(\alpha_v)}(x). \tag{13.21}$$

The two lowest degrees $\hat{L}_n^{(\alpha_v)}(x)$ required to start the recursion are

$$\hat{L}_0^{(\alpha_v)}(x) = e^{-x/2}, \tag{13.22a}$$

$$\hat{L}_1^{(\alpha_v)}(x) = (\alpha + 1 - x)\, e^{-x/2}. \tag{13.22b}$$

Rewritten in terms of the GLFs, the Morse eigenfunction equation (13.11) reads

$$\psi_v^{(M)}(x) = \mathcal{N}_v\, x^{\alpha_v/2}\, \hat{L}_v^{(\alpha_v)}(x). \tag{13.23}$$

The FORTRAN-90 program MorseFCF evaluates eigenfunctions for the upper and lower vibrational states,

$$\psi_v^{(M)}(r) = e^{[\ln(\mathcal{N}_v) + \alpha_v \ln(x)]}\, \hat{L}_v^{(\alpha_v)}(x), \tag{13.24}$$

and uses the vibrational eigenfunctions to compute the Franck–Condon factors and the first three r-centroids. This program is particularly useful when experimental data for the band system are inadequate for accurate determination of the upper and lower potential energy functions. The subroutines in MorseFCF.f90 are potentially useful in other applications of Morse eigenfunctions. The source codes are listed in appendix I.

13.3 Morse eigenfunctions as a vibrational basis

Greenawalt and Dickinson [3] explored the use of Morse eigenfunctions as a basis for numerically expressing the solutions for a potential function $V(r)$ for which an analytic solution is not available. Dissociation of the Morse oscillator limits the maximum number of Morse basis functions, equation (13.15). In contrast, the extent of, for example, a harmonic oscillator basis is limited by the computer on which they are used.

The matrix elements of x, equation (13.8), evaluated in the Morse basis, equation (13.11), are

$$\langle n|x|n\rangle = a - 2n - 1 \tag{13.25a}$$

$$\langle n|x|m\rangle = (-)^{m-n}\frac{\mathcal{N}_n\mathcal{N}_m\Gamma(a-m)}{\beta\, n!} \qquad m \geqslant n \tag{13.25b}$$

$$\langle n|x^2|n\rangle = a(a - 2n - 1) \tag{13.26a}$$

$$\langle n|x^2|m\rangle = (-)^{m-n}\frac{\mathcal{N}_n\mathcal{N}_m\Gamma(a-m)}{\beta\, n!}[a(m-n+1) - (m-n)(m+n+1)] \tag{13.26b}$$

From these matrix elements and equation (13.10), one sees that matrix elements of the Morse potential in the Morse basis are given by

$$V_{nm}^{(M)} = D_e\left(1 - \frac{2\langle n|x|m\rangle}{a} + \frac{\langle n|x^2|m\rangle}{a^2}\right). \tag{13.27}$$

Since the diagonal Morse Hamiltonian matrix is known, equation (13.7), the kinetic energy matrix is simply the Hamiltonian matrix minus the potential matrix,

$$K_{nm}^{(M)} = G_n^{(M)}\delta_{nm} - V_{nm}^{(M)}. \tag{13.28}$$

The solutions for an unknown potential are found by computing the matrix elements of the unknown potential in the Morse basis, adding the Morse kinetic energy $K^{(M)}$ matrix, and diagonalization of the result.

Greenawalt and Dickinson [3] found that the Morse basis is useful for shallow potential wells. However, the matrix of x given in equations (13.25) is a full symmetric matrix, not a tridiagonal matrix. This means that the eigenvalues of the x matrix are not the roots of an orthogonal polynomial satisfying a three-term recurrence formula, and that the eigenvalues of x are not the arguments of a Gaussian integration. When numerical diagonalization of large matrices was made practical by advances in computer hardware and software in the 1960s, matrix solutions of the Schrödinger equation became widespread. Early examples are [4–6], and [7].

Given the orthogonal polynomials of degree $n = 0$ to N having the weight function $w(x)$ over the domain $a \to b$, one can compute the matrix elements of x,

$$x_{nm} = \int_a^b w(x)\, P_n(x)\, xP_m(x)\, dx \tag{13.29}$$

A similarity transformation in which S is an orthogonal matrix (i.e., the inverse S^{-1} equals the transpose $\tilde{S}$) yields the eigenvalues λ_i of the x matrix,

$$\lambda_j = \sum_{n=0}^{N}\sum_{m=0}^{N} \tilde{S}_{jn}\, x_{nm}\, S_{mk}. \tag{13.30}$$

This equation can be inverted to give the matrix elements x_{nm} in terms of the eigenvalues λ_j,

$$x_{nm} = \sum_{i=0}^{N}\sum_{j=0}^{N} S_{ni}\, \lambda_j\, \delta_{ji}\, \tilde{S}_{jm} \tag{13.31a}$$

$$= \sum_{j=0}^{N} S_{nj}\, \lambda_j\, \tilde{S}_{jm} \tag{13.31b}$$

$$= \sum_{j=0}^{N} S_{nj}\, \lambda_j\, S_{mj} \tag{13.31c}$$

By comparing this last result with the conversion of the continuous integral equation (13.29) to a discrete sum using Gaussian integration,

$$x_{nm} = \sum_{j=0}^{N} w(\lambda_j)\, P_n(x_j)\, \lambda_j\, P_m(\lambda_j), \tag{13.32}$$

one sees that the two equations become equivalent if one makes the association

$$S_{nj} = \sqrt{w(\lambda_j)}\, P_n(\lambda_j). \tag{13.33}$$

The practical significance is that equation (13.32) yields exact values for all x_{nm} except x_{NN}.

References

[1] ter Haar D 1946 *Phys. Rev.* **70** 222

[2] Shen Jie, Tang Tao and Wang Li-Lian 2010 Spectral Methods, Algorithms *Analysis and Applications* (Heidelberg: Springer)

[3] Greenawalt E M and Dickinson A S 1969 *J. Mol. Spectrosc.* **30** 427

[4] Chan S I and Stelman D 1963 *J. Chem Phys.* **39** 545

[5] Harris D O, Engerholm G G and Gwinn W D 1965 *J. Chem. Phys.* **43** 1515
[6] Zetik D F and Matsen F A 1967 *J. Mol. Spectrosc.* 122
[7] Dickinson A S and Certain P R 1968 *J. Chem. Phys.* 4209

Part II

Selected applications of diatomic spectroscopy

IOP Publishing

Quantum Mechanics of the Diatomic Molecule (Second Edition)

Christian G Parigger and James O Hornkohl

Chapter 14

Introduction to applications of diatomic spectroscopy

Diatomic spectroscopy encompasses the analysis of the interaction of electromagnetic radiation with diatomic molecules. The applications discussed in this book primarily explore molecular characteristics in and near the visible region of the electromagnetic spectrum, but extending to the near-ultraviolet and near-infrared in the range of 200–800 nm. A variety of experimental studies are designed first for identification of molecular signatures or 'fingerprints' in measured spectra, and second for rigorous analysis of recorded data. Of course, analytical chemistry explores opportunities for the establishment of well-defined diagnostic protocols for diatomic molecules.

Accurate predictions of molecular spectra are essential for analysis. Ultimately, computed diatomic spectra are expected to be readily available together with formalized comparison models for diagnosis purposes. In atomic spectroscopy, species identification relies on the occurrences of specific lines and sequences of lines. For example, measurement of a line near 486.1 nm may not be enough to conclusively identify hydrogen, but usually requires positive checks near 656.2 nm, and equally near 434.0 nm. With atomic signatures of the hydrogen alpha-, beta-, and gamma-lines, one can conclude with confidence the presence of hydrogen due to the occurrence of the first three members of the Balmer series lines in the visible portion of the electromagnetic spectrum. Conversely, when one looks for sodium, one would expect signatures of the Na D_1 and D_2 lines near 589.6 and 589.0 nm. Molecular spectroscopy diagnosis works in an analogous manner because one aims to identify the presence of molecules in survey spectra. Fingerprints of molecular spectra measured with spectral resolutions of the order of 0. 1 nm include location of the band head, shading of the $\Delta v = 0$ sequence, and identification of the progression. For example, molecular OH spectra indicate a well-developed, $\Delta v = 0$ band head near 306 nm, are shaded towards the red, and shows a $\Delta v = 1$ band head near 281 nm. Overlapping or interfering spectral regions may require detailed

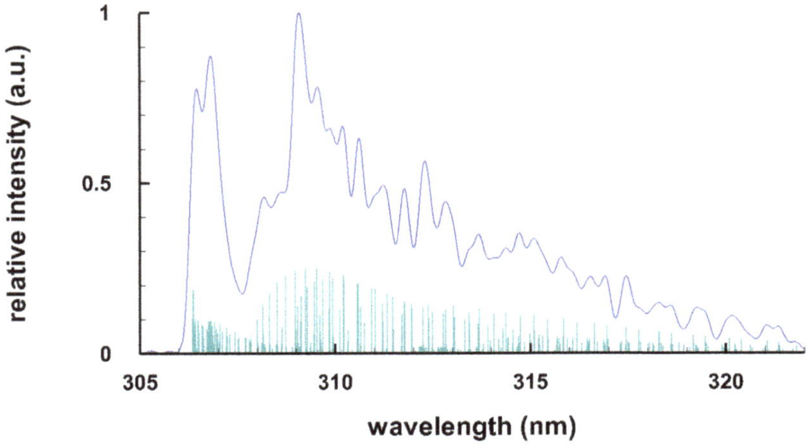

Figure 14.1. Computed spectrum of the $A^2\Sigma \rightarrow X^2\Pi$ UV band of OH, $T = 4\,\mathrm{kK}$: (top panel) spectral resolutions of $\Delta\lambda = 0.32$ nm ($\Delta\tilde{\nu}= 32$ cm^{-1}) and (bottom panel) idealized resolution for the stick spectrum $\Delta\lambda = 0.002$ nm ($\Delta\tilde{\nu}= 0.2$ cm^{-1}) of the $\Delta\nu = 0$ sequence [2].

identification work using well-documented tables [1]. In turn, the presence of carbon C_2 Swan spectra in recorded plasma emission data implies the presence of radiating C_2 diatomic molecules.

Ideally, diatomic molecular spectra should be available in a data base or should be easily computed. To this end, recent publications [2, 3] describe two programs that accomplish prediction of signatures from line strength data for selected transitions of AlO, C_2, CN, and TiO diatomic molecules and fitting of measured spectra; namely, Boltzmann equilibrium spectrum computation and Nelder–Mead temperature programs. However, several other diatomic molecules have been of interest in laser spectroscopy work by the authors, e.g., OH to name but one other example. Figures 14.1–14.7 show selected synthetic spectra for a few diatomic molecules that have been of continued interest: OH, AlO, C_2 Swan band overview (progression) and selected sequences $\Delta\nu = 0, \pm1, \pm2$ of the C_2 Swan bands, and $\Delta\nu = 0, \pm1$ of CN.

Applications of the computed line strengths include analysis of molecular signatures in laser-induced breakdown spectroscopy (LIBS), combustion studies, and analysis of stellar astrophysical spectra. LIBS aims to diagnose elemental composition following generation of a laser spark using analytical chemistry methods analogous to discharge spark spectrometry. A mini-review of diatomic LIBS is presented in appendix H.

Molecular signatures are recognizable in the emitted plasma radiation within the first microsecond after laser-induced optical breakdown, especially when utilizing nominal femto- to several tens of pico-second pulses for generation of laser plasma.

For nominal, 5–10 ns radiation, decaying plasma characteristically reveals relatively large background radiation that masks molecular emissions, but molecular spectra are well-developed in the plume and for time delays of the order of 10–100 μs after initiation of optical breakdown. Combustion studies obviously focus on the

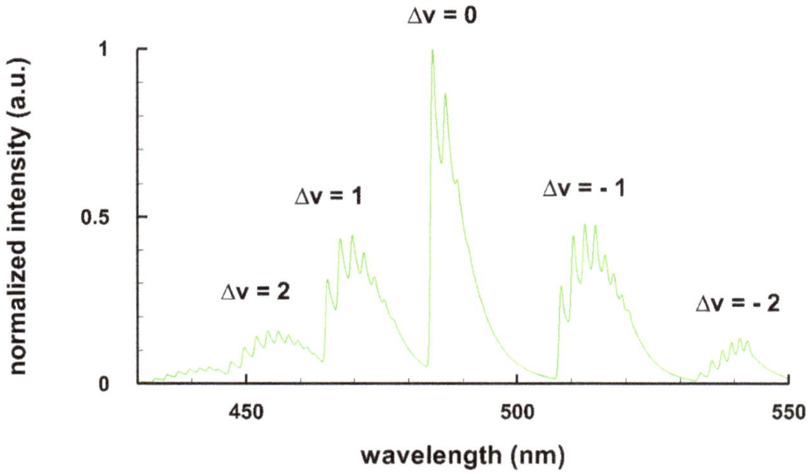

Figure 14.2. Computed spectrum of the AlO $B^2\Sigma^+ \rightarrow X^2\Sigma^+$ band, $T = 4$ kK, for a spectral resolution of $\Delta\lambda = 0.78$ nm ($\Delta\tilde{\nu} = 32$ cm^{-1}). The $\Delta\nu = 0, \pm 1, \pm 2$ sequences of the AlO progression are indicated. Reprinted from [2, 3], copyright (2015), with permission from Elsevier.

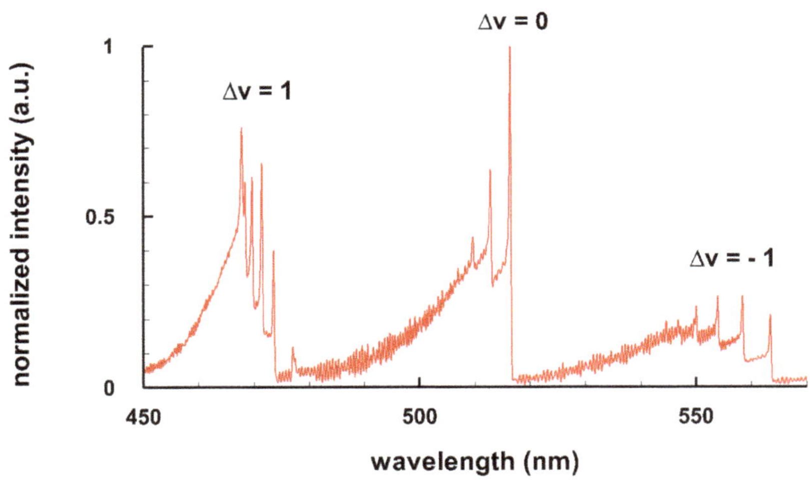

Figure 14.3. Computed spectrum of C_2 Swan $d^3\Pi_g \rightarrow a^3\Pi_u$ band progression, $T = 8$ kK, for a spectral resolution of $\Delta\lambda = 0.32$ nm ($\Delta\tilde{\nu} = 12$ cm^{-1}). The $\Delta\nu = 0, \pm 1$ sequences of the C_2 progression are indicated. Reprinted from [2, 3], copyright (2015), with permission from Elsevier.

presence and dynamics of molecules that would indicate combustion, *viz.* OH in hydrocarbon combustion or NH in ammonia combustion. Stellar astrophysical spectra from white dwarfs at a temperature of the order of 30 kK indicate primarily atomic hydrogen spectra in absorption. As white dwarfs cool to temperatures of the order of 8 kK, molecular spectra are frequently recorded in absorption. The establishment of a connection between astrophysical spectra measured in absorption and laboratory emission spectra measured in the laboratory requires correction for

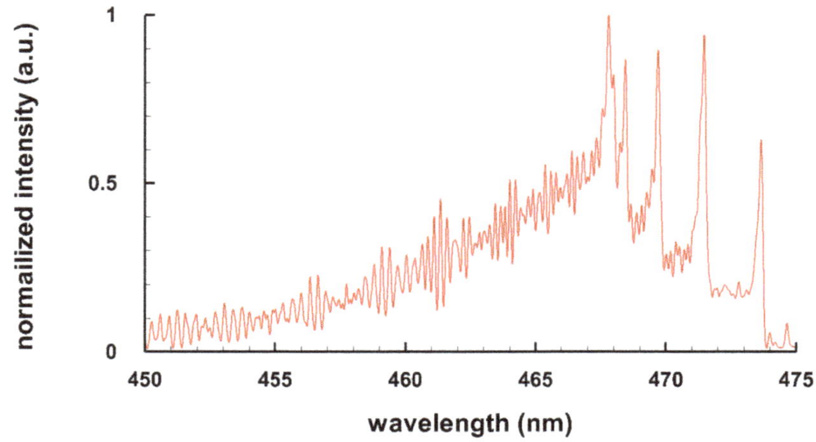

Figure 14.4. C_2 Swan $d^3\Pi_g \rightarrow a^3\Pi_u$ band $\Delta\nu = -1$ sequence, $T = 8$ kK, $\Delta\lambda = 0.13$ nm ($\Delta\tilde{\nu} = 6$ cm^{-1}) [2].

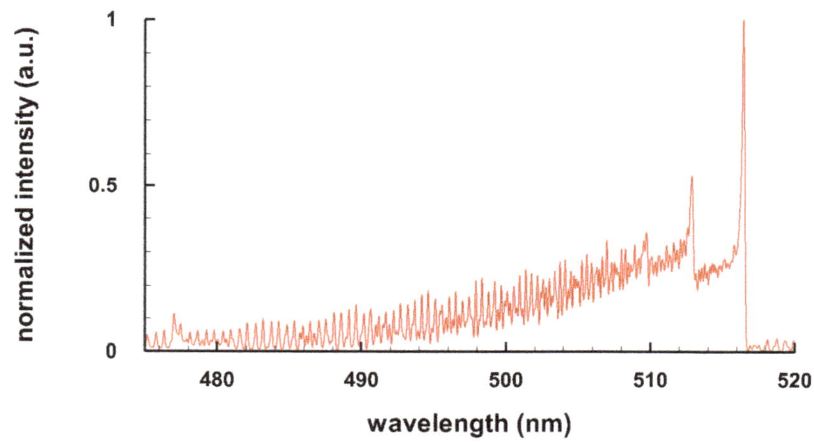

Figure 14.5. C_2 Swan $d^3\Pi_g \rightarrow a^3\Pi_u$ band $\Delta\nu = 0$ sequence, $T = 8$ kK, $\Delta\lambda = 0.15$ nm ($\Delta\tilde{\nu} = 6$ cm^{-1}) [2].

bound-free radiation, but otherwise the established molecular line strengths for molecules such as C_2 and CN are well-suited for astrophysical analysis.

Several laboratory measurements are designed to validate the data sets developed with the theory communicate in part I of this book for selected molecular transitions; equally, the data sets for several molecules are utilized in the analysis of the plasma condition. Moreover, data sets are also utilized in laser-induced fluorescence measurements in combustion studies, yet the primary application comprises analysis of the laser-plasma composition that will serve as diagnosis of the targets, which may be gases, liquids, or solids, including nano-particles.

Typical experimental arrangements are communicated in the next chapter, followed by several chapters on diatomic molecular spectroscopy.

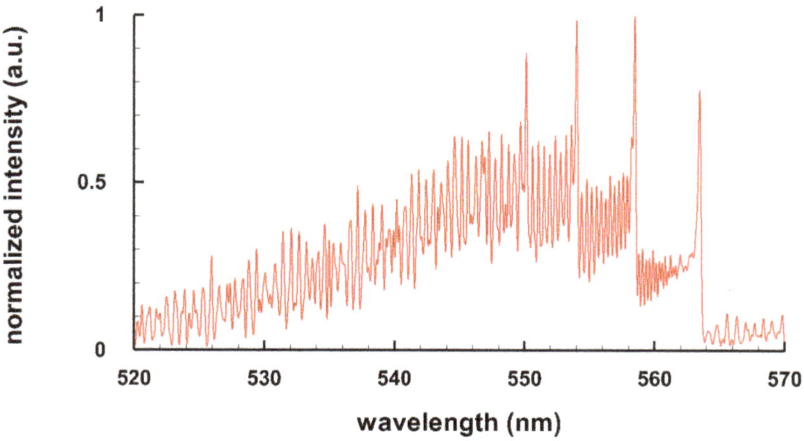

Figure 14.6. C_2 Swan $d^3\Pi_g \rightarrow a^3\Pi_u$ band $\Delta\nu = +1$ sequence, $T = 8$ kK, $\Delta\lambda = 0.18$ nm ($\Delta\tilde{\nu} = 6$ cm^{-1}) [2].

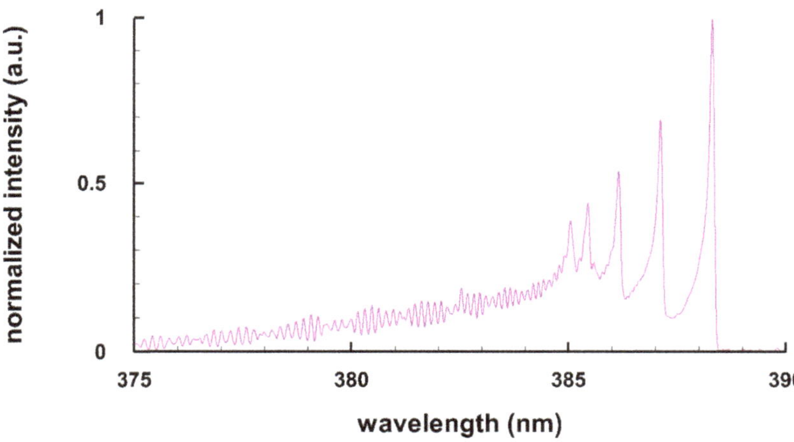

Figure 14.7. Computed spectrum of the CN violet $B^2\Sigma^+ \rightarrow X^2\Sigma^+$ band, $T = 8$ kK, for a spectral resolution of $\Delta\lambda = 0.09$ nm ($\Delta\tilde{\nu} = 6$ cm^{-1}) of the $\Delta\nu = 0$ sequence. Reprinted from [2, 3], copyright (2015), with permission from Elsevier.

References

[1] Wallace L 1962 *Astrophys. J. Suppl. S.* **7** 165

[2] Parigger C G and Hornkohl J O 2010 *Int. Rev. At. Mol. Phys.* **1** 25

[3] Parigger C G, Woods A C, Surmick D M, Gautam G, Witte M J and Hornkohl J O 2015 *Spectrochim. Acta* B **107** 132

IOP Publishing

Quantum Mechanics of the Diatomic Molecule (Second Edition)

Christian G Parigger and James O Hornkohl

Chapter 15

Computation of selected diatomic spectra

15.1 Introduction

This chapter communicates line-strength data and associated scripts for the computation and spectroscopic fitting of selected transitions of diatomic molecules. The scripts for data analysis are designed for inclusion in various software packages or program languages. Selected results demonstrate the applicability of the program for data analysis in laser-induced optical breakdown spectroscopy, primarily at the University of Tennessee Space Institute, Center for Laser Applications. Representative spectra are calculated and referenced to measured data records. Comparisons of experiment data with predictions from other tabulated diatomic molecular databases confirm the accuracy of the communicated line-strength data.

Atomic, molecular, and optical spectroscopy furnishes fundamental insight by decoding light emanating from targets of interest [30–37]. Analytical studies of elements may be straightforward, especially for elements that appear in the first three rows of the periodic table. Balmer-series hydrogen lines or sodium D-lines are usually well separated from spectral interference for low ($\sim$1 eV)-temperature plasma containing sodium as long as reasonable resolving power is available. For example, for the sodium D-lines, a resolving power, R, of $R \simeq$ 1000 is needed to distinguish the two components D1 and D2, separated by $\sim$0.06nm. Resolving individual lines of molecular spectra may require $R >$ 10 000, or at least of the order of one magnitude better resolution than needed for atoms, of course depending on temperature. In molecular spectroscopy, one tends to focus on molecular bands describing electronic transitions. The study of individual atomic or molecular resonances with continuous-wave radiation typically requires GHz scans with nominal MHz or better laser bandwidths. In this work, the focus is on optical spectrometers that measure near-ultraviolet to near-infrared molecular bands with a spectral resolution, $\delta\lambda$, of the order of $\delta\lambda \sim 0.1$ nm.

A collection of molecular diatomic spectroscopy data and expansive literature review and guidance [38] reveals a volley of recent and updated records in the

doi:10.1088/978-0-7503-6204-7ch15

ultraviolet to infrared wavelength range. However, this work's focus is the visible and near-infrared, specific sets of electronic transition data that have been tested in the analysis of experimental records. The mentioned databases [38] predict OH spectra among many others for diatomic molecules, e.g., ExoMol [10] and HITEMP [11], that can be visualized using (for example) PGOPHER [12]—PGOPHER also allows one to model transitions for prediction and comparative analysis. Selected diatomic molecular spectra of AlO, C_2, CN, OH, N_2^+, NO, and TiO, transitions are of interest because these can be observed in laser-induced breakdown spectroscopy (LIBS) [13–15] at standard ambient temperature and pressure (SATP). Diatomic AlO and TiO spectra usually occur following the creation of micro-plasma near or at aluminum and titanium surfaces, respectively. In several cases, molecular spectra may not be of primary interest in elemental analysis with LIBS using nanosecond laser pulses, but molecular spectra are readily observed with femtosecond laser-plasma excitation, or after some time delay (of the order of larger than 100ns for occurrence of CN in CO_2:N_2 gas mixtures) from optical breakdown when using nanosecond laser pulses. Just as for atomic spectra, reasonably accurate molecular spectra are required for analysis [16–20]. The construction of a molecular spectrum relies on: (i) accurate line positions and (ii) reasonably accurate transition strengths [21–24]. For the former, numerical singular value decomposition is employed for upper and lower states of a particular transition. For the latter, Frank–Condon factors and *r*-centroids are computed, and then combined with the rotational factors that usually decouple from the overall molecular line strength due to the symmetry of diatomic molecules.

This work communicates data files and associated scripts for the computation of diatomic molecular spectra, and equally for the fitting of measured data using a nonlinear fitting algorithm. Calculated spectra are presented and references to recorded data sets are provided. Applications comprise fields of chemistry, materials science, astronomy, and finally physics, including astrophysics, e.g., decoding of light from white dwarf stars such as Procyon B. The data are provided as a set of wave numbers, upper-level term values, and line strengths. Originally, FORTRAN/Windows 7 programs computed diatomic molecular spectra [21], but the scripts for the generation of molecular spectra have been redesigned for use with MATLAB [25]. Moreover, this work communicates MATLAB-optimized line-strength files (LSFs) containing three columns, namely wave numbers, upper term values, and line strengths. The codes are operating system independent as long as MATLAB or similar programs are available that allow scripts similar to the ones available in MATLAB. Supplementary data contain programs and nine selected diatomic molecular transitions of AlO, C_2 Swan, CN red, CN violet, OH ultraviolet, N_2^+, NO gamma, TiO γ, and TiO γ'.

15.2 Computation details

The computation of diatomic molecular spectra uses established line-strength data. Programs in FORTRAN accomplish the generation of spectra, coupled with a separate plotting program for visualization, including convenient implementation using the Microsoft-Windows 7 operating system. This work communicates equivalent MATLAB scripts that appear popular with various research groups. First, the Boltzmann equilibrium spectral program (BESP) generates a theoretical spectrum; and second, the Nelder–Mead temperature (NMT) program accomplishes fitting of experimental and theoretical spectra. In principle, BESP can be used to generate maps as a function of temperature and linewidth with subsequent determination of the optimum solution with minimal errors in the least-square sense. In turn, NMT uses nonlinear optimization using geometric constructs, viz. simplicia. The accumulation of experimental spectra in this work is in accord with laser-induced breakdown spectroscopy, or in general laser spectroscopy [26].

15.2.1 MATLAB scripts

The parameter list includes wavelength minimum, maximum, temperature, number of points, normalization factor, and file name. For the BESP.m and NMT.m scripts, the outputs are generated in graphical form. Table 15.1 lists constants that could be used (comment line in the scripts) for the determination of the variation of the refractive index, *n*, of air with wavelength [27],

$$10^6(n - 1) = a_0 + \frac{a_1}{\lambda_N^2} + \frac{a_2}{\lambda_N^4}, \tag{15.1}$$

where λ_N is the wavelength in normal air at 15 °C and 101 325 Pa (760 mm Hg), expressed in terms of micrometers (range 0.2218–0.9000 μm).

Table 15.2 lists constants that are used to account for the variation of the refractive index, r_i, of air at 15 °C, 101 325 Pa, and 0% humidity, with wave number [28],

Table 15.1. Constants for variation of refractive index, *n* (see equation (15.1)).

Parameter	Value
a_0	272.643
a_1	1.2288 (μm^2)
a_2	0.035 55 (μm^4)

Table 15.2. Constants for variation of refractive index (see equation (15.2)).

Parameter	Value (μm^{-2})
k_0 (k0)	238.0185
k_1 (k1)	5 792 105
k_2 (k2)	57.362
k_3 (k3)	167 917

Table 15.3. Constants in BESP.m and NMT.m.

Constant	Value
Planck constant (h)	$6.626\,069\,57 \times 10^{-34}$ (J s)
Speed of light (c)	$2.997\,924\,58 \times 10^{8}$ (m s^{-1})
Boltzmann constant (kB)	$1.380\,6488 \times 10^{-23}$ (J K^{-1})

Table 15.4. Parameters and variables in BESP.m and NMT.m.

Description	Variable
Wavelength minimum	wl_min (cm^{-1})
Wavelength maximum	wl_max (cm^{-1})
Temperature	T (kK)
Full-width at half maximum	FWHM, δλ (nm)
Number of points	N
Normalization	norm
File name	x

$$10^8(r_i - 1) = \frac{k_1}{(k_0 - \sigma^2)} + \frac{k_3}{(k_2 - \sigma^2)}, \qquad (15.2)$$

where σ is the wave number in units of μm^{-1}.

Tables 15.3 and 15.4 summarize script constants and input variables that are important for spectra computations, respectively. However, redesign of BESP.m and NMT.m from the FORTRAN/Windows 7 version [29] was accomplished with extensive discussions [24]. Edited versions of BESP.m and NMT.m are communicated in this work along with nine separate data files.

NMT.m

The details of the NMT script are deferred to appendix J. The adaptation of a previous FORTRAN code with Windows 7 libraries for a Microsoft platform is no longer viable due to support discontinuation of the Windows 7 operating system. However, the NMT.m script delivers spectra-fitting results identical to those obtained with the FORTRAN/Windows 7 implementation.

Data files

This section explains the line-strength data communicated in this work. The LSFs contain wave numbers, upper term values, and the line strengths. Table 15.5 summarizes contents of line-strength data. The air wavelength in the program,

Table 15.5. Line-strength data contents: vacuum wave numbers and upper term values, line strengths.

Description	Variable	Column
Wave number	WN (cm^{-1})	1
Upper term value	Tu (cm^{-1})	2
Line strength	S (stC^2 cm^2) [a]	3

[a] 1 stC = 3.356 10^{-10} C.

Table 15.6. Diatomic molecules, line-strength data files, wavelength range, and number of spectral lines.

Diatomic molecule	LSF	Range (nm)	Number of lines
Aluminum monoxide (AlO)	AlO-BX-LSF.txt	430.72–997.66	33 484
Carbon Swan spectra (C_2)	C2-Swan-LSF.txt	410.93–678.58	29 004
Cyanide red (CNr)	CNr-LSF.txt	499.89–4997.56	40 728
Cyanide violet (CNv)	CNv-LSF.txt	372.88–425.22	7960
Hydroxyl (OH) violet	OH-LSF.txt	278.65–379.72	1683
Nitrogen monoxide (NO) gamma	NO-GAMMA-LSF.txt	200.41–285.95	13 000
Singly ionized nitrogen (N_2^+)	N2p-LSF.txt	319.04–501.46	7302
Titanium monoxide (TiO) γ	TiO-AX-LSF.txt	599.58–945.44	66 962
Titanium monoxide (TiO) γ'	TiO-BX-LSF.txt	582.73–679.12	34 648

WL, is in units of nm. The two programs BESP and NMT convert the vacuum wave numbers to air wavelengths for the analysis of measured data; see equation (15.1). Table 15.6 associates the diatomic molecules and their line-strength data, including the wavelength range.

The LSFs contain significantly more data than illustrated in this communication. Applications of the LSFs include data analysis of laser-induced fluorescence and computation of absorption spectra. Some of these applications are elaborated on in the discussion of C_2 Swan spectra [22].

15.3 Results

This section summarizes the communicated line-strength data. Table 15.7 associates the diatomic molecules and their LSFs. The LSFs contain wave numbers, upper term values and the line strength. The two programs BESP and NMT convert the vacuum wave numbers to air wavelengths for the analysis of the measured data. Table 15.7 displays spectral resolution and temperature, and table 15.7 also communicates but gives one reference each for measurement and fitting selected

Table 15.7. Diatomic molecules, spectral resolution, temperature, and one typical reference each that uses the data.

Diatomic molecule	LSF	$\delta\lambda$ (nm)	T (kK)	Refs.	figure
Aluminum monoxide (AlO)	AlO-BX-LSF.txt	1.0	3.33	[30]	Figure 15.1
Carbon Swan spectra (C_2)	C2-Swan-LSF.txt	0.39	6.75	[31]	Figure 15.2
Cyanide red (CNr)	CNr-LSF.txt	0.38	7.5	[32] [a]	Figure 15.3
Cyanide violet (CNv)	CNv-LSF.txt	0.030	7.94	[33]	Figure 15.4
Singly ionized nitrogen (N_2^+)	N2p-LSF.txt	0.035	5.1	[34]	Figure 15.5
Hydroxyl (OH) ultraviolet	OH-LSF.txt	0.35	3.39	[35]	Figure 15.6
Nitrogen monoxide (NO) gamma	NO-GAMMA-LSF.txt	0.056	6.8	[36]	Figure 15.7
Titanium monoxide (TiO) γ	TiO-AX-LSF.txt	0.10	3.03	[37] [b]	Figure 15.8
Titanium monoxide (TiO) γ'	TiO-BX-LSF.txt	0.40	3.6	[38]	Figure 15.9

[a] Experiments at Johannes Kepler University, Linz, Austria. [b] Experiments in part at Chemical Research Center of the Hungarian Academy of Science, Budapest, Hungary.

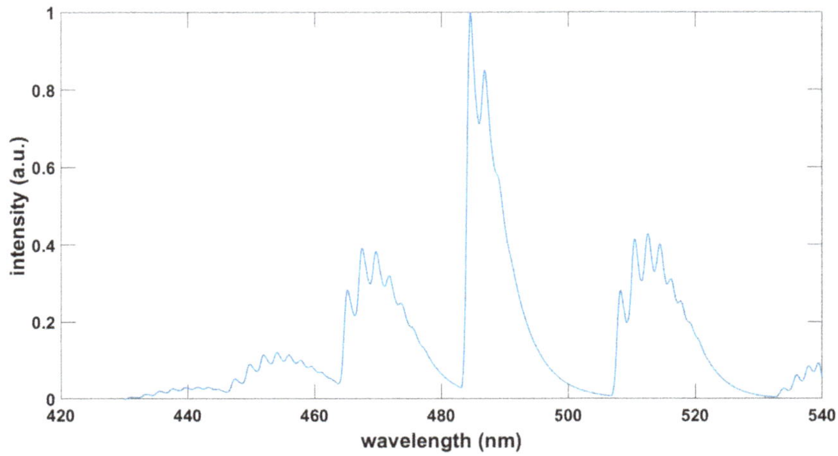

Figure 15.1. Computed AlO spectrum, $\Delta v = 0, \pm 1, \pm 2, +3, \delta\lambda = 1.0$ nm, $T = 3.33$ kK. Reproduced from [39]. CC BY 4.0.

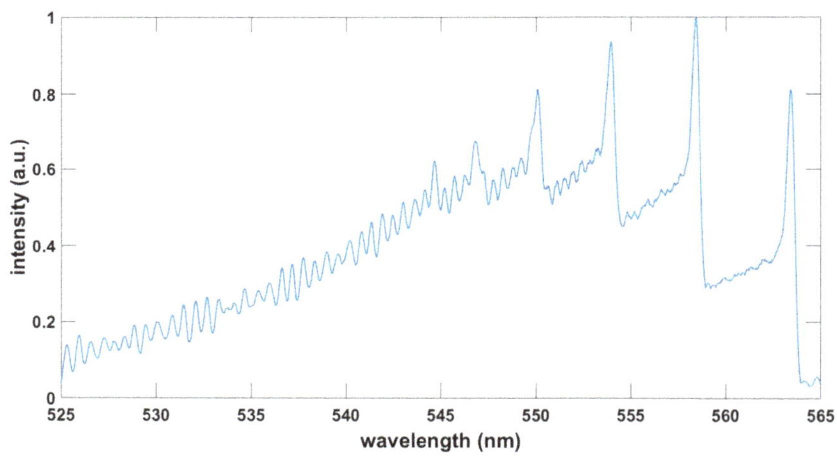

Figure 15.2. Computed C_2 Swan spectrum, $\Delta v = -1$, $\delta\lambda = 0.39$ nm, $T = 6.75$ kK. Reproduced from [39]. CC BY 4.0.

molecular transitions of AlO, C_2, CN, OH, N_2^+, NO, and TiO. Figures 15.1–15.9 illustrate computed spectra that refer to the measured ones in the references.

15.4 Discussion

The accurate prediction of the line positions of diatomic molecules is important for the identification and, of course, for the fitting of measured data. The line positions are usually more accurate than the intensity values. The selected transitions for most of the communicated diatomic molecules, especially AlO, C_2 Swan, CN, and OH,

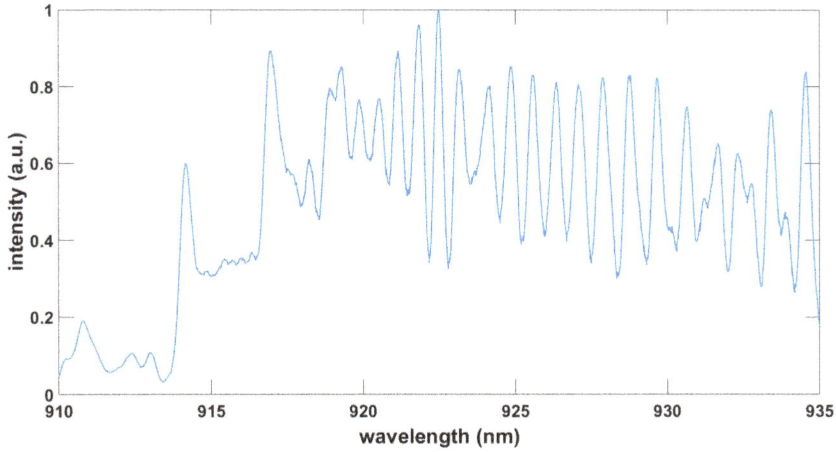

Figure 15.3. Computed CN red spectrum, $\Delta v = +1$, $\delta\lambda = 0.38$ nm, $T = 7.5$ kK. Reproduced from [39]. CC BY 4.0.

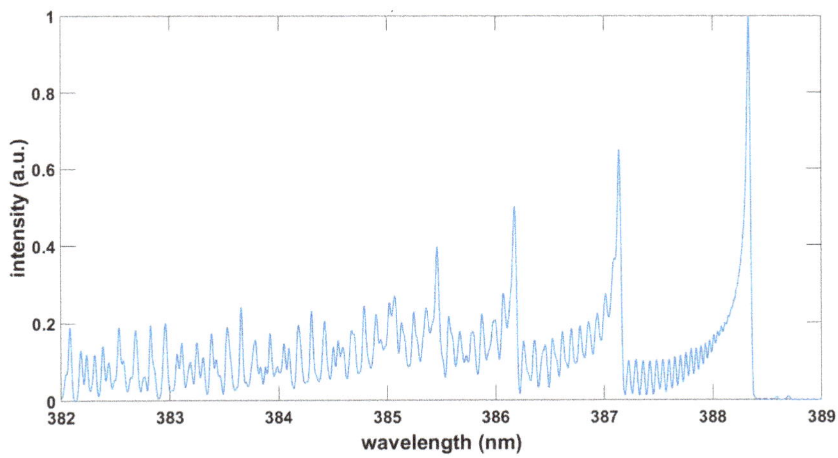

Figure 15.4. Computed CN violet spectrum, $\Delta v = 0$, $\delta\lambda = 0.030$ nm, $T = 7.94$ kK. Reproduced from [39]. CC BY 4.0.

have been extensively tested in the study of laser-induced optical breakdown. Comparisons of the analysis of an experimental OH ultraviolet data record using the communicated OH table and the recent and updated ExoMol diatomic molecular databases reveal agreements of most wavelength positions, of the order of 10% variations of the line strengths, but better than 3% agreement in fitted temperature, spectral resolution, and background. This bodes well for applications of expansive databases such as ExoMol in analytical laser-plasma research for the other diatomic molecules communicated in this work.

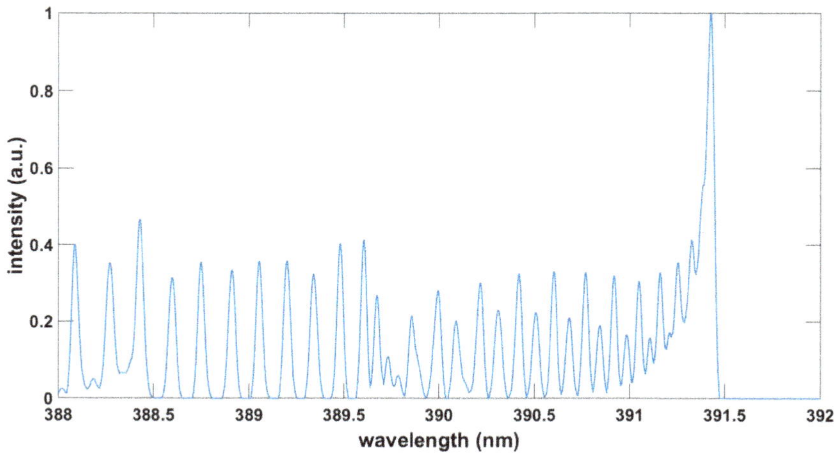

Figure 15.5. Computed N_2^+ spectrum, $\Delta v = 0$, $\delta\lambda = 0.035$ nm, $T = 5.1$ kK. Reproduced from [39]. CC BY 4.0.

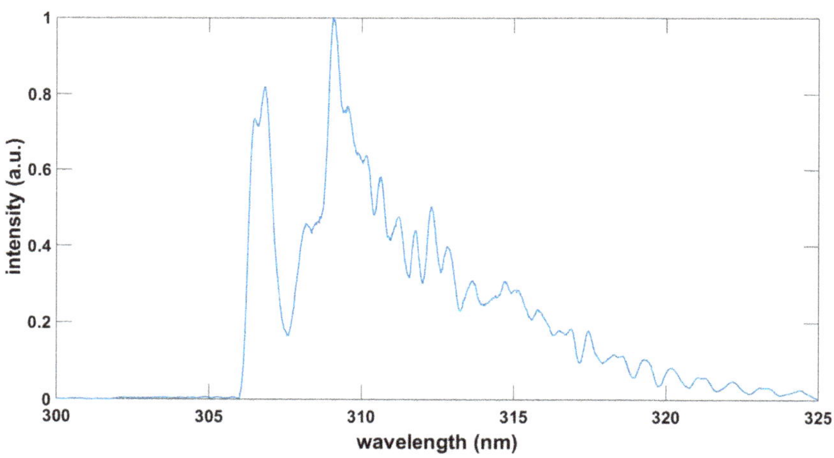

Figure 15.6. Computed OH spectrum, $\Delta v = 0$, $\delta\lambda = 0.35$ nm, $T = 3.39$ kK. Reproduced from [39]. CC BY 4.0.

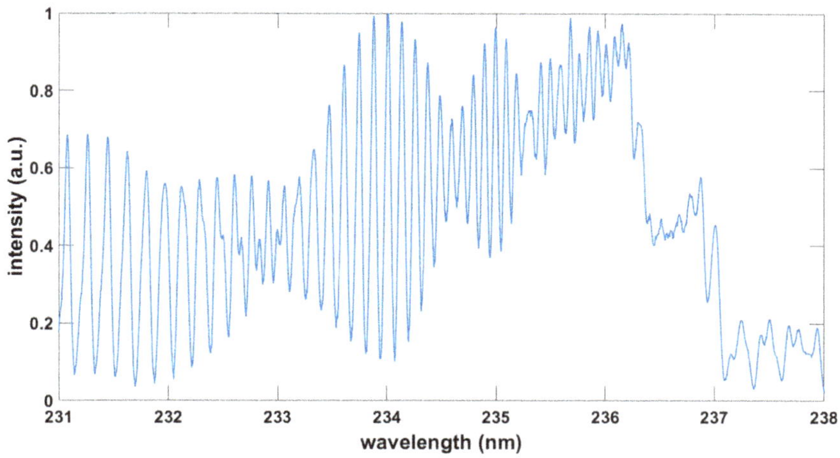

Figure 15.7. Computed NO gamma spectrum, $\Delta v = -1$, $\delta\lambda = 0.056$ nm, $T = 6.80$ kK. Reproduced from [39].

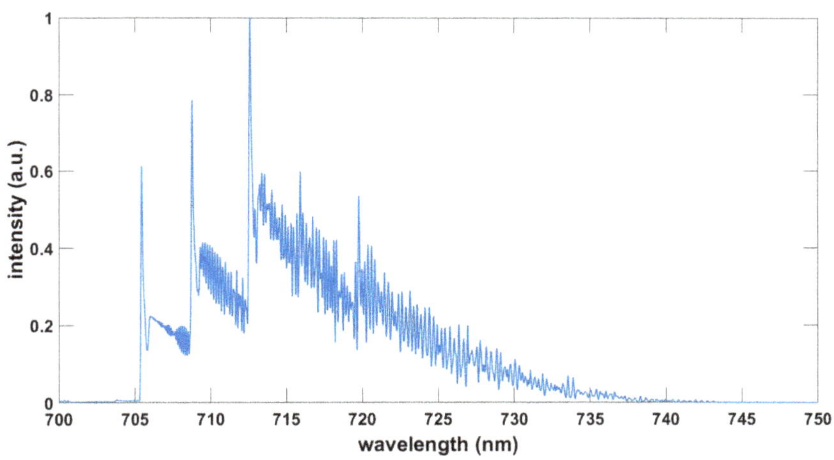

Figure 15.8. Computed TiO γ spectrum, $\Delta v = 0$, $\delta\lambda = 0.10$ nm, $T = 3.03$ kK. Reproduced from [39].

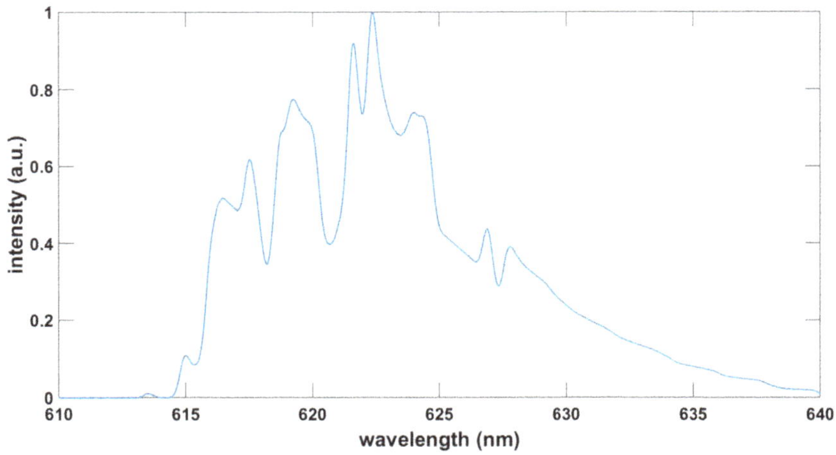

Figure 15.9. Computed TiO γ' spectrum, $\Delta v = 0$, $\delta\lambda = 0.40$ nm, $T = 3.6$ kK. Reproduced from [39]. CC BY 4.0.

References

[1] Kunze H-J 2009 *Introduction to Plasma Spectroscopy* (Heidelberg: Springer)

[2] Fujimoto T 2004 *Plasma Spectroscopy* (Oxford: Clarendon)

[3] Ochkin V N 2009 *Spectroscopy of Low Temperature Plasma* (Weinheim: Wiley)

[4] Omenetto E D 1979 *Analytical Laser Spectroscopy* (New York: Wiley)

[5] Demtröder W 2014 *Laser Spectroscopy 1: Basic Principles* 5th edn (Heidelberg: Springer)

[6] Demtröder W 2015 *Laser Spectroscopy 2: Experimental Techniques* 5th edn (Heidelberg: Springer)

[7] Hertel I V and Schulz C-P 2015 *Atoms, Molecules and Optical Physics 1, Atoms and Spectroscopy.* (Heidelberg: Springer)

[8] Hertel I V and Schulz C-P 2015 *Atoms, Molecules and Optical Physics 2, Molecules and Photons–Spectroscopy and Collisions* (Heidelberg: Springer)

[9] McKemmish L K 2021 *WIREs Comput. Mol. Sci.* **11** e1520

[10] Tennyson J *et al* 2020 *J. Quant. Spectrosc. Radiat. Transf.* **255** 107228

[11] Rothman L S, Gordon I E, Barber R J, Dothe H, Gamache R R, Goldman A, Perevalov V I, Tashkun S A and Tennyson J 2010 *J. Quant. Spectrosc. Radiat. Transf.* **111** 2139

[12] Western C M 2017 *J. Quant. Spectrosc. Radiat. Transfer* **186** 221

[13] Miziolek A W, Palleschi V and Schechter I 2006 *Laser Induced Breakdown Spectroscopy* (New York: Cambridge University Press)

[14] Singh J P and Thakur S N (ed) 2020 *Laser-Induced Breakdown Spectroscopy* 2nd edn (New York: Elsevier)

[15] De Giacomo A and Hermann J 2017 *J. Phys. D Appl. Phys.* **50** 183002

[16] Parigger C G, Surmick D M, Helstern C M, Gautam G, Bol'shakov A A and Russo R 2020 Laser Induced Breakdown *Molecular Laser-Induced Breakdown Spectroscopy. In: Laser Induced Breakdown Spectroscopy* ed J P Singh and S N Thakur 2nd edn (New York: Elsevier)

[17] C. G. Parigger. Laser-induced breakdown in gases: Experiments and simulation, in: Laser-Induced Breakdown Spectroscopy(LIBS), Fundamentals and Applications, A. W. Miziolek, V. Palleschi, I. Schechter, ed, ch 4. (New York: Cambridge University Press) 2006
[18] Parigger C G, Helstern C M, Jordan B S, Surmick D M and Splinter R 2020 *Molecules* **25** 615
[19] Parigger C G, Jordan H B S, Surmick D M and Splinter R 2020 *Molecules* **25** 988
[20] Parigger C G 2020 *Spectrochim. Acta* B **179** 106122
[21] Parigger C G, Woods A C, Surmick D M, Gautam G, Witte M J and Hornkohl J O 2015 *Spectrochim. Acta* B **107** 132
[22] Hornkohl J O, Nemes L and Parigger C G 2009 *Spectroscopy, dynamics and molecular theory of carbon plasmas and vapors. In: Advances in the Understanding of the Most Complex High-Temperature Elemental System, L. Nemes and S. Irle, ed, chapter 4* ed L Nemes and S Irle (Singapore: World Scientific) Chapter 4
[23] Parigger C G and Hornkohl J O 2020 *Quantum Mechanics of the Diatomic Molecule with Applications* (Bristol: IOP Publishing)
[24] Surmick D M and Hornkohl J O 2016 (The University of Tennessee, University of Tennessee Space Institute, Tullahoma, TN, USA) personal communication
[25] *MATLAB Release R2022a Update 5* (MA: Natick)
[26] Parigger C G, Woods A C, Witte M J, Swafford L D and Surmick D M 2014 *J. Vis. Exp.* **84** 51250
[27] Barrell H and Sears J E 1939 *Philos. Trans. Roy. Soc. London* **238** 1
[28] Ciddor P E 1996 *Appl. Opt.* **35** 1567
[29] Parigger C G, Woods A C, Surmick D M, Gautam G, Witte M J and Hornkohl J O 2015 *Spectrochim. Acta* B **107** 132
[30] Dors I G, Parigger C G and Lewis J W L 1998 *Opt. Lett.* **23** 1778
[31] Parigger C G, Plemmons D H, Hornkohl J O and Lewis J W L 1994 *J. Quant. Spectrosc. Radiat. Transf.* **52** 707
[32] Trautner S, Jasik J, Parigger C G, Pedarnig J D, Spendelhofer W, Lackner J and Veis P 2017 and *J. Heitz. Spectrochim. Acta* A **174** 331
[33] Hornkohl J O, Parigger C G and Lewis J W L 1991 *J. Quant. Spectrosc. Radiat. Transf.* **46** 405
[34] Parigger C G, Plemmons D H, Hornkohl J O and Lewis J W L 1995 *Appl. Opt.* **34** 3331
[35] Parigger C G 2022 *Foundations* **2** 934
[36] Hornkohl J O, Fleischmann J P, Surmick D M, Witte M J, Swafford L D, Woods A C and Parigger C G 2014 *J. Phys. Conf. Ser.* **548** 12040
[37] Parigger C G, Woods A C, Keszler A, Nemes L and Hornkohl J O 2012 *AIP Conf. Proc.* 1464 628
[38] Woods A C, Parigger C G and Hornkohl J O 2012 *Opt. Lett.* **37** 5139
[39] Parigger C G 2023 *Foundations* **3** 1

Chapter 16

Experimental arrangement for laser-plasma diagnosis

16.1 Spectroscopy

Experimental investigations of diatomic molecular-emission spectra utilize laboratory, table-top scale laser, spectrometer, and detector equipment. Laser-plasma is generated by focusing pulsed radiation to an irradiance sufficient to achieve optical breakdown at the target. The irradiance threshold for standard ambient temperature and pressure (SATP) dry air is of the order of 0.3 TW cm $^{-}$2, which can be accomplished with f/10 focusing of 10 ns, 100 mJ/pulse, ir radiation at 1.064 μm radiation from a Q-switched Nd:YAG device, e.g., the ratio of a single 10 cm focal-length lens and a 1 cm beam diameter at the lens amounts to 10, labeled as f/10 focusing. The spot-size diameter is approximately the product of wavelength and f-number; therefore, the cross-section focal area is of the order of 10^{-6} cm^2. The peak power of a 100 mJ, 10 ns pulse is 10 MW; consequently, the peak irradiance in focus is of the order of 10 TW cm $^{-}$2. However, a detailed calculation that includes consideration of physical optics and primarily spherical aberration correction would reveal a multiple-peak distribution in the focal volume. The focal volume is defined by the Raleigh-length that extends from either side of focus to an area cross section twice that of the focal area. Typical results are communicated in recent laser-induced breakdown spectroscopy (LIBS) [1, 2].

A recent experimental arrangement for laboratory measurements of laser-induced plasma [3] includes a Q-switched Nd:YAG laser device that generates 6 ns laser pulses with an energy of 850 mJ per pulse at the fundamental wavelength of 1064nm. Radiation is tightly focused to a spot size of the order of 10 μm to achieve a peak irradiance above 1 TW cm $^{-}$2, or well above optical breakdown in SATP laboratory air and gaseous hydrogen at a pressure of the order of one atmosphere. The emanating light from the micro-plasma is recorded with a Czerny-Turner type spectrometer that shows a resolution of 0.02 nm when using a 3600 groves/mm

doi:10.1088/978-0-7503-6204-7ch16

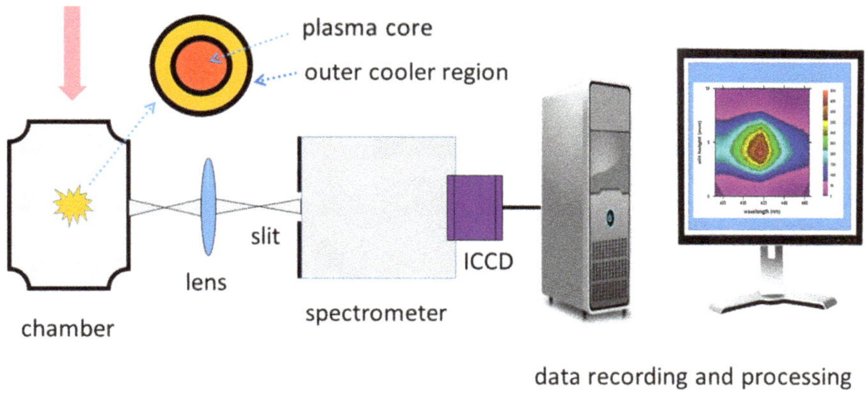

Figure 16.1. Optical breakdown plasma is generated by focusing the laser beam from the top and parallel to the slit. Time-resolved 1:1 images are recorded with the ICCD [3].

holographic grating and an intensified diode array or two-dimensional detector such as an intensified charge-coupled detector (ICCD) for temporally- and spatially-resolved spectroscopy. Figure 16.1 illustrates an experimental schematic of atomic- and molecular-emission spectroscopy following optical breakdown in gases.

Several variants of the experimental arrangement are employed. For example, a cell is not required for studies of air breakdown, laser-ablation of solids and liquids or droplets, or nano-particles. In addition, for studies of laser-induced ignition of combustible mixtures specially designed burners with capabilities of generating a shroud flow are utilized. Other variants of the experiments discussed in this book include different laser devices for plasma generation, including a pico-second excimer laser operating at a wavelength $0.308\,\mu$m [4] and a CO_2 laser operating at $10.6\,\mu$ [5]. A scanning dye laser is employed for planar laser-induced fluorescence experiments.

Figure 16.2 displays a schematic that includes the excitation laser in the experimental arrangement for spectroscopy. For 1:1 imaging of the plasma onto the 100 μm spectrometer slit, a fused silica plano-convex lens, Thorlabs model LA4545, is employed. For the OH experiments, the laser pulse energy is attenuated with beam splitters and apertures from 850–170 mJ/pulse. The laser beam is focused as for experiments in a chamber, but for the OH measurements discussed in chapter 25 the cell in figure 16.2 was removed.

The shadowgraph experiments [7] utilize two separate laser devices, Continuum Surelite model SL I-10, that can be externally operated to deliver laser pulses with a well-defined time delay showing less than ±1 ns trigger-jitter between the pulses. For visualization studies, both lasers are frequency-doubled to operate at the second harmonic, 532 nm wavelength, and both beams are spatially overlapped.

Various details of laser spectroscopy and high peak power laser–matter interactions have been elaborated [8–11]. Wavelength and sensitivity corrections are important for comparison of recorded and computed data due to the wavelength range of molecular spectra, especially when considering rotation-vibration spectra

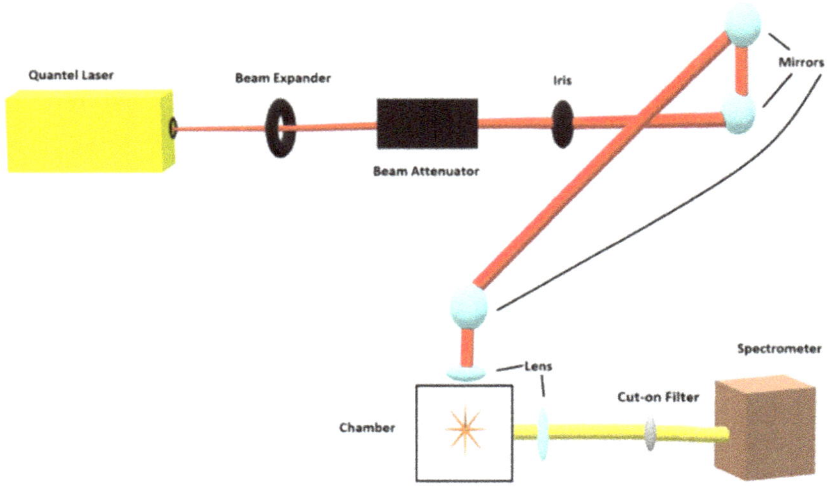

Figure 16.2. Schematic experimental arrangement for spectroscopy. Reproduced from [6]. CC BY 4.0.

sequences and progressions. For summaries of LIBS that show 6–8kK excitation temperature molecular-emission spectra see Miziolek *et al* [12] and Jagdish and Thakur [13].

Detailed equipment lists are communicated in the handbook by Cremers and Radziemski [14] for laser-induced breakdown spectroscopy. The equipment list and performance parameters are of course applicable for molecular plasma-emission spectroscopy. Recent laboratory methods are communicated in a video [15] and are routinely applied in the Center for Laser Application at the University of Tennessee Space Institute. The text associated with [15] conveys a detailed list of equipment for atomic and molecular spectroscopy.

16.2 Shadowgraphy

Shadowgraphs are recorded by external synchronization of the Surelite and Quantel laser devices, and by externally triggering the camera, Silicon Video 9C10, that records the images that are projected onto a screen. Figure 16.3 illustrates the shadowgraph arrangement. Figure 16.4 communicates a photograph of a typical shadowgraph experiment.

16.3 Summary

The analysis of the measured spectra uses the concepts and methods discussed in part I of this book. For isolated molecular transitions, the Franck–Condon factors are found by solving the radial Schrödinger equation numerically, and the Hönl–London strengths are obtained by numerical diagonalization of the rotational and fine-structure Hamiltonian, see Hornkohl *et al* [16]. The computation of a diatomic spectrum is accomplished by specifying the temperature and spectral resolution. In turn, nonlinear fitting routines can be employed to infer the spectroscopic

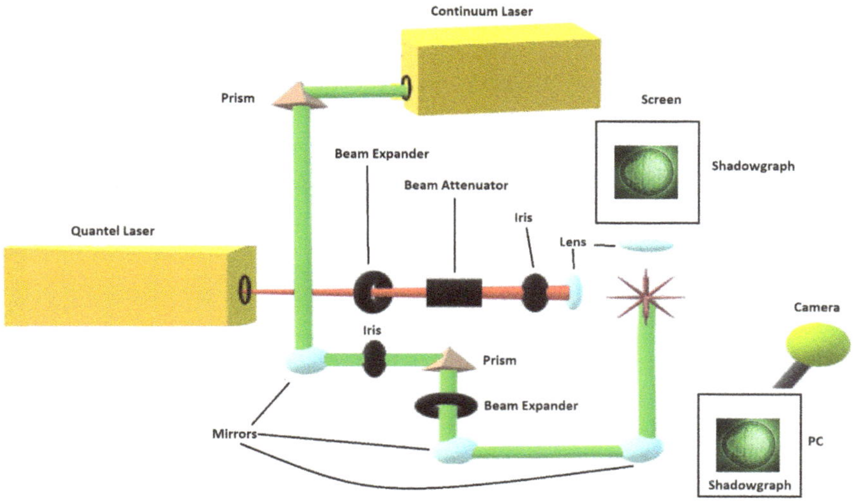

Figure 16.3. Schematic experimental arrangement for shadowgraph recording. Reproduced from [6]. CC BY 4.0.

Figure 16.4. Photograph of the experimental arrangement for air breakdown. Reproduced from [6]. CC BY 4.0.

temperature, spectral resolution for the fitting, and background contributions due to blackbody radiation or contributions from other species [17]. Of interest are diatomic molecules such as AlO, C_2, CH, CN, CrF, N_2^+ 1^{st}Neg, N_2 1^{st}Pos, N_2 2^{nd}Pos, NH, NO, OH. Accurate line-strength files are in use for spectroscopic analysis with the so-called Boltzmann equilibrium spectrum program, see Hornkohl and Parigger [18]. The BESP is used, for example, in the analysis of laser-induced optical breakdown and laser-induced fluorescence. The computed or synthetic spectra have been instrumental in the analysis of molecular laser-induced optical breakdown, and equally for experimental check-points of the computed data-base for diatomic molecular spectroscopy, see, for example, [4, 5, 19–24].

References

[1] Parigger C G 2006 Laser-induced breakdown in gases: experiments and simulation *Laser-Induced Breakdown Spectroscopy (LIBS), Fundamentals and Applications* ed A W Miziolek, V Palleschi and I Schechter (New York: Cambridge University Press) ch 4

[2] Parigger C G, Surmick D M, Helstern C M, Gautam G, Bol'shakov A A and Russo R 2020 Molecular laser-induced breakdown spectrosocpy *Laser-Induced Breakdown Spectroscopy* ed J P Singh and S N Thakur (New York: Elsevier Science) ch 7

[3] Parigger C G, Drake K A, Helstern C M and Gautam G 2018 *Atoms* **6** 36

[4] Hornkohl J O, Parigger C G and Lewis J W L 1991 *J. Quant. Spectrosc. Radiat. Transfer* **46** 405

[5] Parigger C G, Hornkohl J O and Nemes L 2010 *Int. J. Spectrosc.* **2010** 159382

[6] Parigger C G, Helstern C M and Gautam G 2020 *Symmetry* **12** 2116

[7] Parigger C G, Jordan H B S, Surmick D M and Splinter R 2020 *Molecules* **25** 988

[8] Kunze H-J 2009 *Introduction to Plasma Spectroscopy* (Heidelberg: Springer)

[9] Bauer D and Mulser P 2010 *High Power Laser-Matter Interaction* (Heidelberg: Springer)

[10] Hertel I V and Schulz C-P 2015 *Atoms, Molecules and Optical Physics 1, Atoms and Spectroscopy* (Heidelberg: Springer)

[11] Hertel I V and Schulz C-P 2015 *Atoms, Molecules and Optical Physics 2, Molecules and Photons—Spectroscopy and Collisions* (Heidelberg: Springer)

[12] and A W Miziolek, V Palleschi and I Schechter 2006 *Laser-Induced Breakdown Spectroscopy (LIBS)—Fundamentals and Applications* (New York: Cambridge University Press)

[13] Singh J P and Thakur S N (ed) 2020 *Laser-Induced Breakdown Spectroscopy* (New York: Elsevier Science)

[14] Cremers D E and Radziemski L J 2006 *Handbook of Laser-Induced Breakdown Spectroscopy* (New York: Wiley)

[15] Parigger C G, Woods A C, Witte M J, Swafford L D and Surmick D M 2014 *J. Vis. Exp.* **84** 51250

[16] Nemes L, Hornkohl J O and Parigger C G 2005 *Appl. Opt.* **44** 3686

[17] Parigger C G, Woods A C, Surmick D M, Gautam G, Witte M J and Hornkohl J O 2015 *Spectrochim. Acta* B **107** 132

[18] Hornkohl J O and Parigger C G 1996 Boltzmann equilibrium spectrum program (BESP), 1996 (accessed January 19, 2016). http://view.utsi.edu/besp/

[19] Parigger C G, Plemmons D H, Hornkohl J O and Lewis J W L 1994 *J. Quant. Spectrosc. Radiat. Transfer* **52** 707

[20] Hornkohl J O, Nemes L and Parigger C G 2009 Spectroscopy, dynamics and molecular theory of carbon plasmas and vapors *Advances in the Understanding of the Most Complex High-Temperature Elemental System* ed L Nemes and S Irle (Singapore: World Scientific) ch 4

[21] Nemes L, Keszler A M, Hornkohl J O and Parigger C G 2005 *Appl. Opt.* **44** 3661
[22] Dors I G, Parigger C G and Lewis J W L 1998 *Opt. Lett.* **23** 1778
[23] Parigger C G, Guan G and Hornkohl J O 2003 *Appl. Opt.* **42** 5986
[24] Woods A C, Parigger C G and Hornkohl J O 2012 *Opt. Lett.* **37** 5139

IOP Publishing

Quantum Mechanics of the Diatomic Molecule (Second Edition)

Christian G Parigger and James O Hornkohl

Chapter 17

Methylidyne, CH, cavity ring-down spectroscopy in a microwave plasma discharge

17.1 Introduction

This chapter communicates cavity ring-down spectroscopy (CRDS) of methylidyne (CH) in a chemiluminescent plasma that is produced in a microwave cavity. Of interest are the rotational lines of the 0–0 vibrational transition for the A–X band and the 1-0 vibrational transition for the B–X band. The reported investigations originate from CH radical research in 1996 that constituted the first case of applying CRDS to the CH radical. This chapter also includes recent analysis that shows excellent agreement of measured and computed data, and it communicates CH line strength data. The CH radical is an important diatomic molecule in hydrocarbon combustion diagnosis and analysis of stellar plasma emissions, to name just two examples for analytical plasma chemistry.

CRDS was introduced by O'Keefe and Deacon in 1988 [1] and has since been used to an increasing extent for the measurement of weak absorbers or minute amounts of substances in the gaseous phase. Thus, overtone bands [2] and the Herzberg absorption system in molecular oxygen [3] have been analyzed this way. Jet-cooled metal clusters [4, 5] and trace gas components [6] were probed by CRDS. Additionally, CRDS has proved eminently applicable for chemical kinetic system analysis, e.g., see [7, 8], that often involve transient radicals. Free radicals such as oxymethyl (HCO) in hydrocarbon flames [9] or the methyl (CH_3) radical [10] were studied by this technique. We have applied this method in the form of coherent CRDS [11] to the spectroscopic analysis of the CH radical.

This report communicates selected data records from investigations in 1996. Specifically, the CH B–X transition has been the subject of research in subsequent years [12–14]. In addition, this report summarizes recent analysis that utilizes accurate line strength data for CH [15, 16], and provides the CH line strength data for the A–X and B–X transitions. The line strength files (LSFs) for CH can also

doi:10.1088/978-0-7503-6204-7ch17

be applied for analysis of emission spectra that may be collected in laser-induced breakdown spectroscopy [17, 18]. The work in this report may have applications in astrophysics [19–21], combustion studies [22], and diamond film chemical vapor deposition [23].

17.2 Experiment details

A schematic view of the experimental CRDS arrangement is nicely described in [24], including cascade arc plasma source, gas injection, optical cavity, and photo-multiplier/oscilloscope detection, but this work employs a grating spectrometer, as further described in this section. The CH radicals were generated by the oxidation of acetylene (C_2H_2) using excited oxygen atoms produced in an inductively coupled microwave plasma (200 W at 2.45 GHz) in oxygen gas bubbled through water. The discharge was initiated in argon employing a flow inlet a few centimeters from the cavity mirrors. The flow of argon suppressed etching of the coating of the reflective mirrors by the flow of radicals. The chemiluminescent reaction leading to the generation of CH in the CRDS cavity occurred upon mixing the wet oxygen and acetylene via a distributed set of inlet openings, while the cavity was continuously pumped by two Roots pumps of a total capacity of 500 m^3/hour. This source was previously described [25] by Ubachs *et al.* The microwave power source was a resonant cavity powered by a Microtron 200 Microwave Power Generator, Mark III (Electro-Medical Supplies Ltd., England). Under optimum conditions for CH generation this source was operated at the upper 200 W limit, with little reflected power, coupling microwave energy very efficiently to the discharge.

The total pressure of the reactive gas mixture (Ar, O_2, C_2H_2, and water vapor) was kept at 400 Pa (3 Torr) because this provided the optimum setting for the CRDS signals. The CRDS mirrors had reflectivities of $R_1 = 0.993$ and $R_2 = 0.997$ in the 363–430 nm range and a focal length of about 250 mm, consisting of dielectric layers deposited on a Suprasil substrate (Laseroptik GmBH, Garbsen, Germany). For cavity ring-down (CRD) experiments, a Continuum model-TDL60 Nd:YAG-pumped dye laser was employed. The A–X transition access was accomplished with Coumarin 120 dye that shows a gain maximum of 440 nm. The B–X transition was reached by frequency doubling the output using Styryl 7 dye that shows a gain maximum at 720 nm, or frequency doubled at 360 nm. The available output power ranged from 5 to 15 mJ per pulse, but it was attenuated with diaphragms for CRDS. Emission spectra from the reaction zone were recorded using a low resolution Jobin-Yvon grating spectrometer in the spectral range 230-590 nm using a UV-sensitive photomultiplier (EMI) at a resolution of 0.1 nm.

17.3 Diatomic spectra computation details

The computations of the A–X and B–X transitions of CH rely on the establishment of accurate line strengths. For analysis of the measured CH transitions, two sets of line strength files are communicated as a supplement to this work. The development of line strength data is discussed with specific details for the C_2 Swan bands, computation of laser-induced fluorescence, and absorption spectra [15]. The line strengths for diatomic molecules follow recently published procedures [16]. Several applications in the analysis of optical breakdown spectra are communicated [16–18], including data files and the two programs—the Boltzmann equilibrium spectrum program (BESP) and Nelder–Mead temperature (NMT)—for analysis of the selected diatomic molecules [26].

Tables 17.1 and 17.2 communicate excerpts of the set of line strength data applicable for analysis of recorded CRDS data. These data files can be conveniently utilized with BESP and NMT, see reference [26]. For computation of emission spectra in the analysis of laser-plasma, only the wave number, upper term value, and line strengths are needed. For computation of emission spectra [17], MATLAB [27] source code [28] has been made available recently. However, for computation of absorption spectra, the lower term values are required. The collated CH data files in tables 17.1 and 17.2 also show standard designations for diatomic molecules [29].

17.4 Results and discussion

17.4.1 Methylidyne overview spectra

A computed overview emission spectrum for CH A–X illustrates the wavelength range of the provided line strength data. Figure 17.1 shows $\Delta v = 0$ transitions for v' = v" = 0, 1, 2. An instrument resolution, $\delta\lambda$, of $\delta\lambda = 0.05$ nm is selected, and an equilibrium temperature, T, is set to 3.0 kK. Such a spectrum may apply to analysis of laser-plasma emission. Figure 17.2 displays computed CH B–X spectra for $\Delta v = 0, +1$ transition, i.e., $v'' = v = 0$, 1, and $v'' = 1 to v'' = 0$. The spectra displayed in figures 17.1 and 17.2 are normalized separately to the maximum intensity for the A–X and B–X bands.

17.4.2 Emission- and cavity ring-down- spectra of the A–X and B–X bands

CH emission spectra are characterized by a strong band 430 nm, the A–X transition of CH, as well as a weaker emission at 390 nm from the B–X transition of CH. In addition, a medium strong band at 306 nm indicated the presence of OH (the A–X band), probably containing also the C–X transition of CH. There are medium strong vibrational progressions of the C_2 molecule A–X band (Swan band) at 470 nm, 520 nm, and 560 nm, e.g., see reference [26]. The presence of C_2 radicals is evident from the greenish color of the discharge under conditions when the acetylene:oxygen ratio is increased. Optimum conditions for the observation of the A–X and B–X CH bands were accomplished by decreasing the acetylene:oxygen ratio to produce an almost pure blue color, which is well known from flame emission studies. Upon comparing the microwave discharge to the oxy-acetylene flame, the former appears to be a much neater source of CH.

The CRD spectra obtained in the 429-432 nm turned out to correspond to a pure $A^2\Delta(v = 0) \longleftarrow X^2\Pi(v = 0)$ band of CH. The LIFBASE program [31, 32], based on CH A–X and CH B–X research [33, 34], was used to simulate this spectral region where weak features belonging to the excited vibrational transitions 1–1 and 2–2 can also be seen in regions not overlapped by the strong 0–0 features. In this work, accurate line strengths [16, 18] are employed for analysis. Figure 17.3 illustrates a

Table 17.1. First two dozen of 1384 lines for CH $A\ ^2\Delta \leftrightarrow X\ ^2\Pi$.

J''	J''		v''	v''	p''	p''	N''	N''	$F_{J''}$	$F_{J''}$	$\tilde{\nu}$	$S_{J''J''}$	$S_{S_{n''v''J''n''v''J''}}$
1.5	2.5	P_{22}	0	0	+f	-f	2	3	24 663.5612	1569.6083	23 093.9531	0.2013	0.8026
1.5	2.5	P_{21}	0	0	+f	-f	2	2	24 663.5612	1489.0759	23 174.4844	0.1996	0.7973
1.5	2.5	P_{11}	0	0	-e	+e	1	2	24 663.5612	1489.2381	23 174.3223	0.2004	0.8001
1.5	2.5	P_{12}	0	0	-e	+e	1	3	24 663.5612	1569.1156	23 094.4453	0.2005	0.7986
1.5	1.5	Q_{21}	0	0	+f	-e	2	1	24 663.5612	1433.8288	23 229.7324	0.7998	3.200
1.5	1.5	Q_{22}	0	0	+f	-e	2	2	24 663.5612	1482.8608	23 180.7012	0.8038	3.211
1.5	1.5	Q_{12}	0	0	-e	+f	1	2	24 663.5612	1483.1056	23 180.4551	0.8075	3.224
1.5	1.5	Q_{11}	0	0	-e	+f	1	1	24 663.5612	1433.8051	23 229.7559	0.7963	3.184
1.5	0.5	R_{22}	0	0	+f	-f	2	1	24 663.5612	1416.0299	23 247.5312	2.005	8.020
1.5	0.5	R_{11}	0	0	-e	+e	1	0	24 663.5612	1415.9191	23 247.6426	2.005	8.021
2.5	3.5	P_{12}	0	0	-f	+f	2	4	24 661.8291	1683.5813	22 978.2480	0.007 0179	0.027 893
2.5	3.5	P_{11}	0	0	-f	+f	2	3	24 661.8291	1573.3187	23 088.5098	0.3709	1.478
2.5	3.5	P_{21}	0	0	+e	-e	3	3	24 750.4863	1573.6950	23 176.7910	0.2022	0.8070
2.5	3.5	P_{22}	0	0	+e	-e	3	4	24 750.4863	1682.7661	23 067.7207	0.5651	2.248
2.5	3.5	P_{22}	0	0	-f	+f	3	4	24 750.4863	1683.5813	23 066.9043	0.5666	2.254
2.5	3.5	P_{21}	0	0	-f	+f	3	3	24 750.4863	1573.3187	23 177.1680	0.2009	0.8017
2.5	3.5	P_{11}	0	0	+e	-e	2	3	24 661.8291	1573.6950	23 088.1348	0.3707	1.478
2.5	3.5	P_{12}	0	0	+e	-e	2	4	24 661.8291	1682.7661	22 979.0625	0.007 2268	0.0287 23
2.5	2.5	Q_{11}	0	0	-f	+e	2	2	24 661.8291	1489.2381	23 172.5918	1.738	6.944
2.5	2.5	Q_{12}	0	0	-f	+e	2	3	24 661.8291	1569.1156	23 092.7129	0.1448	0.5773
2.5	2.5	Q_{22}	0	0	+e	-f	3	3	24 750.4863	1569.6083	23 180.8789	2.093	8.354
2.5	2.5	Q_{21}	0	0	+e	-f	3	2	24 750.4863	1489.0759	23 261.4102	0.4911	1.964
2.5	2.5	Q_{21}	0	0	-f	+e	3	2	24 750.4863	1489.2381	23 261.2480	0.4952	1.980
2.5	2.5	Q_{22}	0	0	-f	+e	3	3	24 750.4863	1569.1156	23 181.3711	2.089	8.335

Line strength table with column headings: J'' upper and J'' lower total angular momentum quantum number (nuclear spin not included); P_{ij}, or Q_{ij}, or R_{ij}, line designation based on J'', J'', $F_{J''}$, $F_{J''}$; v'' upper and v'' lower vibrational quantum number; p'' upper and p'' lower parity designations, the ± total parity eigenvalue is followed by the *e*/*f* parity; N'' upper and N'' lower total orbital angular momentum quantum number; $F_{J''}$ upper and $F_{J''}$ lower term value computed from model Hamiltonian, cm^{-1}; $\tilde{\nu}$ vacuum wave number, $\tilde{\nu} = F_{J''} - F_{J''}$, cm^{-1}; $S_{J''J''}$ Hönl–London term, unitless; and $S_{n''v''J''n''v''J''}$ line strength, $stC^2\ cm^2$ (1 stC $= 3.356 \times 10^{-10}$ C)

Table 17.2. First two dozen of 261 lines for CH $B\ ^2\Sigma^- \leftrightarrow X\ ^2\Pi$.

J''	J''		v''	v''	p''	p''	N''	N''	$F_{J''}$	$F_{J''}$	$\tilde{\nu}$	$S_{J''J''}$	$S_{n''v''J''n''v''J''}$
0.5	1.5	P_{11}	0	0	-f	+f	0	1	27 114.2564	1433.9116	25 680.3457	2.498	0.2572
0.5	1.5	P_{12}	0	0	-f	+f	0	2	27 114.2564	1483.2126	25 631.0430	0.1750	0.018 018
0.5	1.5	P_{22}	0	0	+e	-e	1	2	27 139.5581	1482.9686	25 656.5898	2.493	0.2567
0.5	1.5	P_{21}	0	0	+e	-e	1	1	27 139.5581	1433.9356	25 705.6230	0.1802	0.018 552
0.5	0.5	Q_{11}	0	0	-f	+e	0	0	27 114.2564	1416.0057	25 698.2500	1.337	0.1376
0.5	0.5	Q_{21}	0	0	+e	-f	1	0	27 139.5581	1416.1159	25 723.4414	1.337	0.1376
1.5	2.5	P_{11}	0	0	+f	-f	1	2	27 139.5166	1489.1826	25 650.3340	3.508	0.3612
1.5	2.5	P_{12}	0	0	+f	-f	1	3	27 139.5166	1569.7157	25 569.8008	0.1008	0.010 374
1.5	2.5	P_{22}	0	0	-e	+e	2	3	27 190.0681	1569.2245	25 620.8438	3.505	0.3609
1.5	2.5	P_{21}	0	0	-e	+e	2	2	27 190.0681	1489.3449	25 700.7227	0.1032	0.010 624
1.5	1.5	Q_{12}	0	0	+f	-e	1	2	27 139.5166	1482.9686	25 656.5488	0.019 638	0.002 0218
1.5	1.5	Q_{11}	0	0	+f	-e	1	1	27 139.5166	1433.9356	25 705.5801	3.723	0.3833
1.5	1.5	Q_{21}	0	0	-e	+f	2	1	27 190.0681	1433.9116	25 756.1562	0.017 592	0.001 8112
1.5	1.5	Q_{22}	0	0	-e	+f	2	2	27 190.0681	1483.2126	25 706.8555	3.725	0.3835
1.5	0.5	R_{11}	0	0	+f	-f	1	0	27 139.5166	1416.1159	25 723.4004	0.6683	0.068 806
1.5	0.5	R_{21}	0	0	-e	+e	2	0	27 190.0681	1416.0057	25 774.0625	0.6683	0.068 803
2.5	3.5	P_{11}	0	0	-f	+f	2	3	27 189.9989	1573.4256	25 616.5742	4.511	0.4645
2.5	3.5	P_{12}	0	0	-f	+f	2	4	27 189.9989	1683.6892	25 506.3105	0.070 992	0.007 3091
2.5	3.5	P_{22}	0	0	+e	-e	3	4	27 265.6964	1682.8762	25 582.8203	4.510	0.4643
2.5	3.5	P_{21}	0	0	+e	-e	3	3	27 265.6964	1573.8017	25 691.8945	0.072 257	0.007 4393
2.5	2.5	Q_{11}	0	0	-f	+e	2	2	27 189.9989	1489.3449	25 700.6543	5.838	0.6010
2.5	2.5	Q_{22}	0	0	+e	-f	3	3	27 265.6964	1569.7157	25 695.9805	5.838	0.6011
2.5	1.5	R_{11}	0	0	-f	+f	2	1	27 189.9989	1433.9116	25 756.0879	1.494	0.1538
2.5	1.5	R_{12}	0	0	-f	+f	2	2	27 189.9989	1483.2126	25 706.7871	0.1099	0.011 316

Line strength table with column headings (identical to the ones in table 17.1)

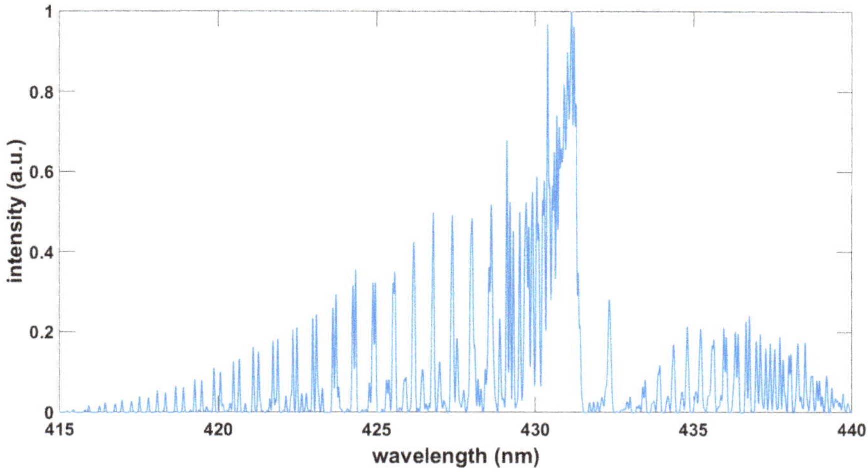

Figure 17.1. Computed CH A–X spectrum, $\Delta v = 0$, $\delta\lambda = 0.05$ nm, $T = 3.0$ kK. Reproduced from [30].

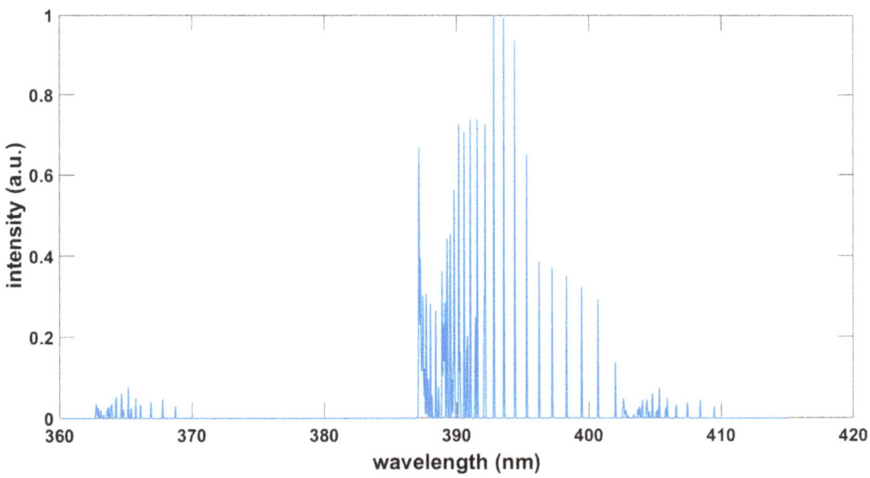

Figure 17.2. Computed CH B–X spectrum, $\Delta v = 0, +1$, $\delta\lambda = 0.05$ nm, $T = 3.0$ kK. Reproduced from [30].

comparison of measured and fitted absorption spectra. The absorption spectra comparisons illustrated in figure 17.3 are determined by calculating the absorption spectra as outlined in reference [15]. The experimental spectrum displayed in figure 17.3 is normalized the maximum intensity in the indicated wavelength range for the A–X band. For completeness, figure 17.4 illustrates computed emission spectra in the same wavelength range as for figure 17.3. As expected, there are subtle

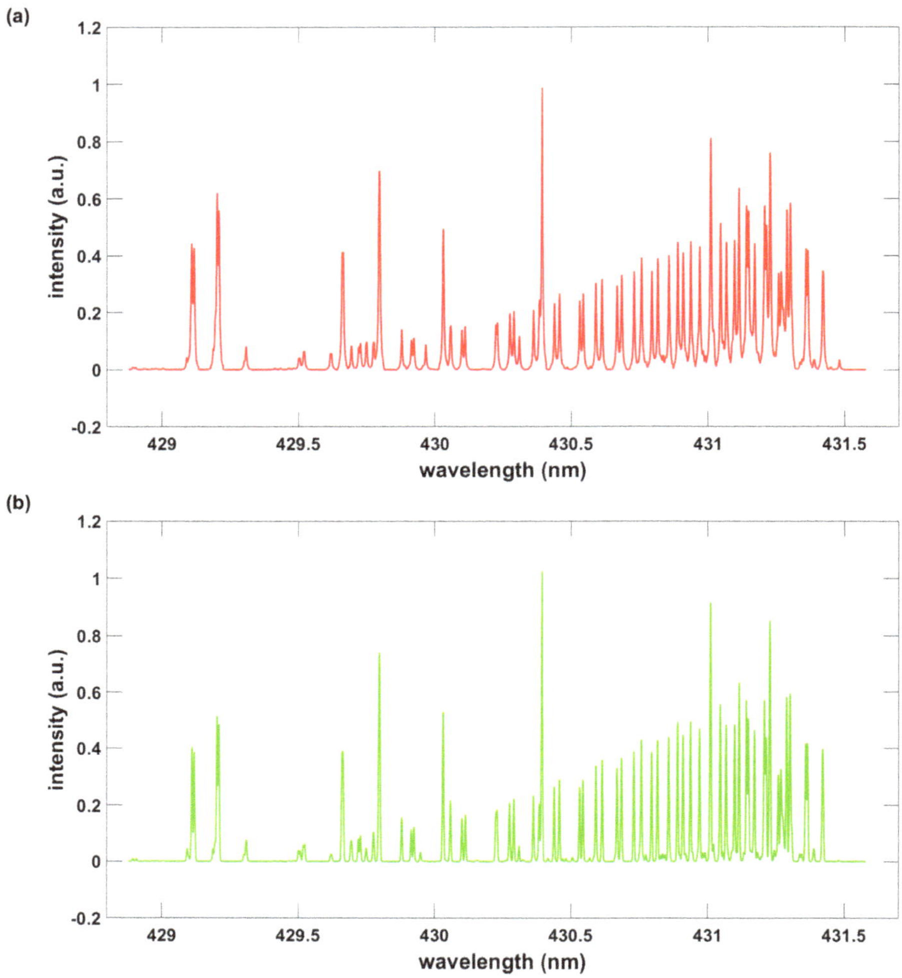

Figure 17.3. Comparison of measured **(a)** and fitted **(b)** CH A–X spectra, $\delta\lambda = 0.005$ nm, $T = 1.47$ kK. Reproduced from [30]. CC BY 4.0.

differences for comparisons of absorption spectra, see figure 17.3(a), with emission spectra in figure 17.4. Frequently, emission spectra from plasma are of interest, e.g., in laser-induced breakdown spectroscopy [18]. Computation of absorption spectra requires knowledge of lower state term values, whereas emission spectra rely on upper state term values. The CH line strength data and MATLAB scripts [17] are provided for computation of emission spectra as illustrated in figure 17.4.

Figure 17.5 displays a recorded B–X CH spectrum, and figure 17.6 shows an emission spectrum for the same wavelength range of 364.964 nm (27400cm^{-1}) to 364.033 nm (27470cm^{-1}).

Recent advances and updates of accurate molecular data for application in astrophysics include the ExoMol [35] database. Comparisons with the provided CH

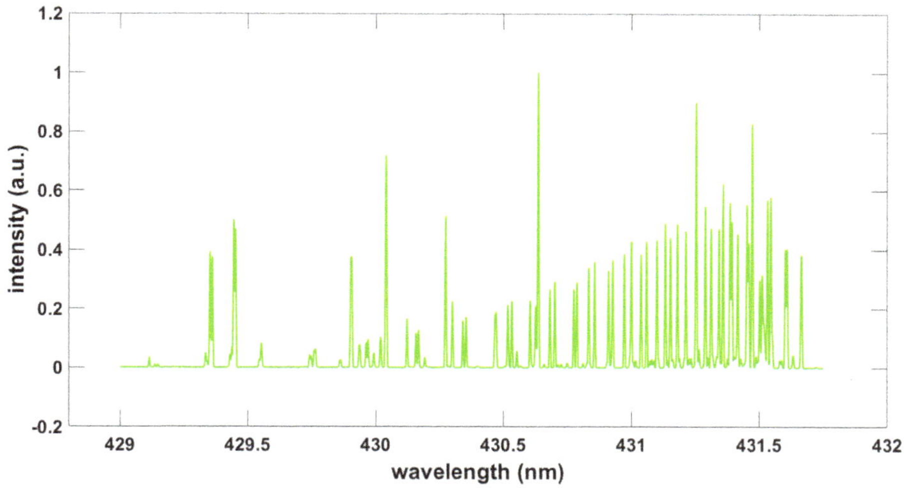

Figure 17.4. Computed CH A–X emission spectrum, $\delta\lambda = 0.005$ nm, $T = 1.5$ kK. Reproduced from [30]. CC BY 4.0.

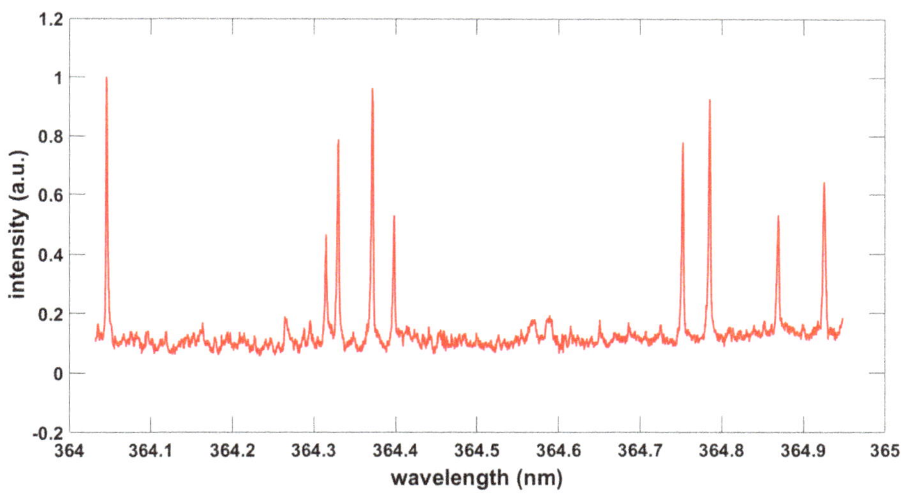

Figure 17.5. Recorded CH B–X spectrum using CRDS. Reproduced from [30]. CC BY 4.0.

A–X and B–X line strengths reveal identical lines with an emission spectrum as displayed in figure 17.6. However, there are also two lines that would correspond to the measured CRD lines (see figure 17.5) at 364.568nm (27 429.73 cm^{-1}) and 364.589 nm (27 428.11 cm^{-1}), yet with little effect on the temperature (0.63 kK versus 0.65 kK). The temperature is inferred using the nine prominent lines in figure 17.5, and figure 17.6 illustrates the result. Noteworthy, the spectrum in figure 17.5 is identically reproduced in the latest LIFBASE version 2.1.1 [32].

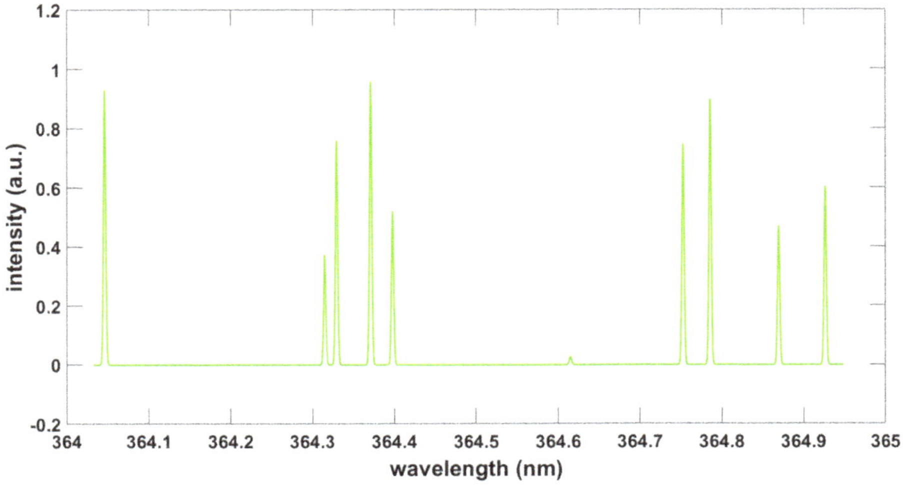

Figure 17.6. Computed CH B–X emission spectrum, $\delta\lambda = 0.0033$ nm, $T = 0.65$ kK. Reproduced from [30].

The two lines near 365.6nm are reproduced by resorting to the ExoMol [35] referenced molecular line lists, intensities, and spectra (MoLLIST) [36] for the $^{12}C^{1}H$ isotopologue of CH. However, analysis and establishment of high-resolution line lists of ^{12}CH continues to be of interest [37].

17.5 Conclusions

This work communicates a convincing comparison of recorded cavity ring-down spectra and of computed CH A–X absorption spectra using line strength data. Furthermore, these comparisons agree well with advances in spectral simulation for diatomic molecules (LIFBASE), and database advances in exoplanet and other hot atmospheres modeling (ExoMol). Higher resolution for the investigated CH B–X transition than for the CH A–X transition also confirms the reasonable accuracy of measured and computed line positions. Emission spectra of CH A–X and B–X were observed but the focus of this work was the application of CRDS for the CH radical characterizations. However, the provided line strength data are expected to continue to be useful in absorption and emission spectroscopy of plasma that contains hydrocarbons.

C.G.P. acknowledges support in part from the State of Tennessee funded Center for Laser Applications at the University of Tennessee Space Institute. L.N. acknowledges a short term visiting scientist support from the Nederlandse Organistie voor Wetenschappelijk (NWO), as well as support from the Hungarian Research Fund (OTK #3079) for computational equipment in Hungary during the initial research analysis in 1996. Finally, L.N. and C.G.P. acknowledge the consent of Professor Hans J.J. ter Meulen (University of Nijmegen, The Netherlands) to publish the so far unpublished original material and CRDS data.

References

[1] O'Keefe A and Deacon D A G 1988 *Rev. Sci. Instrum* **59** 2544
[2] Romanini D and Lehmann K K 1993 *J. Chem. Phys.* **99** 6287
[3] Huestis D L, Copeland R A, Knutsen K, Slanger T G, Jongma R T, Boogaarts M G H and Meijer G 1994 *Can. J. Phys.* **72** 1109
[4] Scherer J J, Paul J B, Collier C P and Saykally R J 1995 *Chem. Phys. Lett.* **102** 5190
[5] O'Keefe A, Scherer J J, Cooksy A L, Sheeks R, Heath J and Saykally R J 1990 *Chem. Phys. Lett.* **172** 214
[6] Jongma R T, Boogaarts M G H, Holleman I and Meijer G 1995 *Rev. Sci. Instrum.* **66** 2821
[7] Yu T and Lin M C 1993 *J. Am. Chem. Soc.* **115** 4371
[8] Yu T and Lin M C 1994 *J. Phys. Chem.* **98** 9697
[9] Cheskis S 1995 *J. Chem. Phys.* **102** 1851
[10] Zalicki P, Ma Y, Zare R N, Wahl E H, Dadamino J R, Owano T G and Kruger C H 1995 *Chem. Phys. Lett.* **234** 269
[11] Meijer G, Boogaarts M G H, Jongma R T, Parker D H and Wodtke A M 1994 *Chem. Phys. Lett.* **217** 112
[12] Wang C-C, Nemes L and Lin K-C 1995 *Chem. Phys. Lett.* **245** 585
[13] Nemes L and Szalay P G 1999 *Models Chem.* **136** 205
[14] Szalay P G and Nemes L 1999 *Molec. Phys.* **96** 359
[15] Hornkohl J O, Nemes L and Parigger C G 2009 Spectroscopy, dynamics and molecular theory of carbon plasmas and vapors *Advances in the Understanding of the Most Complex High-Temperature Elemental* ed L Nemes and S Irle (Singapore: World Scientific) ch 4
[16] Parigger C G and Hornkohl J O 2020 *Quantum Mechanics of the Diatomic Molecule with Applications* (Bristol: IOP Publishing)
[17] Parigger C G 2023 *Foundations* **3** 1
[18] Parigger C G, Surmick D M, Helstern C M, Gautam G, Bol'shakov A A and Russo R 2020 Molecular laser-induced breakdown spectroscopy *Laser Induced Breakdown Spectroscopy* ed J P Singh and S N Thakur 2nd edn (New York: Elsevier)
[19] Brzozowksi J, Bunker P, Elander N and Erman P 1976 *Astrophys. J.* **207** 414
[20] Erman P 1979 Astrophysical applications of time resolved spectroscopy of small molecules *Molecular Spectroscopy Volume 6: A Review of the Literature published in 1977 and 1978* ed R F Barrow, D A Long and J Sheridian (London: The Royal Society Chemistry)
[21] Erman P 1979 *Phys. Scr.* **20** 575
[22] Warnatz J 1984 *Combustion Chemistry* (New York: Springer)
[23] Raiche G A and Jeffries J B 1993 *Appl. Opt.* **32** 4629
[24] Engeln R, Letourneur K G Y, Boogarts M G H, van den Sanden M C M and Schram D C 1999 *Chem. Phys. Lett.* **310** 405
[25] Ubachs W, Meijer G, ter Meulen J J and Dymanus A 1986 *J. Chem. Phys.* **84** 3032
[26] Parigger C G, Woods A C, Surmick D M, Gautam G, Witte M J and Hornkohl J O 2015 *Spectrochim. Acta* B **107** 132
[27] *MATLAB Release R2022a Update 5* (Natick, MA: The MathWorks, Inc)
[28] Surmick D M and Hornkohl J O 2016 (The University of Tennessee, University of Tennessee Space Institute, Tullahoma, TN, USA) personal communication
[29] Hornkohl J O 2004 Private communication, The University of Tennessee, University of Tennessee Space Institute, Tullahoma, TN, USA
[30] Nemes L and Parigger C G 2023 *Foundations* **3** 16

[31] Luque J and Crosley D R 1999 *LIFBASE, Database and spectral simulation for diatomic molecules (v 1.6)* (Menlo Park, CA: SRI International) SRI International Report MP-99-009
[32] Luque J and Crosley D R 2021 *LIFBASE: Database and Spectral Simulation for Diatomic Molecules* (Menlo Park, CA: SRI International)
[33] Luque J and Crosley D R 1996 *J. Chem. Phys.* **104** 2146
[34] Luque J and Crosley D R 1996 *J. Chem. Phys.* **104** 3907
[35] Tennyson J *et al* 2020 *J. Quant. Spectrosc. Radiat. Transf.* **255** 107228
[36] Masseron T, Plez B, Van Eck S, Colin R, Daoutidis I, Godefroid M, Coheurand P-F, Bernath P, Jorissen A and Christlieb N 2014 *Astron. Astrophys.* **571**
[37] Furtenbacher T, Hegedus S T, Tennyson T and Császár A G 2022 *Phys. Chem. Chem. Phys.* **24** 19287

IOP Publishing

Quantum Mechanics of the Diatomic Molecule (Second Edition)

Christian G Parigger and James O Hornkohl

Chapter 18

Cyanide, CN

Diagnosis of cyanide (CN) is important in a variety of fields of science. CN readily occurs as a results of recombination in laser-induced plasma in air. The measured signal strength is largely increased over that in air when 1:1 mole ratio mixtures of CO_2:N_2 are investigated [1, 2]. This chapter also summarizes recent research communicated at various national and international conferences.

The mole ratio and density of CN can be predicted for thermodynamic equilibrium using the so-called chemical equilibrium and applications code [3, 4]. During the plasma expansion in the first microsecond after initiation of optical breakdown, shock wave phenomena affect the distribution of atomic and molecular species. CN can be measured within the first 0.1 μs, but is well developed after several dozens of microseconds. Diatomic molecular spectra of CN with excitation temperature in the range of 6–9 kK are also routinely measured in hydrocarbon combustion in air, the fuel providing carbon and the air providing nitrogen for the formation of CN due to combustion or due to recombination radiation in laser-induced breakdown plasma.

18.1 Analysis of CO_2 laser plasma

Experimental investigations for measurements of CN utilize a standard laser-induced breakdown spectroscopy (LIBS) arrangement: laser device, spectrometer, detector-system, and a sample. For studies that uses a CO_2 laser as the radiation source [5], the experiment details are as follows: the CO_2 laser is the transversely excited atmospheric-pressure (TEA) model TEA-820 Lumonics, the spectrometer is the 1/2m model 500 SpectraPro Acton Research Corporation, and the detector is the intensified model 1460 EG&G Princeton Applied Research detector/controller optical multichannel analyzer. The laser plasma is first generated for alignment purposes and initial diagnostic studies in laboratory air using f/4 focusing. The plasma is then generated in CO_2 gas that expands into the focal volume at a rate of

doi:10.1088/978-0-7503-6204-7ch18

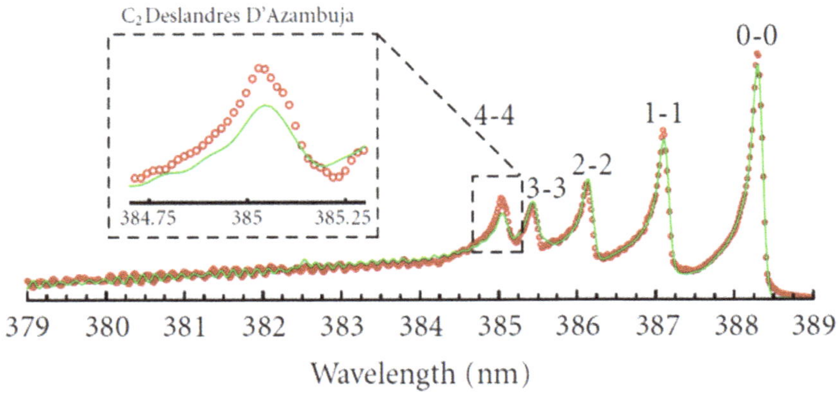

Figure 18.1. Measured (circles) and fitted (solid line) spectrum of the CN $\Delta v = 0$ sequence of the CN $B^2\Sigma^+ \longrightarrow X^2\Sigma^+$ violet system. The insert indicates emission from the C_2 Deslandres DAzambuja system [5].

10 L/m into the laboratory. Survey spectra from laser-induced optical breakdown (LIOB) in CO gas are reported here.

Typical operating parameters of our CO_2 TEA laser are (i) pulse width of 70–100 ns full width at half maximum (FWHM), or 10 ×longer than Nd:YAG pulses, (ii) pulse energy of 2000–3000mJ, or typically 10–100 ×larger than nominal Nd:YAG laser devices, and (iii) wavelength of 10.6 μm, or 10 ×longer than the 1.064 μm fundamental Nd:YAG laser radiation. Figure 18.1 shows the recorded and fitted spectrum of the CN $\Delta v = 0$ sequence of the CN $B^2\Sigma^+ \longrightarrow X^2\Sigma^+$ violet system. The spectral resolution amounted to 0.12 nm when using the 2400 grooves/mm grating. Particularly strong CN emissions occur and dominate the molecular emissions in the visible region when using CO_2 flow expanding into laboratory air. We used a gate width of 20 μs and a time delay of 20 μs. The inferred temperature amounts to 7.22 kK, using our fitting routines for diatomic spectra [1, 5–7]. Figure 18.1 also shows an insert near the wavelength region of 4–4 band of CN that we excluded in the fitting routine. The additional spectroscopic feature indicates presence of the C_2 Deslandres D'Azambuja system [8].

18.2 Analysis of CN in Nd:YAG laser plasma

Various applications, including LIBS, utilize Q-switched Nd:YAG laser pulses to generate plasma [1]. Figure 18.2 illustrates measured and fitted spectra of the CN violet system recorded with a spectral resolution of $\Delta\lambda = 0.03$nm (2 cm^{-1}). Individual band heads of the $\Delta v = 0$ sequence are clearly discernible. The fitting of the experimental spectrum utilizes the line-strength data for CN and the NMT program as elaborated in the supplement of [9]. The computed spectra are based on the provided line-strength data, and the fitting routine uses nonlinear fitting based on simplices in the Nelder–Mead algorithm [10, 11].

The Nelder–Mead method [11] minimizes the difference between a fitting function and a set of data by iteratively adjusting the parameters of the fitting function.

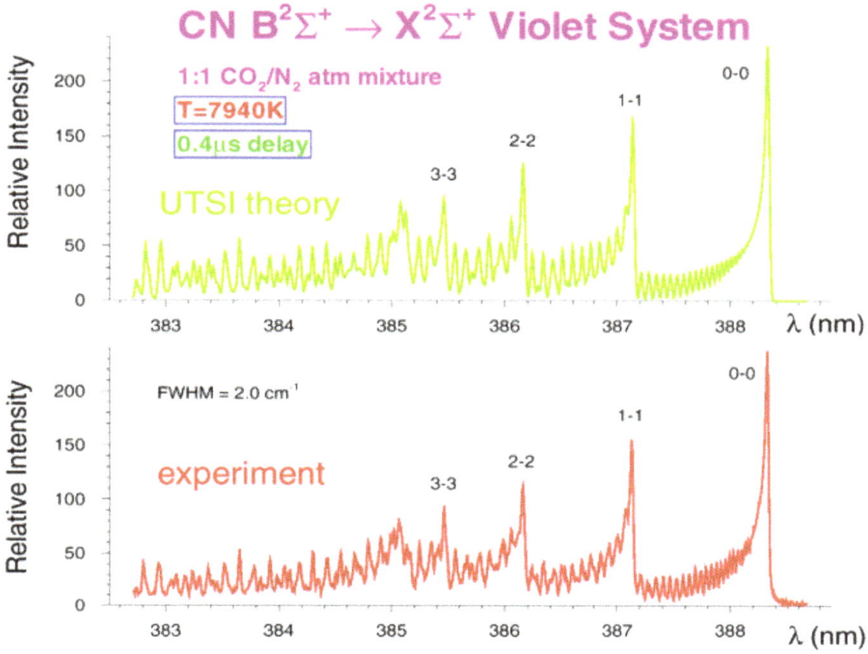

Figure 18.2. Measured (bottom panel) and fitted (top panel) spectrum of the CN $\Delta v = 0$ sequence of the CN $B^2\Sigma^+ \longrightarrow X^2\Sigma^+$ violet system. Reprinted from [1], copyright (1991), with permission from Elsevier.

The method does not require that the derivative of the fitting function to be known. For a set of N parameters, the Nelder–Mead method begins with $N + 1$ trial values of the parameters and $N + 1$ values of the difference between the fitting function and the data, and the method operates by moving away from the maximum difference. Each of the $N + 1$ trial values of the parameters and the $N + 1$ function differences is called a vertex. Each vertex is a point in a N-dimensional space called a simplex. In essence, the volume of the simplex is minimized, thus the name *downhill simplex*. Figure 18.3 displays measured, fitted, and difference spectra at a spectral resolution of 0.093 nm.

Alternate programs that allow analysis of measured diatomic spectra include LIFBASE [12] and PGOPHER [13]. The former appears to work well for lower excitation temperatures, and the latter has enjoyed significant updates and improvements for CN. The current (March 2019) pgo-file for the BAX CN transitions [14] delivers results that are consistent with the NMT fitting in terms of inferred temperature and fitted spectral resolution, see figure 18.3. Figure 18.4 compares experimental and computed spectra, but fitting algorithms in PGOPHER utilize vacuum wavenumbers, therefore requiring conversion from air wavelengths [15] in the experiment. Moreover, relative intensities are used in the PGOPHER program; in other words, computation of partition functions is avoided for plasma excitation temperatures of the order of 8 kK. The spectral resolution for PGOPHER fitting (version 10.1 of December 2018) is accomplished with 5.2 cm^{-1} Gaussian and 0.6 cm^{-1} Lorentzian widths.

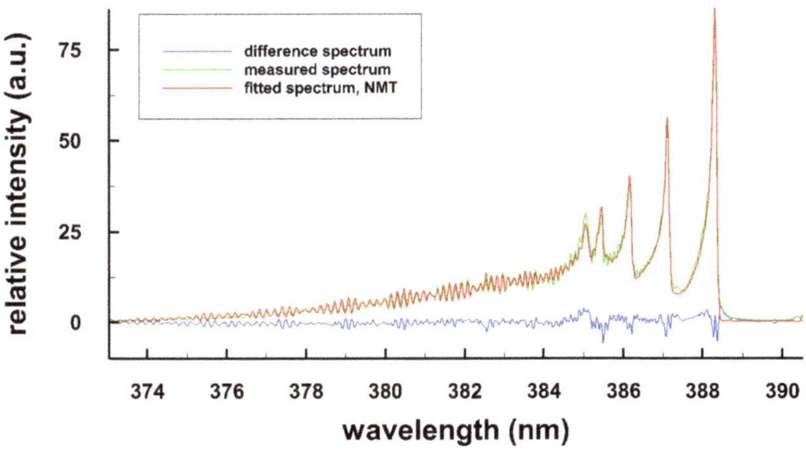

Figure 18.3. Measured, fitted, and difference spectra of the CN $\Delta v = 0$ sequence of the CN $B^2\Sigma^+ \longrightarrow X^2\Sigma^+$ violet system at a temperature of $T = 6.84$ kK [1, 9].

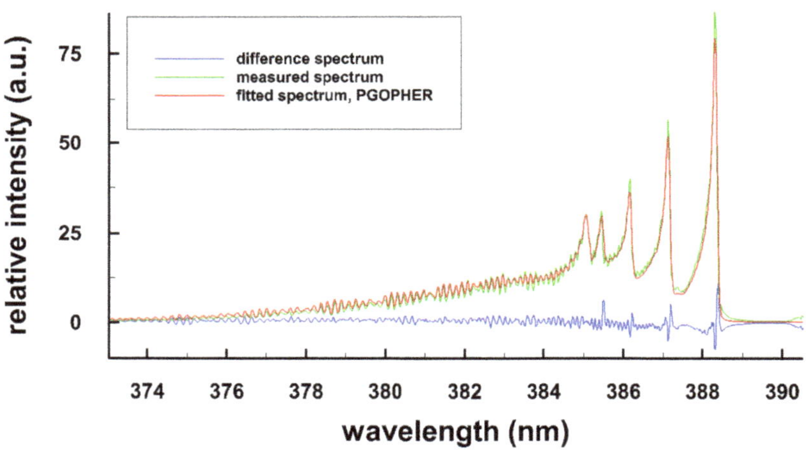

Figure 18.4. Results of fitting the experimental data of figure 18.3 with the PGOPHER program. Difference, measured, and fitted spectra for the CN $\Delta v = 0$ sequence of the CN $B^2\Sigma^+ \longrightarrow X^2\Sigma^+$ violet system yield a temperature of $T = 6.83$ kK.

18.3 Spatially and temporally resolved CN spectra

Recent experimental investigations explore the CN violet system using an intensified charge-coupled detector (ICCD) that is arranged to spatially resolve spectra along the slit dimension [2]. Laser-induced optical breakdown is generated by propagating the infrared laser beam from the top and parallel to the spectrometer slit, see the experimental arrangement in figure 16.1 in chapter 16. Initial experimental studies address details of the focal volume intensity distribution and effects on optical breakdown in laboratory air at standard ambient pressure and temperature (SATP).

18.3.1 Laser-beam focusing

A laser beam with 850 mJ/pulse is employed for plasma generation in SATP laboratory air, and another synchronized beam of 13 ns duration intersects the plasma to obtain shadowgraphs on a screen. A standard charge-coupled device (CCD) is utilized for the recording of subsequent shadowgraphs. The focal distribution is obtained by solving diffraction integrals for a beam from the laser device that is expanded to a diameter of 10 mm. A fused silica, 100 mm, plano-convex lens (Thorlabs model LA4545) focuses the laser beam with f/10 optics. The f-number (f/#) is computed as the ratio of beam focal length and the beam diameter at the lens. Figure 18.5 illustrates computed distributions including aberrations [2, 16] with an aspect ratio of 50:1 for the Thorlabs LA4545 lens for focusing with f/10 optics.

The peak irradiance distributions are computed for 850 mJ, 6 ns, 1064nm radiation. Tighter f/5 focusing for a perfect lens would show about one order of magnitude (or by a factor of 2^3) smaller focal volume than that obtained for f/10 focusing. The optical breakdown threshold is 0.28 TW cm $^{-2}$ [17, 18] at 1 atm in dry air and for 1064nm radiation. The yellow pseudo-color range near 1 TW cm $^{-2}$ (10^0 in the figure) indicates an irradiance that is a factor of three to four higher than optical breakdown threshold.

Several regions along the optical axis indicate above-threshold irradiance as the laser beam reaches focus near 100 mm. Consequently, optical breakdown will occur at multiple spots, as indicated in the figure with subsequent absorption of the remainder of the 6 ns pulse. As the energy/pulse is reduced to about 8.5 mJ/pulse, optical breakdown would occur at focus, causing a single spherical expansion.

The irradiance along the optical axis (x = 0) exhibits undulations above and below the perfect, aberration-free lens. Figure 18.6 shows optical-axis variations of

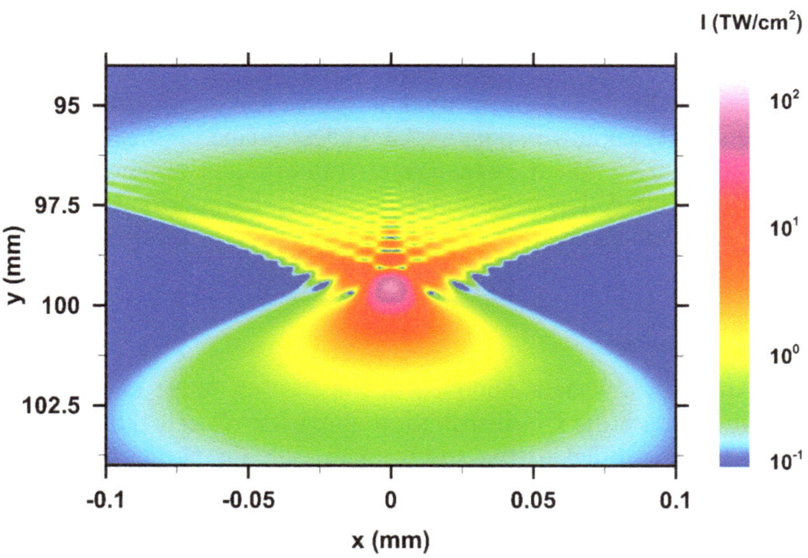

Figure 18.5. Computed focal distribution for f/10, 1064nm focusing from the top.

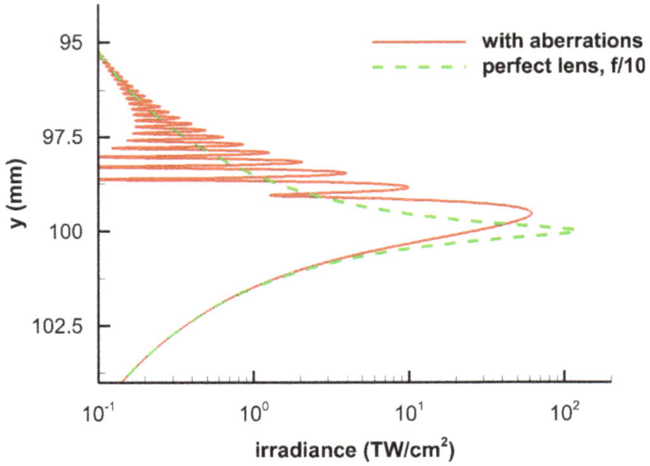

Figure 18.6. Irradiance along the optical axis for f/10 focusing.

the f/10 with aberrations. There are about 10 maxima and minima for an irradiance larger than the clean air optical breakdown threshold of 0.28 TW cm $^{-}$2 for clean air.

18.3.2 Shadowgraphs

Optical breakdown visualizations utilize the shadowgraph method, or visualization based on the second derivative of the index of refraction. The shadowgraphs are recorded using a second laser beam that is externally synchronized to the pulse that generates optical breakdown [19]. Figure 18.7 illustrates optical breakdown for 0.05, 0.6, 1.5, and 3 μs time delays. The representative images are taken from a set of 124 recorded shadowgraphs from different breakdown events; in other words, figure 18.7 is put together from different experiments. The background shows the particular mode structure of the second beam used as source for the shadowgraph. Clearly, there are multiple breakdown spots in the 0.05 μs image, and each of these spots indicates a spherically symmetric expansion. At threshold irradiance, only one laser-spot would be measured. The deviation from spherical expansion is indicated by the forward cone toward the bottom of the images, which is nicely developed in the 0.06 μs image.

18.3.3 Raw CN spectra

Figure 18.8 displays raw data obtained in the spectral range of the CN $\Delta v = 0$ sequence and illustrate the CN distribution along the plasma that is generated by focusing the beam parallel to the spectrometer slit. Of particular interest is the spatial distribution of the recorded spectra that are subjected to wavelength and detector-sensitivity calibrations. For diatomic spectra, detector-sensitivity corrections are significant because it is not unexpected to see a variation across several nanometer, or across the vibrational and rotational peaks indicated in the figure.

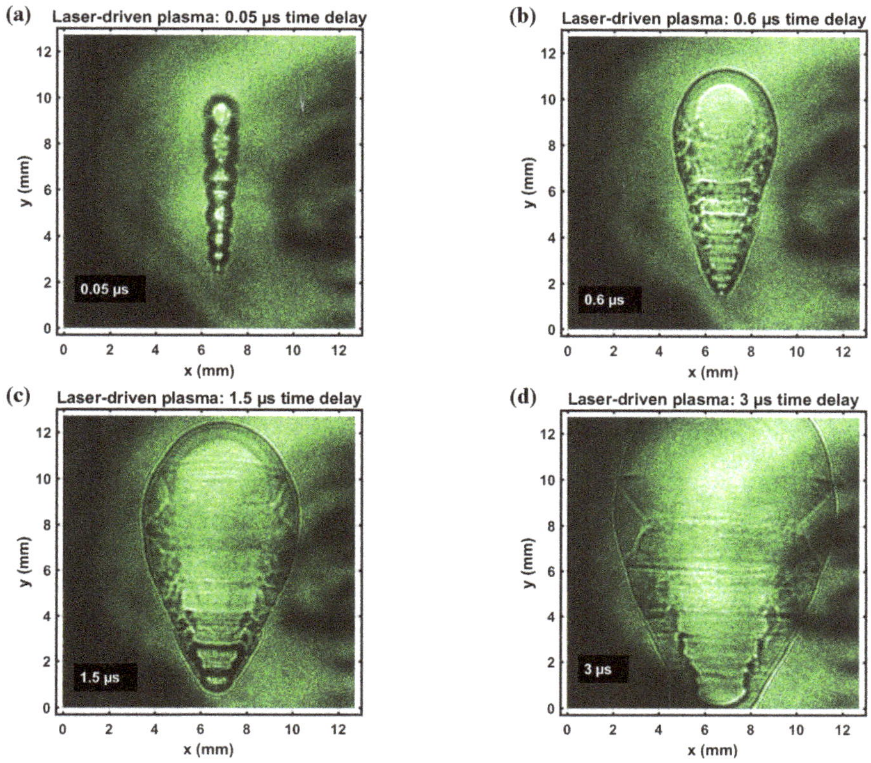

Figure 18.7. Shadow-graphs. Time delays (a) 0.05 μs, (b) 0.6 μs, (c) 1.5 μs, and (d) 3 μs. Multiple breakdown spots lead to stagnation regions. Absorption of the laser beam near the top of the images leads to near spherically- or spheroidally-symmetric expansion. The central vertical line is imaged on the spectrometer slit for capture of spectra along the vertical direction [19].

The intensity distribution along the slit height near pixel 512 displays the variation of the rotational spectra. The vertical distributions along the slit are measures for the excitation temperature of CN in the plasma. For time delays within the first microsecond, the vertical CN spectra indicate expansion dynamics associated with laser-induced optical breakdown. This is further elaborated in the next section.

18.3.4 Abel-inverted CN spectra

Recent experiments [2, 20, 21] explore expansion dynamics of laser-induced optical breakdown in a 1:1 mole ratio CO_2:N_2 mixture at 1 atm. Abel-inversion of the CN data [22] is aimed to obtain spatial distribution reminiscent of the expanding shock wave. In this section, analysis of the line-of-sight data is elaborated first, and then results are presented of the Abel-inverted data.

The experimental arrangement consists of a set of components typical for time-resolved, laser-induced optical emission spectroscopy [9, 20, 23], or nanosecond LIBS. The experimental series for the separate measurements of CN molecular distribution after optical breakdown includes evacuating the cell to a nominal

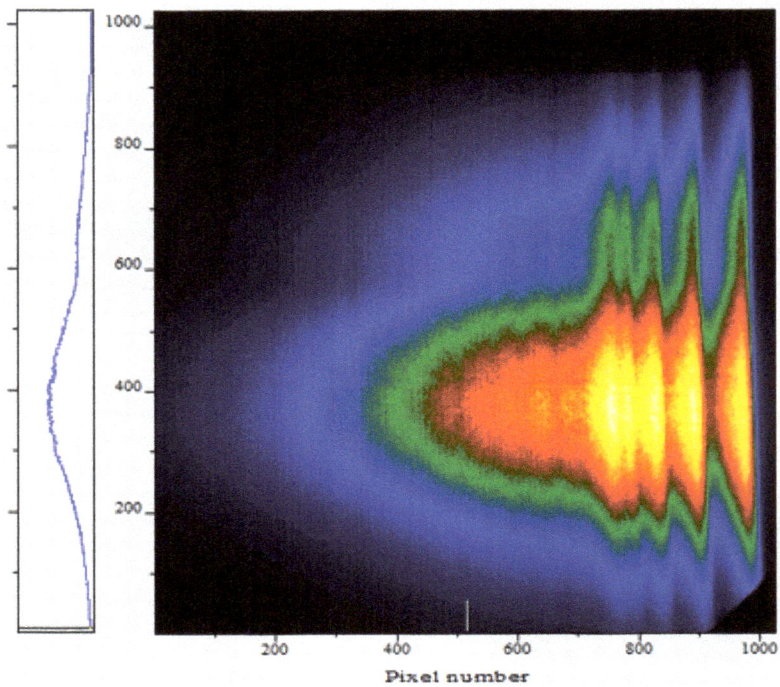

Figure 18.8. Raw CN rotational-vibrational spectra recorded with an ICCD. The horizontal axis corresponds to wavelength, and the vertical axis to the slit height. The four band heads of the CN $\Delta v = 0$ are well developed between pixels 700 and 1000, including undulations between pixels 500 and 700 due to overlapping rotational spectra.

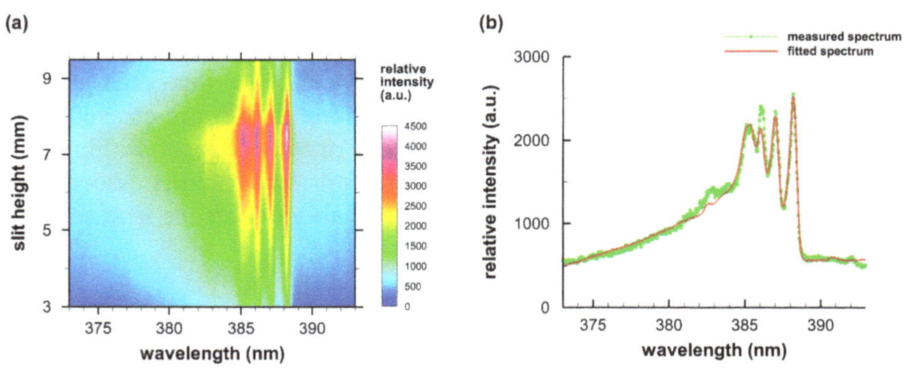

Figure 18.9. Time delay 0.95 μs; (a) pseudo-colored contour map and (b) measured and fitted spectrum at a slit height of 7 mm, inferred temperature: $T = 9.04$ kK [21].

mercury pump vacuum of 10^{-4} Pa (10^{-6} Torr) and then introducing the N_2:CO_2 mixture. Figure 18.9 illustrates the wavelength-calibrated, sensitivity-corrected, and background-subtracted maps of captured time-resolved data following optical breakdown in the ultra-high pure N_2 and research grade CO_2. The figure shows a single measured [20] and computed [9] spectrum at the center of the plasma. The fitted spectral resolution, $\Delta\lambda$, amounts to $\Delta\lambda = 0.44$ nm.

Signatures of the 0–0, 1–1, 2–2, 3–3, and 4–4 band heads begin to emanate for time delays of the order of 100 ns from optical breakdown. Moreover, the plasma typically propagates towards the top or laser side. The recorded data indicate a ~0.8 mm upward CN-signal propagation in the 370–393.5 nm spectral, ~7 mm object window during the first 5 microsecond, from a delay of 0.2–5.2 μs.

The detector pixels are binned in four tracks along the slit direction, resulting in obtaining 256 spectra for each time delay. The average of accumulated data from 100 consecutive optical breakdown events is recorded for 21 different time delays using a gate width of 125 ns. In figure 18.9, the vertical axis indicates the slit height. With 1:2 imaging, and a pixel resolution of 13.6 μm, the clearly discernable plasma size in the cell amounts to ~3 mm. The figure shows that the CN band heads of the $\Delta v = 0$ sequence are well developed, and it also shows an atomic line near 386.2 nm that likely is the carbon CI 193.09 nm atomic line recorded in second order. The electron density, n_e, can be determined from the Stark FWHM, $\Delta\lambda_{\rm Stark}$, or from the Stark shift, $\delta\lambda_{\rm Stark}$, of the carbon line [24],

$$\Delta\lambda_{\rm Stark} = 2w\, n_e[10^{17}\ {\rm cm}^{-3}], \quad {\rm or}\ \delta\lambda_{\rm Stark} = d\, n_e[10^{17}\ {\rm cm}^{-3}], \tag{18.1}$$

where the width parameter, w, and shift parameter, d, amount to w = 0.002 91 nm and d = 0.002 94 nm [24], respectively, at an electron temperature of 40 kK. At 200 ns time delay, the full width of ~0.4 ± 0.1nm in second order reveals an electron density of $\sim 34 \times 10^{17}$ cm^{-3}. The shift of 0.2 ± 0.05 nm in second order yields a consistent result for the electron density. The accuracy of the Stark width parameter is listed at 30% [25].

Figure 18.10 displays measured and fitted spectra at the center and at the top of the spectra at a time delay of 2.2 μs. The plasma shows a spatial temperature variation with the temperature at the center at 8.42 kK and at the top at 9.03 kK. The higher temperature near the edge versus the center indicates higher electron density. Additionally, the presence of the CI 193.09 nm atomic line recorded in

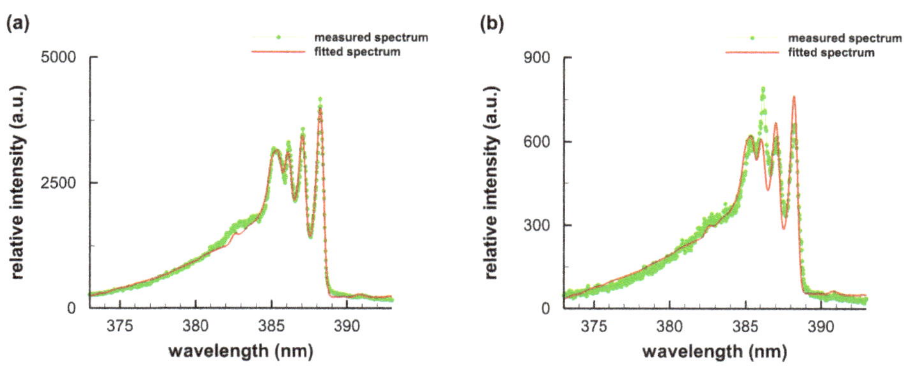

Figure 18.10. Time delay 2.2 μs; (a) center, $T = 8.42$ kK, and (b) top, $T = 9.03$ kK [21].

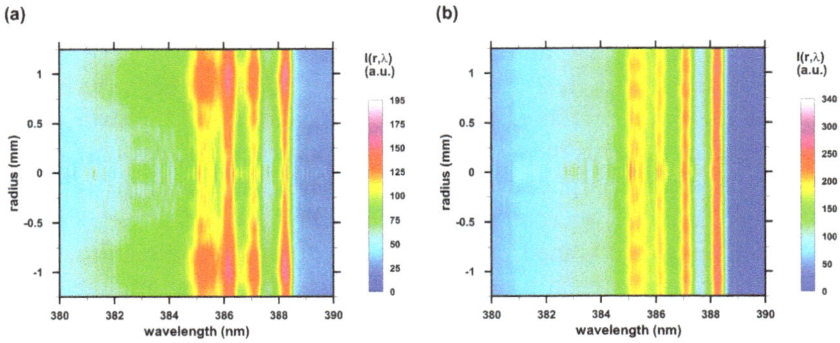

Figure 18.11. Abel-inverted spectra versus radius. Time delays (a) 1.2 μs [2] and (b) 3.2 μs.

second order as displayed in figures 18.9(b) and 18.10(b) indicates an electron density that is larger near the edges than at the center of the plasma.

Near the edges of the expanding plasma, the CN signals are lower than near the center. Computations of equilibrium CN distributions versus temperature [20] indicate about a factor of 3 smaller CN mass fractions at 9 kK than at 8 kK. Moreover, application of Abel inverse integral transforms [20] reveal slightly lower CN temperature in the plasma center for time delays of the order of 1 μs. Near the plasma edges at the top and the bottom (see figure 18.9(a)), the signals of the likely atomic carbon line are spectrally wider, show larger shifts, and are stronger than the CN 0–0 band head.

The line-of-sight recorded spectra are further analyzed using Abel-inversion techniques. In view of the shadowgraphs for optical air breakdown with 850 mJ/pulse 6 ns excitation, spherical symmetry may only be indicated near the top of the images, see figure 18.7. However, as the energy per pulse is reduced by about one order of magnitude, a smaller number of breakdown sites occurs. The optical breakdown is initiated near focus of a perfect lens, of course with residual aberration effects, but with diminished forward-cone structure. Hence, the expanding breakdown plasma can be viewed as spherically symmetric. Slight deviation from spherical symmetry, indicated by a forward cone, reflect the amount of plasma generation with above-threshold irradiance. Figure 18.11 displays Abel-inverted spectra for time delays of 1.2 and 3.2 μs. Clearly, the 1.2 μs spectra indicate diminished signals of CN near the center. The 193.09 nm carbon line, measured in second order, shows a larger width near the edges than at center. This implies larger electron density near the edges than at center, and thus higher temperature for larger electron density due to isentropic expansion. In other words, hotter CN spectra are measured near the expanding shock wave. In turn, for the 3.2 μs time-delay map, the CN distribution appears flat, but perhaps indicating slightly higher temperature at center than at the edges. There are no signatures from the carbon line for the 3.2 μs data.

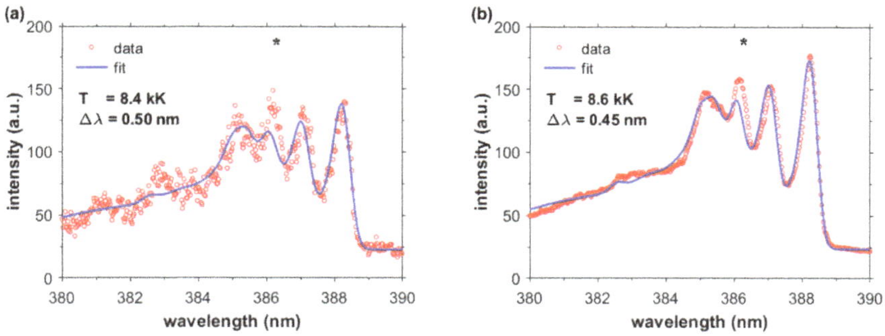

Figure 18.12. Inferred CN excitation temperature (a) at the center, (b) at a radius of 0.85 mm, 1.2 μs time delay, gate 0.125 μs. The asterisks indicate the 386.2 nm position for the 193.1 nm carbon line measured in second order [2].

Figure 18.12 illustrates measured and fitted CN spectra at the center and at a radius of 0.85 mm for data recorded at 1.2 μs time delay. A modified Boltzmann-plot method [1] is utilized in the recently published program [9] for the determination of the best-fit FWHM, $\Delta\lambda$, and temperature, T, from the entire spectrum. Appendix C further discusses the modified Boltzmann-plot method. The determination of the temperature and density variation of CN and carbon line-width along the slit is a topic of ongoing research. The variation is expected to be analogous to recently reported hydrogen-nitrogen laser plasma work [26, 27]. Recorded spectra following optical breakdown in the carbon dioxide-nitrogen mixture at a time delay of 3.7 μs and at 1 mm show lower temperature [20] than that reported here for the 1.2 μs time delay.

The temperature differences in figure 18.12 appear to be rather small; however, a systematic investigation of both the line-of-sight and Abel-inverted CN spectra is expected to reveal similar, relative temperature profiles as obtained in hydrogen-nitrogen laser-plasma experiments [26, 27], i.e., larger density and temperature near the edges than at center, including a temperature drop just inside the shock wave, and minute variations near the center. This trends appears to also be indicated by the carbon line that is measured in second order.

Figure 18.13 displays the hydrogen-atoms temperature variation along the plasma. The shock wave radius, $R(\tau)$ as function of time delay, τ, is evaluated with the Taylor–Sedov [19] formula for spherical expansion,

$$R(\tau) = (E_p/\rho\, \tau^2)^{1/5}. \tag{18.2}$$

The laser-pulse energy, E_p, for the hydrogen-nitrogen experiments [27] amounts to 0.15 J, and the density of the 1:1 nitrogen hydrogen mixture, ρ, equals 0.37kg m^{-3}. For $\tau = 3$ μs, one finds for the radius R = 5.2 mm.

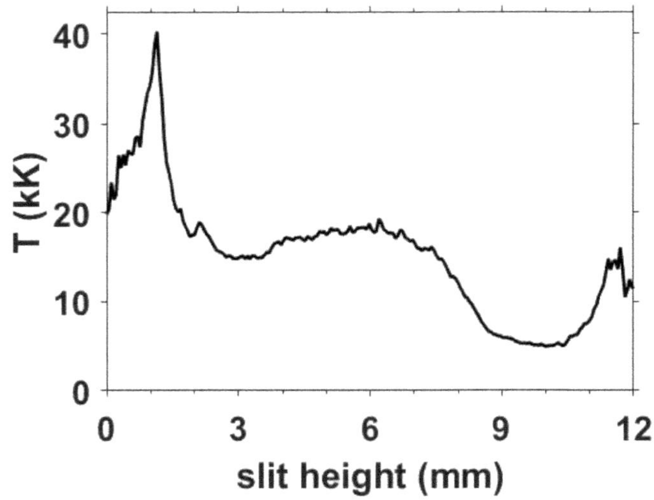

Figure 18.13. Electron temperature from line-of-sight data, 3 μs time delay, 0.025 μs gate. Average temperature: 15.4 kK [27].

References

[1] Hornkohl J O, Parigger C G and Lewis J W L 1991 *J. Quant. Spectrosc. Radiat. Transfer* **46** 405

[2] Parigger C G, Helstern C M and Gautam G 2019 *Atoms* **7** 7030074

[3] Gordon S and McBride B J 1976 Computer program for calculation of complex equilibrium compositions, rocket performance, incident and reflected shocks, and chapman- jouguet detonations *NASA Lewis Research Center, Interim Revision, NASA Report SP-273* **273**

[4] McBride B J and Gordon S 2005 *Computer Program for Calculating and Fitting Thermodynamic Functions, NASA RP-1271* **1271** 1271

[5] Parigger C G, Hornkohl J O and Nemes L 2010 *Int. J. Spectrosc.* **2010** 159382

[6] Hornkohl J O and Parigger C G 1996 Boltzmann equilibrium spectrum program (BESP), 1996 (accessed January 19, 2016). http://view.utsi.edu/besp/.

[7] Nemes L, Keszler A M, Hornkohl J O and Parigger C G 2005 *Appl. Opt.* **44** 3661

[8] Park H S, Nam S H and Park S M 2005 *J. Appl. Phys.* **2005** 113103

[9] Parigger C G, Woods A C, Surmick D M, Gautam G, Witte M J and Hornkohl J O 2015 *Spectrochim. Acta* B **107** 132

[10] Nelder J A and Mead R 1965 *Comput. J.* **7** 308

[11] Salieri F, Quarteroni A and Sacco R 2000 *Numerical Mathematics.* (New York: Springer)

[12] Luque J and Crosley D R 1999 *LIFBASE, Database and spectral simulation for diatomic molecules (v 1.6* (Menlo Park, CA: SRI International) SRI International Report MP-99-009

[13] Western C M 2017 *J. Quant. Spectrosc. Radiat. Transfer* **186** 221

[14] Western C M 2019 Private communication

[15] Downs M J and Birch K B 1994 *Metrologia* **31** 315

[16] Parigger C G 2006 Laser-induced breakdown in gases: experiments and simulation *Laser-Induced Breakdown Spectroscopy (LIBS), Fundamentals and Applications* ed A W Miziolek, V Palleschi and I Schechter (New York: Cambridge University Press) ch 4

[17] Kroll N and Watson K M 1972 *Phys. Rev.* A **5** 1883
[18] Thiyagarajan M and Thompson S 2012 *J. Appl. Phys.* **111** 073302
[19] Gautam G, Helstern C M, Drake K A and Parigger C G 2016 *Int. Rev. At. Mol. Phys.* **7** 45
[20] Parigger C G, Helstern C M, Drake K A and Gautam G 2017 *Int. Rev. At. Mol. Phys.* **8** 53
[21] Helstern C M and Parigger C G 2019 *J. Phys.: Conf. Ser.* **1289** 012016
[22] Parigger C G, Surmick D M, Helstern C M, Gautam G, Bol'shakov A A and Russo R 2020 Molecular laser-induced breakdown spectrosocpy *Laser-Induced Breakdown Spectroscopy* ed J P Singh and S N Thakur (New York: Elsevier Science) ch 7
[23] Parigger C G, Woods A C, Witte M J, Swafford L D and Surmick D M 2014 *J. Vis. Exp.* **84** 51250
[24] Griem H 1974 *Spectral Line Broadening by Plasma* (New York: Academic)
[25] Konjević N 2002 *J. Phys. Chem. Ref. Data* **31** 819
[26] Parigger C G 2019 *Atoms* **7** 7030061
[27] Parigger C G 2019 *Contr. Astron. Obs. Skalnaté Pleso* **50** accepted

IOP Publishing

Quantum Mechanics of the Diatomic Molecule (Second Edition)

Christian G Parigger and James O Hornkohl

Chapter 19

Cyanide molecular laser-induced breakdown spectroscopy with current databases

19.1 Introduction

This chapter discusses diatomic molecular spectroscopy of laser-induced plasma and analysis of data records, specifically signatures of cyanide (CN). Line strength data from various databases are compared for simulation of the CN, B $^2\Sigma^+ \longrightarrow$ X $^2\Sigma^+$, Δv $= 0$ sequence. Of interest are recent predictions using an astrophysical database, i.e., ExoMol, a laser-induced fluorescence database, i.e., LIFBASE, and a program for simulating rotational, vibrational, and electronic spectra, i.e., PGOPHER. Cyanide spectra that are predicted from these databases are compared with line-strength data that have been in use by the author for the last three decades in the analysis of laser–plasma emission spectra. Comparisons with experimental laser–plasma records are also communicated for spectral resolutions of 33 and 110 pm. The accuracy of the CN line-strength data is better than one picometer. Laboratory experiments utilize 308 nm, 35 ps bursts within an overall 1 ns pulse-width, and 1064nm, 6 ns pulse-width radiation. Experimental results are compared with predictions. Differences of the databases are elaborated for equilibrium of rotational and vibrational modes and at an internal, molecular temperature of the order of 8000 K. Applications of accurate CN data include, for example, combustion diagnosis, chemistry, and supersonic and hypersonic expansion diagnosis. The cyanide molecule is also of interest in the study of astrophysical phenomena.

The diatomic molecule cyanide, CN, occurs in various forms in nature. The toxicity of cyanide motivates health studies, in particular those of hydrogen cyanide gas that kills by inhalation. Cyanide radical measurements characterize interstellar clouds, including temperature inferences of the cosmic microwave background radiation [1, 2]. The primary interest in this work is in the CN violet $B^2\Sigma^+ - X\ ^2\Sigma^+$ band systems. Extensive experimental and theoretical details convey the complexity of the CN radical [3, 4]. In particular, the CN red systems are

doi:10.1088/978-0-7503-6204-7ch19

mentioned in a study of laboratory spectroscopy of astrophysically interesting molecules [5].

Spectroscopy [6–10] of laser-induced air plasma reveals 'fingerprints' of CN early (within the first microsecond) in the plasma decay [11]. Recent developments of laser-induced breakdown spectroscopy indicate interest in the diagnosis of molecular species. This work investigates existing databases for diatomic molecules, in particular the ones for CN, for analysis of measured emission spectra. Laser-induced breakdown with 6 ns, 1064nm radiation and associated diagnosis [12] discusses the determination of CN distribution and shock-wave phenomena in hypersonic and supersonic expansion.

For cyanide spectroscopy, one can employ the ExoMol database [13], the LIFBASE laser-induced fluorescence database [14], and the PGOPHER program for simulating rotational, vibrational, and electronic spectra [15]. There are, of course, other databases that can be accessed [16] for diatomic molecules, including HITEMP, that, for example, show hydroxyl, OH, data [17]. The ExoMol, LIFBASE, and PGOPHER predictions of the CN violet $\Delta v = 0$ sequence are compared with experimental data. The laser–plasma recombination spectra that are utilized in the comparisons were captured following optical breakdown with picosecond laser pulses in a 1:1 molar CO_2:N_2 atmospheric gas mixture, and in standard ambient temperature and pressure laboratory conditions. Astrophysical ExoMol databases are expected to work well for optical spectroscopy of laser–plasma. The measurement of molecular spectra may be accomplished with nanosecond laser-induced optical breakdown—molecular spectra are readily observed with femtosecond or picosecond laser–plasma excitation—after some time delay (of the order larger than 100ns for the occurrence of CN in CO_2:N_2 gas mixtures) from optical breakdown when using nanosecond laser pulses. In addition, analysis is discussed with previously established line-strength data that are freely available along with MATLAB [18] scripts for a subset of transitions associated with the CN violet and red band systems [19, 20].

19.2 Computation of diatomic spectra

The computation of optical molecular spectra relies on sets of wavelength positions and line strengths for molecular band systems with, in principle, theoretically resolved vibrational and rotational transitions. The experimental spectral resolutions discussed in this work are 0.033–0.11 nm. Consequently, there are a multitude of lines for each wavelength-bin of a digital array detector that captures time-resolved data following individual laser–plasma events. For example, there are well over 2000 lines in the CN violet $\Delta v = 0$ spectral region of 370 nm to 390 nm, and most lines are bunched together, including overlapped vibrational band heads. The measurements utilize a typical laser-induced optical breakdown experimental arrangement [21]; namely, (i) laser device, (ii) focusing lens, (iii) cell that contains the atmospheric mixture, (iv) spectrometer, (v) linear diode detector and optical multichannel analyzer, (vi) electronic timing, and (vii) monitor devices and equipment for subsidiary measurements capturing time-resolved CN B–X data

Table 19.1. Spectroscopic constants for the A–X and B–X transitions of $^{12}C^{14}N$.

$Y_{k,l}$	Constant	$X^2\Sigma^+$ (cm^{-1})	$A^2\Pi$ (cm^{-1})	$B^2\Sigma^+$ (cm^{-1})
$Y_{0,0}$	T_e	0.0	9245.28	25 752.0
$Y_{1,0}$	ω_e	2068.59	1812.5	2163.9
$-Y_{2,0}$	$\omega_e x_e$	13.087	12.60	20.2
$Y_{3,0}$	$\omega_e y_e$	−0.009 09	−0.0118	—
$Y_{0,1}$	B_e	1.8997	1.7151	1.973
$-Y_{1,1}$	α_e	0.017 36	0.017 08	0.023
$Y_{2,1}$	γ_e	-3.10×10^{-5}	-3.6×10^{-5}	—
$-Y_{0,2}$	D_e	6.40×10^{-6}	5.93×10^{-6}	[6.6×10^{-6}]
$-Y_{1,2}$	β_e	12×10^{-9}	42×10^{-9}	—

19.2.1 Traditional simulation of diatomic molecular spectra

The traditional approach for the generation of simulated spectra is based on molecular constants. The energy levels can be evaluated by employing a Dunham expansion [22],

$$E(\mathrm{v}, J) = \sum_{k,l} Y_{k,l}(\mathrm{v} + 1/2)^k[J(J + 1)]^l. \tag{19.1}$$

The constant coefficients $Y_{k,l}$ are called Dunham parameters, with the indices k and l corresponding to vibrational and rotational contributions, respectively. The $k = 0$ and $l = 0$ parameter $Y_{0,0}$ represents the minimum electronic energy, T_e. Table 19.1 lists the molecular constants in units of cm^{-1} for the A–X and B–X transitions for $^{12}C^{14}N$ [23].

The application of traditional molecular constants (see table 19.1) is expected to reveal line-position and intensity inconsistencies. For example, air emission spectra predictions with the NEQAIR code [24, 25] reveal slightly different line positions leading to difficulties in air-plasma analysis of superposition spectra composed of several species. The laser plasma that is generated in dry air shows a variety of species [26–28]; consequently, accurate predictions of individual components are desirable. Commercial SPECAIR software is available that focuses on calculating emission or absorption spectra of air-plasma radiation [29]. Conversely, one can utilize published molecular line positions and strengths for individual species for the spectra simulation of the CN diatomic molecule.

19.2.2 Line positions and strengths of diatomic spectra

In this book, the computation of diatomic molecular spectra utilizes line-strength data. The Boltzmann equilibrium spectral program (BESP) and the Nelder–Mead temperature (NMT) program allow one to, respectively, compute an emission spectrum and fit theoretical to experimental spectra, see chapter 15. The construction of the communicated molecular CN line strengths 'CNv-lsf' [20] first makes use of Wigner–Witmer eigenfunctions and a diatomic line-position fitting program; second, it computes Frank–Condon factors and *r*-centroids; and third, it combines these factors with the rotational factors that usually decouple from the overall molecular line strength due to the symmetry of diatomic molecules. In turn, the ExoMol states and transition files for CN [4, 30, 31] and the PGOPHER data file [32] are examined in order to generate line-strength data that can be used with BESP and NMT. The LIFBASE program is utilized for visual comparisons of CN B–X simulated and recorded data.

The ExoMol and the PGOPHER data show Einstein A-coefficients that are converted to line strengths [33–35], S, for electric dipole transitions, using

$$A_{ul} = \frac{16\pi^3}{3g_u h\epsilon_0 \lambda^3}(e\ a_0)^2 S_{ul}, \qquad g_u = 2(2J_u + 1). \tag{19.2}$$

Here, A_{ul} denotes the Einstein A-coefficient for a transition from an upper, u, to a lower, l, state, and h and ε_0 are Planck's constant and vacuum permittivity, respectively. The elementary charge is e, the Bohr radius is a_0, and the transition strength is S_{ul}. The line strength, S, that is used in the MATLAB scripts is expressed in traditional spectroscopy units (stC^2 cm^2). The wavelength of the transition is λ, g_u is the upper state degeneracy, and J_u is the total angular momentum of the upper state.

19.3 Results

This section elaborates analysis of recorded CN spectra of the B $^2\Sigma^+ \longrightarrow$ X $^2\Sigma^+$, $\Delta v = 0$ sequence. The spectra were captured using an intensified 1024-diode array detector and a laboratory-type Czerny–Turner (Jobin Yvon model HR-640, Fr) spectrometer. The spectral resolutions of the experimental arrangement amount to 0.11 nm ($\simeq$7 cm^{-1}) and 0.033 nm ($\simeq$2 cm^{-1}) for the employed 1200 and 3600 groves/mm gratings, respectively. Detector sensitivity corrections and wavelength calibrations were accomplished with standard spectroscopic light sources.

The time-resolved measurement gate-delays were 400 ns (ns), with gate-open times of 500 ns and 600 ns for the higher and lower resolution data, respectively. An average of 40 spectra were collected; consequently, the inferred temperature is an average of the line-of-sight data during the gate-open duration.

The analysis discusses the 0.033 nm and then the 0.11 nm resolution data. For both data sets, we discuss the fitting results obtained with

- The PGOPHER program using PGOPHER data,
- The NMT program using PGOPHER- and ExoMol line strengths,
- The NMT program using CNv-lsf data, comparisons,
- The LIFBASE program.

One of the primary interests is the use of updated ExoMol sets of line-strength data that appear to be in use for extragalactic studies [31]. Availability of an extensive line list obviously would alleviate computation of specific transitions that are investigated in laser–plasma laboratory experiments, including CN B–X violet and CN A–X red systems. The ExoMol database shows 2285 103 transitions up to 60 000 cm^{-1} between the 3 lowest electronic states, $X\ ^2\Sigma^+$, $A\ ^2\Pi$, and $B\ ^2\Sigma^+$. The PGOPHER data set for CN A–X and B–X includes 191 109 transitions.

The CNv-lsf and CNr-lsf data contain 7960 and 40 728 transitions, respectively. The differences in number of transitions are in part due to the number of rotational states (up to $\simeq$120 in ExoMol), the cutoffs for Einstein A-coefficients, and associated line strengths (see equation (19.2)), or the establishment of sets of computed molecular parameters that fit data from high-resolution, Fourier-transform spectroscopy. The line positions are determined from high-resolution data with a standard deviation comparable to the estimated experimental errors. The obtained, simulated line strengths are typically better than 0.05 cm^{-1}.

19.3.1 Analysis of the 0.033 nm spectral resolution data

19.3.1.1 PGOPHER program using PGOPHER data

The full CN B–X and A–X transition data are available for the construction of simulated spectra. Figure 19.1 displays measured and fitted CN violet spectra. The fitted spectrum also shows the baseline correction and difference between experimental and simulated spectra.

The synthetic spectrum is fitted using the PGOPHER program and an overlay of the experimental record of recorded optical multichannel analyzer (OMA) counts versus vacuum wave numbers that are computed from the recorded air wavelengths. Table 19.2 lists results for parameters that were selected for fitting. Normalized intensity units were used, i.e., the partition function is implemented. Following a separate fit for determination of the wave number offset of −0.31 cm^{-1} (−0.005 nm-the Doppler width at $\simeq$ 8000 K for CN equals $\simeq$ 0.005 nm), the selected parameters included Gaussian width, temperature, scaling, and T_{vib} that assess deviation from thermodynamic equilibrium. The T_{vib} results confirms Boltzmann distribution equilibrium within the standard deviation.

In this work, a Gaussian profile models the spectrometer and intensified linear-array detector transfer function. However, a measured system transfer function or a Voigt function can replace the selected Gaussian profile provided that changes are implemented in the MATLAB source scripts for the recently communicated BESP and NMT scripts [20]. The PGOPHER program allows one to accomplish Voigt profile fits.

Investigations of fitting Gaussian, w_G, and Lorentzian, w_L, widths, while leaving all other parameters constant, lead to the PGOPHER results $w_G = 1.6\ cm^{-1}$ (0.024 nm) and $w_L = 0.64\ cm^{-1}$ (0.0095nm). Application of an empirical Voigt width, w_V, approximation [37],

$$w_V \approx w_L/2 + \sqrt{w_L^2/4 + w_G^2}, \tag{19.3}$$

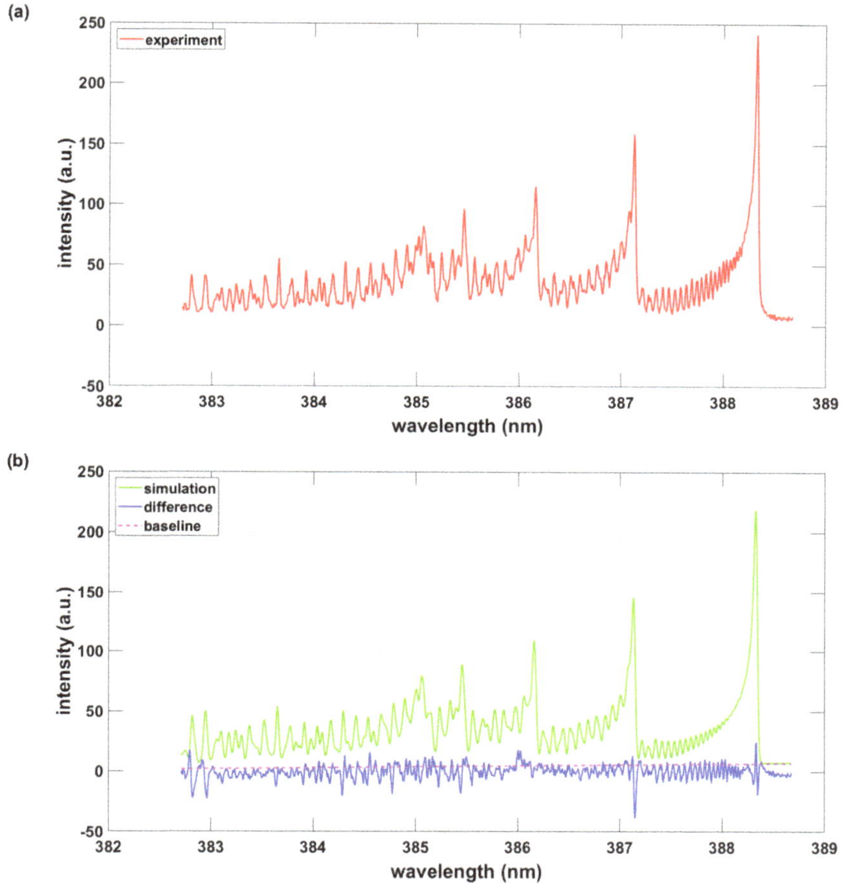

Figure 19.1. (**a**) Experimental spectrum; (**b**) simulation with PGOPHER, T = 8103 K, $\Delta\lambda = 0.033$ nm. Reproduced from [36]. CC BY 4.0.

Table 19.2. PGOPHER fitting parameters for the 0.033 nm data and results.

Parameter	Value	Standard deviation
Temperature (K)	8103	187
Gaussian width (cm^{-1})	2.2	0.03
Scale (au)	43 960	790
T_{vib} (K)	8108	146

results in $w_V \approx 2.0\ \mathrm{cm}^{-1}$ (0.029 nm). The widths in units of nm that are indicated in brackets are calculated using $\Delta\lambda = \Delta\tilde{\nu} \times \lambda^2$ at a wavelength of 386 nm (central wavelength of the 0.033 nm spectral resolution data). The Lorentzian contribution to the Voigt profile amounts to $\approx$ 30%.

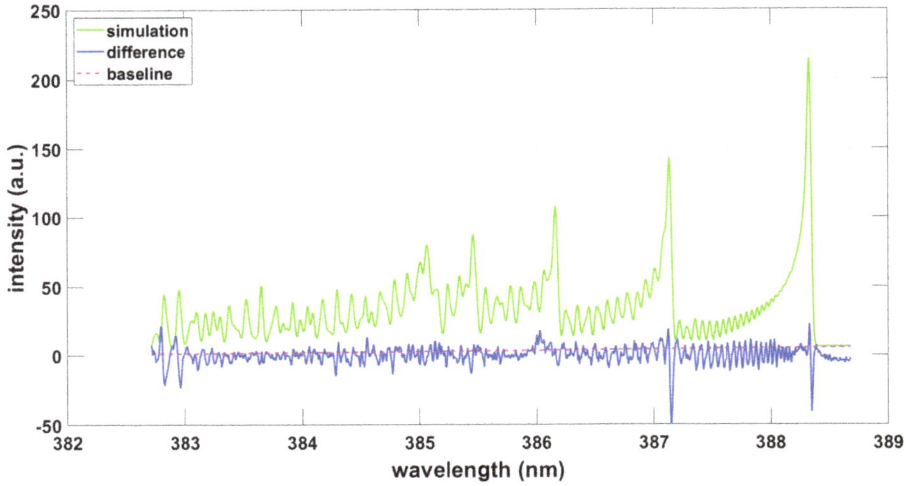

Figure 19.2. Simulation using PGOPHER data and the NMT program, $T = 8240$ K, $\Delta\lambda = 0.034$ nm. Reproduced from [36]. CC BY 4.0.

19.3.1.2 NMT program using PGOPHER and ExoMol line strengths

In addition, CN B–X line positions and Einstein A-coefficients (that are converted to line strengths) are collected in a data file that is compatible with the mentioned NMT spectral fitting program. Figure 19.2 illustrates spectra determined from temperature and Gaussian line-width fitting. The results are consistent with those obtained from the versatile PGOPHER program. The inclusion of CN A–X data in the analysis would change the inferred temperature by $\simeq 0.1\%$.

Similarly, the recommended ExoMol databases for states and transitions were utilized to generate a subset of lines and strengths for analysis with the NMT program and for the 0.033 nm resolution data.

The results displayed in figure 19.3 indicate a Gaussian full width at half maximum (FWHM) of 0.038 nm, indicative of line-position differences in the ExoMol data. However, the inferred temperature appears consistent with the results illustrated in figures 19.1 and 19.2.

19.3.1.3 NMT program using CNv-lsf data, comparisons

Analysis of the measured data with the CNv-lsf data reveals a temperature of $T = 8140$ K and a fitted FWHM of 0.038 nm. Figure 19.4 shows the fitting results.

The simulated spectra in figures 19.1–19.4 display superposition spectra from quite a few individual rotational–vibrational transitions of the CN B–X $\Delta \mathrm{v} = 0$ sequence. Actually, small contributions from the CN A–X transition are also included in the PGOPHER and ExoMol databases. Tables 19.3 and 19.4 summarize the number of lines in the data files.

Comparisons of CN B–X line-position accuracies in the PGOPHER and ExoMol databases with those of the CNv-lsf database are further elaborated for the range of the 0.033 nm resolution experiments (25 725 cm^{-1} to 26 125 cm^{-1}) and for Einstein

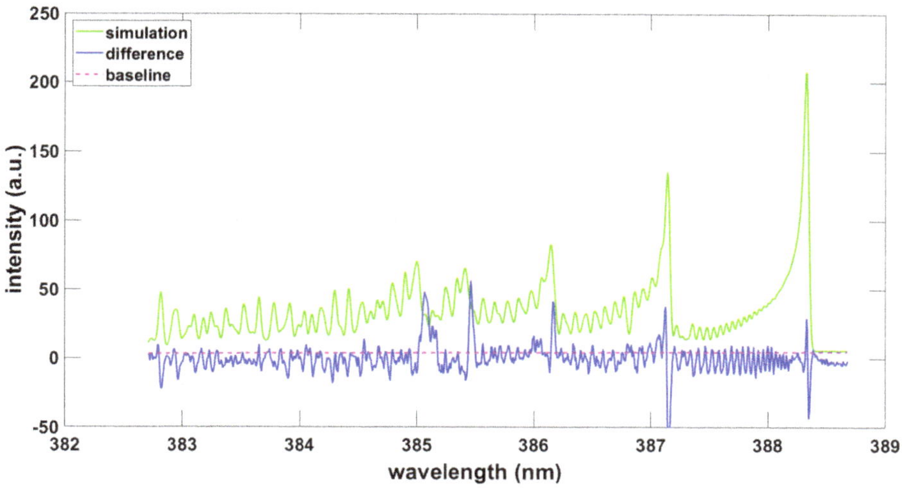

Figure 19.3. Simulation using ExoMol data and the NMT program, T = 8150 K, $\Delta\lambda = 0.038$ nm. Reproduced from [36]. CC BY 4.0.

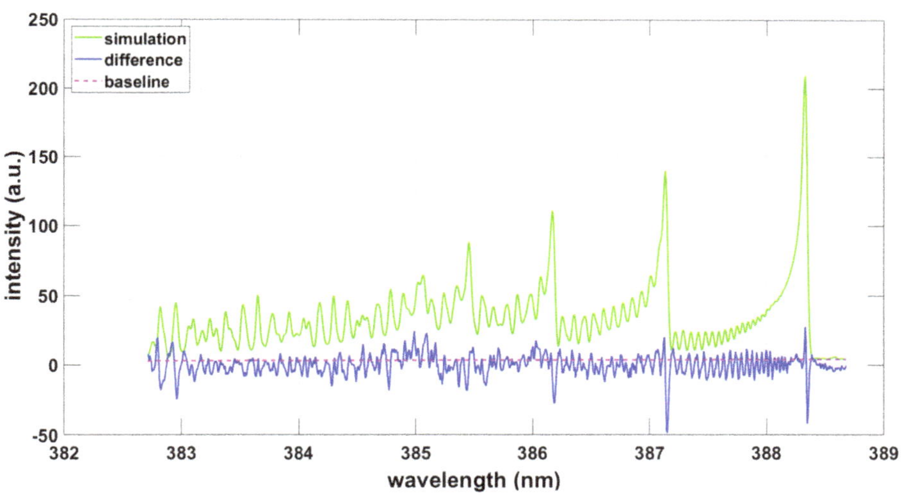

Figure 19.4. Simulation using CNv-lsf data and the NMT program, T = 8140 K, $\Delta\lambda = 0.036$ nm. Reproduced from [36]. CC BY 4.0.

Table 19.3. Number of transitions of the simulated spectra in the measured experimental ranges for the 0.033 nm (25 725 cm^{-1} to 26 125 cm^{-1}) data.

Database	0.033 nm range CN B–X	0.033 nm range CN B–X and A–X
PGOPHER	3598	5631
ExoMol	4302	17 181
CNv-lsf	2461	2461

Table 19.4. Number of transitions of the simulated spectra in the measured experimental ranges for the 0.033 nm data (see table 19.3) with Einstein A-coefficients larger than 10^3 s^{-1}.

Database	0.033 nm range CN B–X	0.033 nm range CN B–X and A–X
PGOPHER	3205	4532
ExoMol	3625	8270
CNv-lsf	2461	2461

Table 19.5. Subset CN B–X lines of the PGOPHER and ExoMol data that agree within $\Delta\tilde{\nu}$ of 2461 CN B–X transitions in the CNv-lsf data for the 0.033 nm range (25 725 cm^{-1} to 26 125 cm^{-1}) [36].

Database	$\Delta\tilde{\nu}$ <0.05 cm^{-1}	$\Delta\tilde{\nu}$ <0.2 cm^{-1}	$\Delta\tilde{\nu}$ <1.0 cm^{-1}	$\Delta\tilde{\nu}$ <2.0 cm^{-1}
PGOPHER	1200	1422	1823	2012
ExoMol	158	463	1266	1935

A-coefficients larger than 10^3 s^{-1} (see first column in table 19.4). Table 19.5 displays agreements of lines within the indicated wave number range and otherwise the same angular momentum values for upper and lower levels of the transitions.

Various aspects of the accuracy of the CNv-lsf database have been extensively tested [38], including analysis of laser-induced fluorescence and nominal 300 K temperature Fourier-transform spectra. The differences in predictions for line positions appear larger for levels with higher angular momenta, J' and J. The accuracy of line positions in the CNv-lsf database is better than 0.05 cm^{-1}. Table 19.5 reveals that PGOPHER line positions compare favorably with those in the CNv-lsf database. The ExoMol database appears acceptable within $\simeq$2 wave numbers that correspond to a spectral resolution of $\simeq$ 0.033 nm. However, high-temperature ($\simeq$ 8000 K) inferences from CN B–X spectra at a 0.033 nm experimental spectral resolution appear only minimally affected (see figures 19.2 and 19.3).

19.3.1.4 LIFBASE program

The LIFBASE program is also used for simulation of the measured spectrum. A Voigt profile of width 0.033 nm with a 30% Lorentzian contribution is selected. Figure 19.5 illustrates the results. It is noteworthy that the PGOPHER program is also capable of fitting Voigt profiles. However, Gaussian profiles are usually selected in the NMT program for modeling of the spectrometer and detector transfer function. The choice of a 30% Lorentzian contribution is based on PGOPHER-program investigations, see equation (19.3).

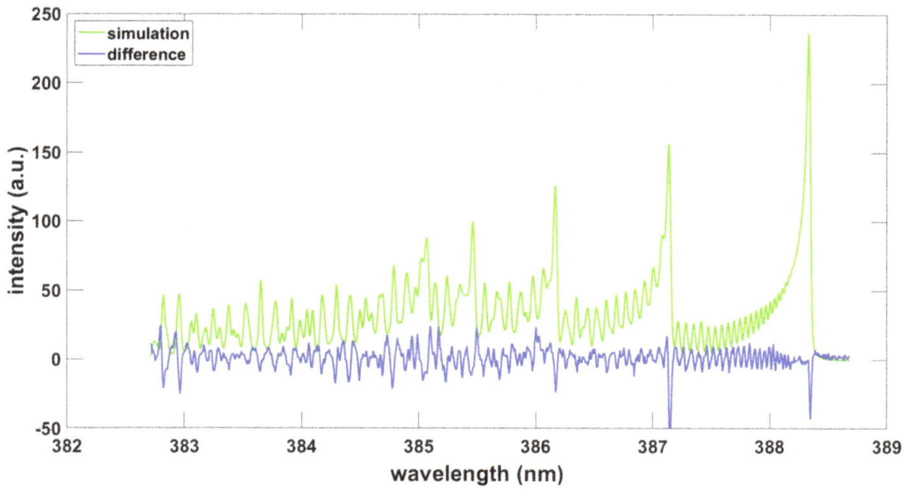

Figure 19.5. Simulation using LIFBASE with fixed $T = 8103$ K, $\Delta\lambda = 0.033$ nm, Voigt profile with 30% Lorentzian contribution. Reproduced from [36]. CC BY 4.0.

19.3.2 Analysis of the 0.11 nm spectral resolution data

The lower resolution 0.11 nm data are analyzed using a similar approach as that for the 0.033 nm data. It is noteworthy that the recorded average spectra for both lower and higher resolution show coincidentally similar detector counts (or intensity in arbitrary units) for the two different gratings, respectively.

19.3.2.1 PGOPHER program using PGOPHER data

Following a separate fit for determination of the wave number offset of 0.13 cm^{-1} (0.002 nm), the selected parameters included Gaussian width, temperature, and scaling. Figure 19.6 illustrates the results obtained by employing the PGOPHER fitting program and the available full CN B–X and CN A–X data set [32]. The effect of the CN A–X transition lines is minimal for temperature inference, viz. 0.1% for the higher spectral resolution data (see figure 19.2).

Table 19.6 summarizes the results of PGOPHER fitting including the standard deviations.

19.3.2.2 NMT program using PGOPHER and ExoMol line strengths

Figure 19.7 displays the results of the NMT program using PGOPHER data. As before, the NMT program resorts to fitting of relative intensities. In turn, the scales that are indicated in tables 19.2 and 19.6 can serve as a calibration of the detector counts in the laboratory-recorded CN spectra.

The fitting of the experimental data using the PGOPHER-extracted wave numbers, upper term values, and Einstein A-coefficients (that are converted to line strengths) with the NMT program shows similar results as obtained for exclusive PGOPHER fitting.

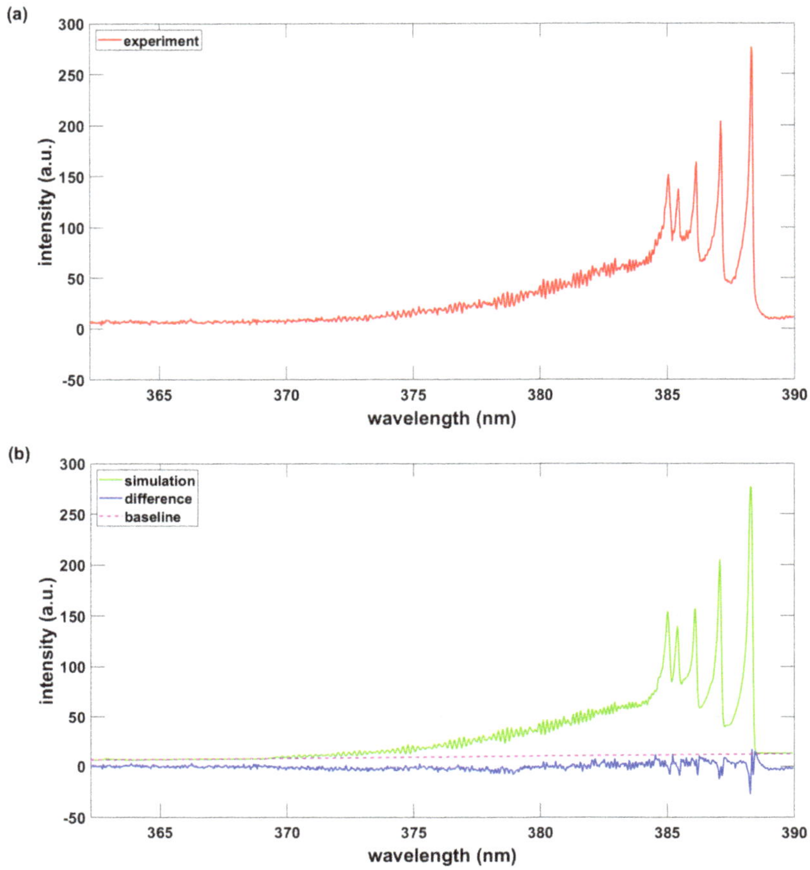

Figure 19.6. (**a**) Experimental spectrum; (**b**) simulation with PGOPHER, $T = 8340$ K, $\Delta\lambda = 0.11$ nm. Reproduced from [36]. CC BY 4.0.

Table 19.6. PGOPHER fitting parameters for the 0.11 nm data and results.

Parameter	Value	Standard deviation
Temperature (K)	8340	43
Gaussian width (cm^{-1})	7.1	0.07
Scale (au)	107 300	850

Fitting the ExoMol data using the NMT program yields a temperature of $T = 9090$ K for a spectral resolution of $\Delta\lambda = 0.11$ nm. Figure 19.8 illustrates the simulated spectrum, difference, and baseline.

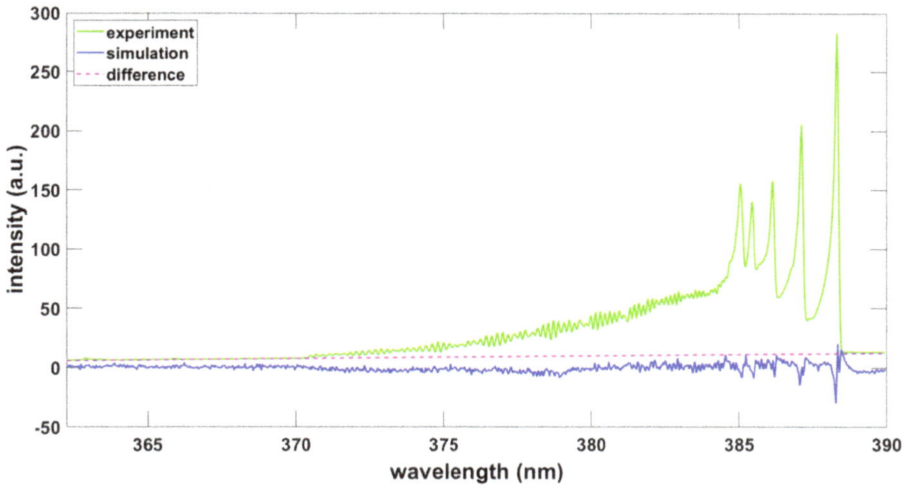

Figure 19.7. Simulation using PGOPHER data and the NMT program, T = 8380 K, $\Delta\lambda = 0.11$ nm. Reproduced from [36]. CC BY 4.0.

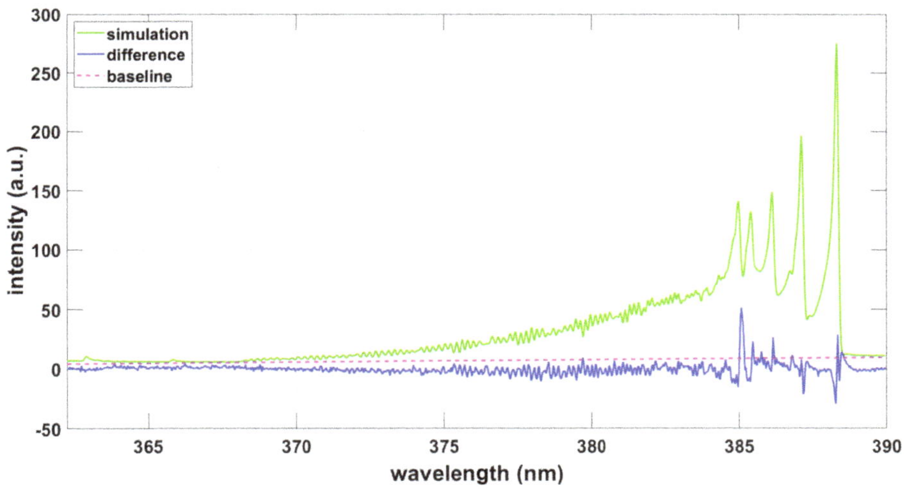

Figure 19.8. Simulation using ExoMol data and the NMT program, $T = 9090$ K, $\Delta\lambda = 0.11$ nm. Reproduced from [36]. CC BY 4.0.

19.3.2.3 NMT program using CNv-lsf data, comparisons

A temperature of $T = 8940$ K is obtained when using CNv-lsf in the NMT program. Figure 19.9 shows the results. Tables 19.7 and 19.8 summarize the number of lines in the data files.

19.3.2.4 LIFBASE program

Similar to the higher resolution data, the LIFBASE program is utilized for the generation of a simulated spectrum. Again, a 30% Lorentzian contribution is

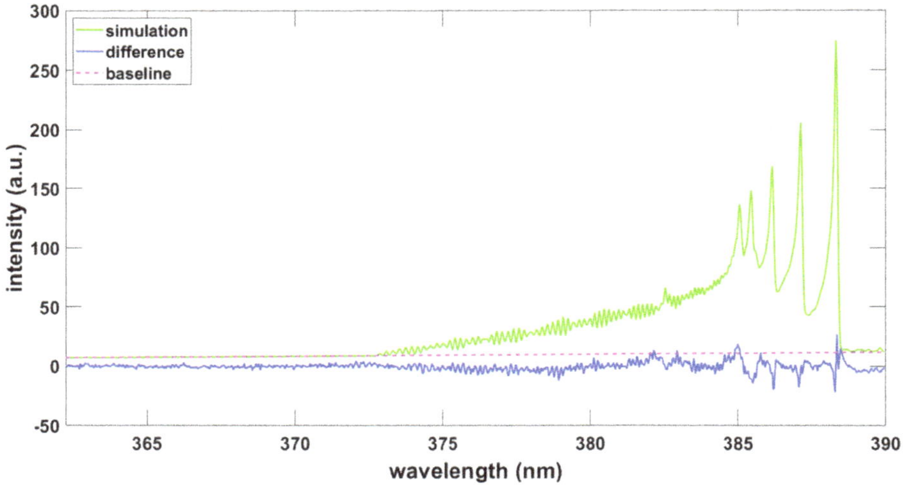

Figure 19.9. Simulation using CNv-lsf data and the NMT program, $T = 8940$ K, $\Delta\lambda = 0.11$ nm. Reproduced from [36]. CC BY 4.0.

Table 19.7. Number of transitions of the simulated spectra in the measured experimental ranges for the 0.11 nm (25 600 cm^{-1} to 27 600 cm^{-1}) data.

Database	0.11 nm range CN B–X	0.11 nm range CN B–X and A–X
PGOPHER	7115	14 349
ExoMol	10 499	78 079
CNv-lsf	3313	3313

Table 19.8. Number of transitions of the simulated spectra in the measured experimental ranges for the 0.11 nm data (see Table 19.7) with Einstein A-coefficients larger than 10^3 s^{-1}.

Database	0.11 nm range CN B–X	0.11 nm range CN B–X and A–X
PGOPHER	6187	9958
ExoMol	8259	28 274
CNv-lsf	3313	3313

selected based on the investigations with Gaussian and Lorentzian profile contributions, see equation (19.3). Figure 19.10 displays the results.

19.4 Discussion

The CN violet $B\ ^2\Sigma^+ - X\ ^2\Sigma^+$, $\Delta v = 0$ sequence reveals a multitude of vibrational and rotational transitions that are usually not individually resolved in the study of

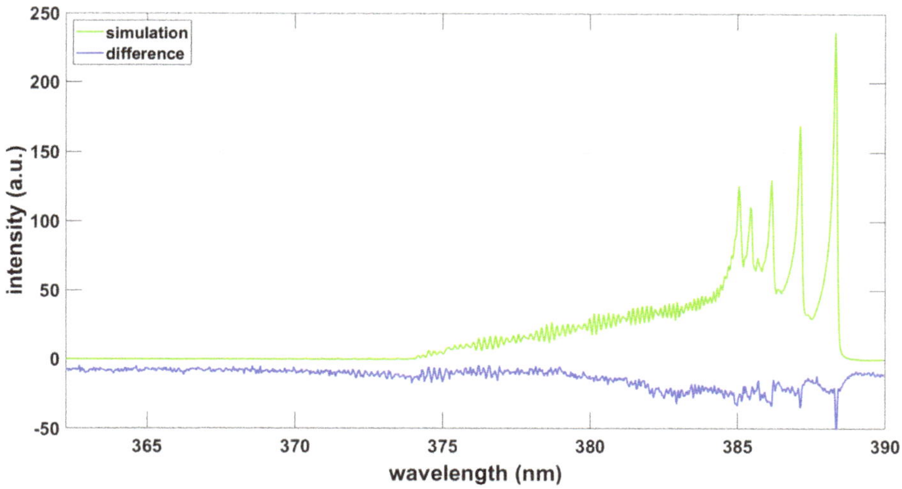

Figure 19.10. Simulation using LIFBASE with fixed $T = 8340$ K, $\Delta\lambda = 0.11$ nm, Voigt profile with 30% Lorentzian contribution. Reproduced from [36]. CC BY 4.0.

laser-induced plasma emissions in the spectral range of 383 nm to 389 nm. Analysis of the 0.033 nm spectral resolution experimental emission spectrum with the PGOPHER program and use of a full CN B–X data file yields CN excitation temperature of $\simeq$ 8100 K and a standard deviation of $\simeq$ 200 K.

Similar results were obtained when using the CN B–X data in conjunction with a Nelder–Mead spectral fitting program. The most recent ExoMol CN data also predict a temperature within the standard deviation of the PGOPHER prediction, but there appear to be discrepancies in the ExoMol predictions near the 2–2, 3–3, and 4–4 band heads or in the range of 385–386 nm. Analogous results are noted when using the CNv-lsf line strengths; however, with slightly better agreement of experimental and simulated spectra than those for ExoMol. Comparative spectra obtained from the LIFBASE program are also largely in agreement with the measured data.

The lower 0.11 nm resolution data in the range of 362–390 nm appear to be well predicted by the PGOPHER data. About 6% higher temperatures are predicted with the ExoMol and the CNv-lsf data. The CN violet spectra have been extensively tested in a variety of experimental studies of laser plasma.

For comprehensive comparisons of experimental and simulated spectra, however, the higher the spectral resolution the better. The expansive PGOPHER and ExoMol line lists allow one to predict simulated spectra consistent with laboratory laser–plasma experiments.

The author (C.G.P.) acknowledges the support in part by the Center for Laser Applications at the University of Tennessee Space Institute. Furthermore, deep appreciation goes to departed J.O. Hornkohl and C.M. Western for their meticulous work in quantum mechanics of the diatomic molecule and contributions toward simulating rotational, vibrational, and electronic spectra, and their contributions, respectively, to the development of BESP, NMT, and PGOPHER programs.

References

[1] Roth K C, Meyer D M and Hawkins I 1993 *Astrophys. J.* **413** L67
[2] Leach S 2004 *Can. J. Chem.* **82** 730
[3] Ram R S, Davis S P, Wallace L, Englman R, Appadoo D R T and Bernath P F 2006 *J. Mol. Spectrosc.* **237** 225
[4] Brooke J S A, Ram R S, Western C M, Li G, Schwenke D W and Bernath P F 2014 *Astrophys. J. Suppl. Ser.* **210** 1
[5] Davis S P 1987 *Publ. Astron. Soc. Pac.* **99** 1105
[6] Kunze H-J 2009 *Introduction to Plasma Spectroscopy* (Heidelberg: Springer)
[7] Fujimoto T 2004 *Plasma Spectroscopy* (Oxford: Clarendon)
[8] Ochkin V N 2009 *Spectroscopy of Low Temperature Plasma* (Weinheim: Wiley)
[9] Demtröder W 2014 *Laser Spectroscopy 1: Basic Principles* 5th edn (Heidelberg: Springer)
[10] Demtröder W 2015 *Laser Spectroscopy 2: Experimental Techniques* 5th edn (Heidelberg: Springer)
[11] Miziolek A W, Palleschi V and Schechter I 2006 *Laser Induced Breakdown Spectroscopy* (New York: Cambridge University Press)
[12] Singh J P and Thakur S N (ed) 2007 *Laser-Induced Breakdown Spectroscopy* (Amsterdam: Elsevier Science)
[13] Tennyson J *et al* 2020 *J. Quant. Spectrosc. Radiat. Transf.* **255** 107228
[14] Luque J and Crosley D R 2021 *LIFBASE: Database and Spectral Simulation for Diatomic Molecules* (Menlo Park, CA: SRI International)
[15] Western C M 2017 *J. Quant. Spectrosc. Radiat. Transfer* **186** 221
[16] McKemmish L K 2021 *WIREs Comput. Mol. Sci.* **11** e1520
[17] Rothman L S, Gordon I E, Barber R J, Dothe H, Gamache R R, Goldman A, Perevalov V I, Tashkun S A and Tennyson J 2010 *J. Quant. Spectrosc. Radiat. Transf.* **111** 2139
[18] *MATLAB Release R2022a Update 5* (MA: Natick)
[19] Parigger C G and Hornkohl J O 2020 *Quantum Mechanics of the Diatomic Molecule with Applications* (Bristol: IOP Publishing)
[20] Parigger C G 2023 *Foundations* **3** 1
[21] Hornkohl J O, Parigger C G and Lewis J W L 1991 *J. Quant. Spectrosc. Radiat. Transfer* **46** 405
[22] Dunham J L 1932 *Phys. Rev.* **41** 721
[23] *National Institute of Standards and Technology (NIST) Chemistry WebBook, SRD 69, for the Cyano Radical, Constants of Diatomic Molecules.* 2021
[24] Whiting E E 1995 Private communication
[25] Whiting E E, Park C, Liu Y, Arnold J and Paterson J 1996 *NEQAIR96, Nonequilibrium and Equilibrium Radiative Transport and Spectra Program: User's Manual* (CA: NASA Ames Research Center) Technical Report NASA RP-1389
[26] Boulous P M I and Pfender E 1994 *Thermal Plasmas–Fundamentals and Applications* (New York: Plenum)
[27] McBride B and Gordan S 1994 *Interim Revision NASA Report RP-1311 Part* I (Cleveland, OH: NASA Lewis Research Center)

[28] McBride B and Gordan S 1996 *Interim Revision NASA Report RP-1311 Part II* (Cleveland, OH: NASA Lewis Research Center)
[29] Laux C O 2002 Radiation and nonequilibrium collisional-radiative models, von Karman Institute Lecture Series 2002-07 *Physico-Chemical Modeling of High Enthalpy and Plasma Flows* ed D Fletcher, J M Charbonnier, G S R Sarma and T Magin (Flanders: Rhode-Saint-Genèse))
[30] Syme A-M and McKemmish L K 2020 *Mon. Not. R. Astron. Soc.* **499** 25
[31] Syme A-M and McKemmish L K 2021 *Mon. Not. R. Astron. Soc.* **505** 4383
[32] Western C M 2019 Private communication
[33] Condon E U and Shortley G 1953 *The Theory of Atomic Spectra* (Cambridge: Cambridge University Press)
[34] Hilborn R C 1982 *Am. J. Phys.* **50** 982 http://arxiv.org/ftp/physics/papers/0202/0202029.pdf
[35] Thorne A P 1988 *Spectrophysics* 2nd edn (London: Chapman and Hall)
[36] Parigger C G 2023 *Atoms* **11** 62
[37] Whiting E E 1968 *J. Quant. Spectrosc. Radiat. Transf.* **8** 1379
[38] Parigger C G and Hornkohl J O 2020 *Quantum Mechanics of the Diatomic Molecule with Applications* (Bristol: IOP Publishing)

Chapter 20

Diatomic carbon, C_2

The prediction of diatomic molecular carbon spectra has been a significant goal [1] in the research efforts that are discussed in this book. Plasma emission spectra serve a gauge for the accuracy of the computed spectra. This chapter focuses on aspects of diatomic carbon molecular spectroscopy that were communicated at two recent line shape conferences [2, 3].

In the laboratory, laser-induced breakdown spectroscopy (LIBS) is a technique utilized for analyzing a substance's composition [4]. It is a noninvasive and noncontact technique that only requires a small amount of the substance to be useful while still allowing for accurate determination of unknown material [5]. This becomes especially useful when the substance itself or its environment is hazardous in nature. However, in order to be of use as a diagnostic tool, well-tested models are required in conjunction with interpretation of well-developed molecular C_2 spectra, such as recombination spectra recorded following laser-induced optical breakdown [6]. Recent investigations [2, 7, 8, 9] carbon Swan spectra emissions from laser-induced plasma.

20.1 Analysis of C_2 in Nd:YAG laser-plasma

Carbon Swan spectra are observed following laser ablation of graphene in laboratory air [2] or on solid targets and in gas-phase optical breakdown [10, 11]. Previous experiments showed temperatures that ranged from 4.5 to 7.5kK for the $\Delta v = 0$ transition and 4.2–4.5 kK for the $\Delta v = -1$ transition for time delays of the order of 1.6 μs to 70 μ s. This experiment explored in greater detail time delays $> 10\ \mu$s for both molecular bands. Temperatures were found to be similar, ranging from 4.5 to 6.7 kK for the $\Delta v = 1$ transition and 3.2–5.5 kK for the $\Delta v = -1$ transition. Figure 20.1 illustrates C_2, Δv = -1 emission spectra recorded at a delay of 30 μs in CO gas at 1 atm.

doi:10.1088/978-0-7503-6204-7ch20

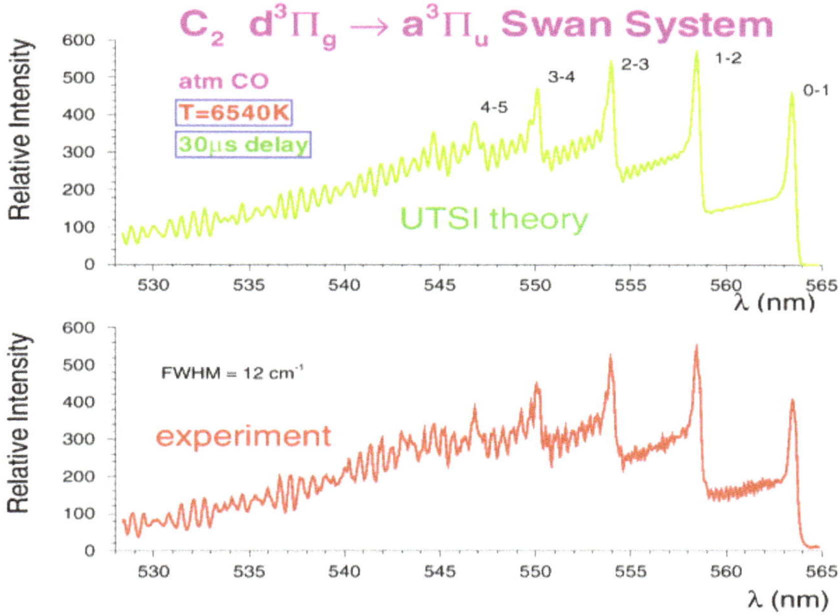

Figure 20.1. Measured and computes C_2, $\Delta v = -1$ sequence. Time delay 30 μ s, temperature 6.54 kK, spectral resolution 0.36 nm (12 cm^{-1}). The spectrum is measured following optical breakdown in CO gas at 1 atm. Reprinted from [6], with permission of Elsevier.

20.2 Detailed fitting of C_2 spectra

Efforts are communicated to investigate the applicability of the local thermodynamic equilibrium (LTE) assumption. Comparisons are discussed in view of previous work that utilized Stark broadening of the H_β line, confirming LTE for delays < 10 μ s, yet further research is needed for longer time delays. Recent experiment utilizes a Nd:YAG pulsed laser operating at 1064nm with 13 ns, 190 mJ/pulse to generate laser-induced breakdown. The emitted spectra were dispersed with an 1800 grooves/mm grating in a HR640 Jobin-Yvon 0.6 m spectrometer and then recorded with an Andor iSstar ICCD. Measurements were time-resolved with varying time delays (10 μs to 100 μ s) and corresponding gate widths (1 μs to 20 μ s). Spectra were sensitivity and wavelength calibrated with a standard tungsten and mercury lamp, respectively.

For diatomic molecules, spectral predictions require a temperature along with a set of line-strengths [1], for the allowed spectral transitions. In order to fit the spectra, the Nelder–Mead algorithm is utilized [12] through the so-called Nelder–Mead-Temperature (NMT) program [13]. The NMT program generates a single temperature spectral fit for the recorded data. This fit, within the accuracy and precision bounds, is also used to refine the calibration of the data. This is repeated several times, with each run increasing from a linear, quadratic, to finally a cubic calibration. After that, the program is run three times in succession to ensure accuracy of the computed temperature. Figures 20.2 and 20.3 are representative of

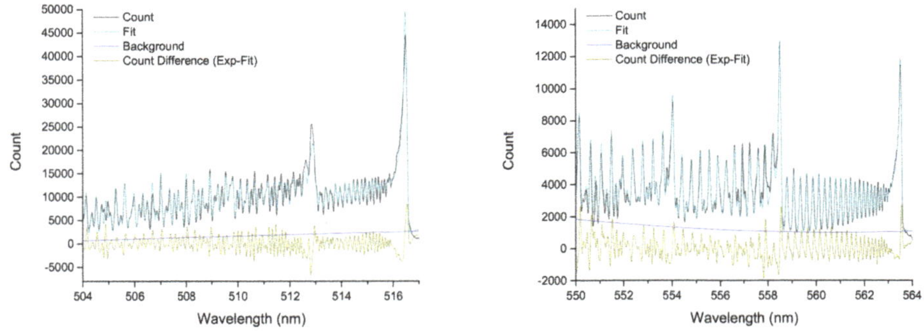

Figure 20.2. Measured and fitted C_2 spectra: 40 μs delay, 5 μs gate width. Temperatures $T = 5.5$ kK and $T = 4.4$ kK, left-hand and right-hand panels, respectively. Reproduced from [2].

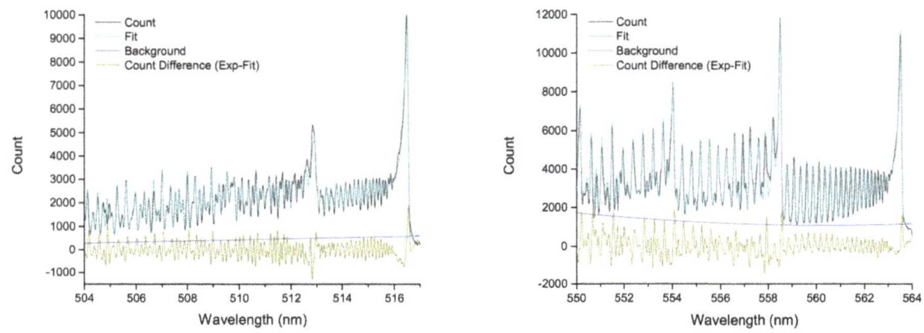

Figure 20.3. 70 μs delay, 10 μs gate width. $T = 5.8$ kK and 3.2 kK, left-hand and right-hand panels, respectively. Reproduced from [2].

the results from the NMT program for carbon Swan spectra of the Δv = 0 and Δv = - 1 bands. Seen in both are the clear vibrational peaks and underlying rotational structure. No atomic lines are present. Note, there is good matching between the experimental and theory spectra because the count difference between both is generally within the computed background line. However, the discrepancy in temperature is in part due to the different backgrounds that were used in the analysis of the Δv = 0 and Δv = -1 analysis. Moreover, the data were collected in separate experimental runs.

Figure 20.4 shows a two-dimensional image of the C_2 spectra for the $\Delta v = 0$ band. Calculated temperatures range from 6600–7700 K, increasing towards 7 mm, with a dip just under it at 6 mm. Also of note is the slight vertical symmetry.

For the theory models, a single temperature is used for fitting with assumption of LTE. For early time delays of the order of 1 μs with laser energy per pulse of the order of 100 mJ, time-delayed spectra indicate LTE due to the presence of H_β [8]. Past that, there is some indication for LTE at later delays. The experimental spectra show good agreement with computed theory spectra within background variations.

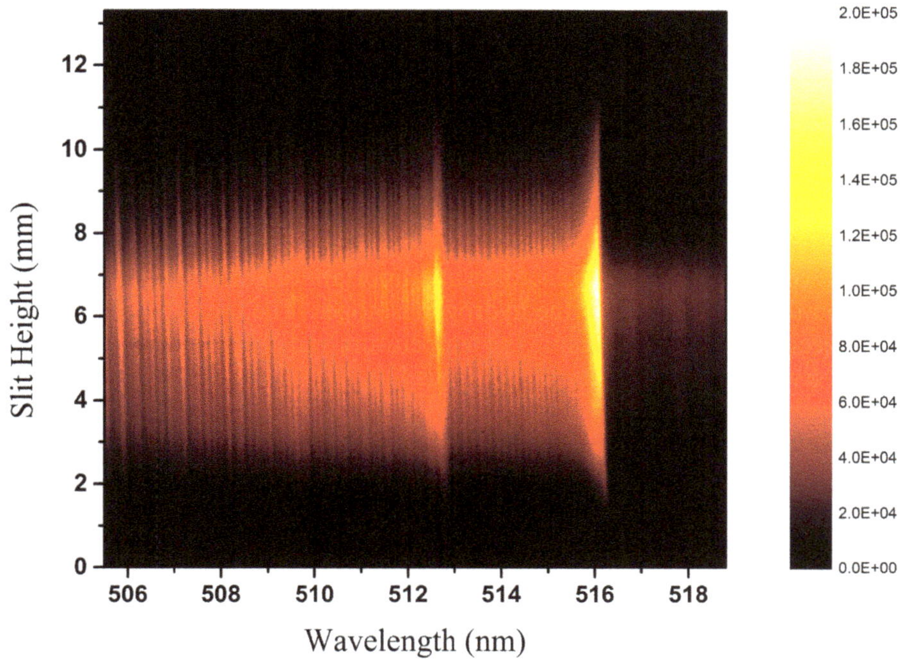

Figure 20.4. Raw image of two-dimensional C_2 $\Delta v = 0$ spectra at 15 μs delay and 5 μs gate width. The full width at half maximum averaged to 0.13 nm for each of the individual spectral lines. Reproduced from [2].

However, there is no definite indication of LTE at later delays. As the electron density diminishes, H_β emissions disappear in the measured spectra past 10 μ s. There is some indication that LTE is absent because there appear to be different temperature ranges for the different wavelength bands, as seen in comparing the $\Delta v = 0$ and $\Delta v = -1$ sequences that were collected from separate graphene laser ablation events.

From the recorded two-dimensional data, the plasma shows a higher temperature towards the top of the recorded emission spectra near 7 mm but is cooler away from it. Just below that region, there appear to be significant variations in inferred temperature. These temperature differences could indicate nonLTE conditions. Consequently, exploration and utilization of Abel and Radon techniques [14] in better analyzing the plasma is called for.

20.3 Superposition spectra of hydrogen and carbon

Superposition spectra of hydrogen and carbon Swan bands [3] occur in hydrogen-carbon plasma generated by high intensity laser radiation. Investigated are the H_α, H_β, and also H_γ lines and their characteristics, as well as broadening and shifts. This article elaborates on previously communicated emission spectroscopy data of the hydrogen Balmer series [15–19]. LIBS historically embraces elemental analysis, or atomic spectroscopy, and to a lesser extent molecular spectroscopy. Time-resolved diagnostics allow us to record atomic emission profiles separate from molecular recombination and/or excitation; however, atomic and molecular species can occur simultaneously within a selected spectral window, for our expanding methane experiments typically on the order of 1–2 μ microseconds after optical breakdown.

Following the invention of laser devices, LIBS evolved into one of the favorite topics not only in atomic and chemical physics but also in chemistry, biology, environmental science, material engineering, and recently space exploration. Modeling of the hydrogen line shapes for the diagnostic for the International Thermonuclear Experimental Reactor is an important ongoing investigation, and it is based on the Balmer lines of the hydrogen isotopes. The use of hydrogen Balmer lines has been employed to measure the temperature and electron density in plasmas, including several theoretical approximations [20–22]. In typical LIBS experiments, the main broadening mechanism of spectral lines is the Stark effect. Line broadening has been studied comprehensively by Griem [23]. A large, linear Stark effect in hydrogen makes its lines attractive for recent research [24]. According to theory, the shapes and shifts of plasma-broadened isolated lines are mainly determined by electron impact with the radiating atom or ion, and a smaller contribution arises from the electric micro-fields generated by essentially static plasma ions. In addition, at a given electron number density of the plasma, the hydrogen lines are broader with respect to the lines of the other elements. Doppler broadening has information about velocity distribution and Stark broadening gives information on electron density [25]. These features make their use particularly interesting in LIBS experiments performed with typically 0.1nm spectral resolution spectrometers where the instrumental broadening introduced by the spectrometer might be comparable, or even larger, than the broadening of the emission lines of elements different from hydrogen. Moreover, the Stark broadening of the emission lines is independent from the fulfillment of LTE conditions [26], and this aspect makes this approach for the electron density measurement an interesting and powerful tool. Measurements of line shapes are important in plasma diagnostics allowing us to investigate fundamental atomic structure, as well as inferring plasma temperature and electron density. In particular, the H_β line is one of the most studied lines, and its use has become a standard technique in plasma spectroscopy [27].

In previous studies by Parigger *et al* [28], analysis of Stark broadening and shifts in measured H_α and H_β spectra, combined with Boltzmann plots from H_α, H_β and H_γ lines to infer the temperature, have been discussed for the electron densities in the range of 10^{16}–10^{19}cm^{-3}, and for the temperatures in the range of 6–100 kK. Lines H_α, H_β, and H_γ were recorded at different delay times from optical breakdown. The electron density was evaluated using different theoretical approaches that are commonly employed for plasma diagnostics [17]. We previously reported measurements [15–19] and comprehensive studies of the diagnostic application of Balmer lines.

For generation of a micro-plasma, a laboratory Nd:YAG laser operated at the fundamental wavelength of 1064nm was used. For the hydrogen plasma investigations, a Continuum YG680S-10 Nd:YAG with 150 mJ energy per pulse and 7.5 ns pulse duration, was focused to typically 1.4 TW cm $^{-2}$ in a pressure cell that was filled with gaseous hydrogen to a pressure of 810 ± 25 Torr (1.07 $\times 10^5$ Pa) and 1010 ± 25 Torr (1.33 $\times 10^5$ Pa), subsequent to evacuation of the cell with a diffusion pump. Spectrometer and detector arrangements are discussed in the Parigger and Oks review article [15]. Recording of shadowgraphs (not illustrated here) was accomplished with Coherent Infinity 40-100 Nd:YAG laser with 50 mJ and 300 mJ pulse energy and 3.5 ns pulse width was focused in laboratory air to an irradiance of typically 10 TW cm $^{-2}$. Perpendicular to the laser beam generating LIB, we typically employed pulsed laser radiation from a 308 nm XeCl excimer laser.

In the expanding methane experiments [18], we used a 1/2 m model 500 SpectraPro Acton Research Corporation spectrometer together with an intensified linear diode array (model 1460 Princeton Applied Research detector/controller optical multichannel analyzer). Averaged over 100 individual LIB events, the captured time-resolved data (measurement window of 0.1 μ s) from the methane breakdown events were detector-noise/background corrected and wavelength and detector sensitivity calibrated using standard calibration lamps.

Measurements of individual profiles of the Balmer series lines H_α, H_β, and H_γ allows us to infer the electron excitation temperature. However, the inferred temperature will show error-bar contributions due to the broadened atomic emissions. The presence of the molecular emissions will add to the background for the hydrogen Balmer series diagnostic. We treat the H_β and H_γ lines as background contribution to extract molecular temperature information for analysis of molecular excitations [19]. Time-resolved spectroscopy measurements of hydrogen- alpha, -beta, and -gamma emissions were performed in an expanding flow of methane gas. Hydrogen Stark widths were used to characterize the plasma decay during the first 2 microseconds after laser-induced optical breakdown [18]. Analysis of the recorded hydrogen spectra include measurements of full width at half maximum (FWHM) and full width half area (FWHA) to infer electron density N_e [16]. Temperature was determined from incomplete hydrogen Balmer Series lines [17]. Yet, for time delays on the order of 1 μ s, molecular emissions can be recognized from the C_2 Swan system, superposed to the H_β and H_γ Balmer series lines. C_2 emissions are clearly recognizable for time delays of the order of 2 μs [3]. Figures 20.5 and 20.6 show these results.

Figure 20.6 indicates the presence of C_2 molecular emissions for a delay time of $\Delta\tau = 2.1\ \mu$ s. Temperature inferred from the fitting of the C_2 'fingerprints' revealed molecular temperatures in the range of 4–6 kK [19]. The H_β profile at a delay time of $\Delta\tau = 2.1\ \mu$s shows an electron density of 0.5 $\times 10^{17}$cm^{-3}. The electron excitation temperature for delay times on the order of 1.1–2.1 μs amounts to 13–11 kK [17], while the molecular temperature of the residual carbon, see figure 14.2, is determined to be 5.2 kK.

The H_β profiles typically show a double-peak structure consistent with computed line profiles that appear asymmetric early in the plasma decay. The major broadening mechanism here is due to the Stark effect, which should cause a Lorentzian profile, and the dip at the line center is due to absence of computed H_β shifts [29, 30] and in part due to electrostatic interactions with ions [31]. The theory for widths of

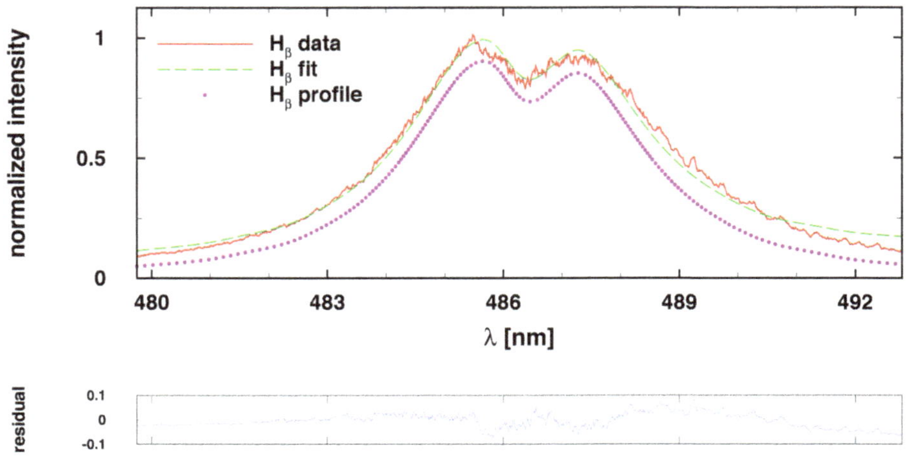

Figure 20.5. H_β recorded and fitted spectra. Delay time $\Delta\tau = 1.1$ μ s. $N_e = 1.0 \times 10^{17}$ cm^{-3}. Reproduced from [3], © IOP Publishing Ltd. CC BY 3.0.

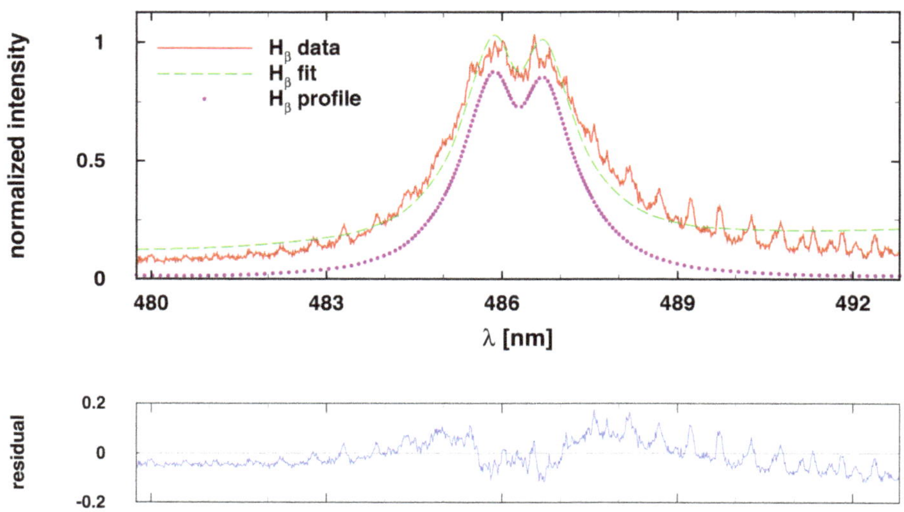

Figure 20.6. H_β recorded and fitted spectra. Delay time $\Delta\tau = 2.1$ μ s. $N_e = 0.5 \times 10^{17}$ cm^{-3}. Reproduced from [3], © IOP Publishing Ltd. CC BY 3.0.

the hydrogenic lines is mostly accurate and best agrees for H_β, where the overall error is about 5%. For hydrogenic systems, a relatively large portion of the broadening is due to quasi-static effects.

Electron density in plasma can be found by measuring Stark broadening, and this is the most precise technique for determining the electron density N_e in plasma. This broadening will become significant for $N_e > 10^{15}$ cm^{-3}. Stark broadening is due to collision between radiating species with charged species, such as electrons and ions. Not only the H_β Stark width but also the separation of the H_β double-peak structure is a measure for the free electron density [28]. Oks's theory of Stark widths incorporates several major theoretical advances and new phenomena in the Stark broadening of hydrogen lines in plasmas [32]. The Stark line width $\Delta\lambda_{FWHM}$ at FWHM can be extracted from the measured line width $\Delta\lambda_{observed}$ by subtracting instrumental $\Delta\lambda_{instrumental}$ and Doppler line broadening $\Delta\lambda_{Doppler}$ [33]. A large electron density ($N_e > 10^{17}\text{cm}^{-3}$) causes Stark broadening that is typically much larger than employed instrumental resolution and Doppler broadening. Experimental studies of aluminum laser ablation [34] within an evacuated cell reveals the limitation of Balmer series diagnostic for electron density and temperature. Early in the plasma decay, the H_α line becomes discernable at delays on the order of 0.025–0.050 μs after LIB, showing an electron density of 10^{18}cm^{-3}. The H_β line appears recognizable at delays on the order of 0.075–0.100 μ s, showing an electron density of 4×10^{17} cm^{-3}.

A plasma's temperature and density are important for its stability. For high-temperature nuclear fusion plasma, noninvasive plasma diagnostics (e.g., optical emission spectroscopy) are necessary [35]. A typical method to infer plasma temperature is to construct a so-called Boltzmann plot [36] including consideration of nonuniformity of laser-induced plasma [37]. Relative ion populations in LTE can be computed using the Saha (or Saha-Eggert) equation. However, when applying the Saha equations to obtain temperature, called the temperature of the ionization equilibrium, relative intensities of lines from different ion stages of the same atom (occasionally different atoms) need to be measured [38].

The recorded emission spectra of the hydrogen Balmer series allow us to infer electron number density, using previously tabulated values. The displayed hydrogen emission profiles are also convolved with the Doppler width that amounts to 0.05 nm at 10 kK and 0.15 nm at 100 kK. There is a substantial red shift of the H_α line profile from its unperturbed position for N_e of the order of 10^{18} cm^{-3}. Larger red shifts in the forward region of the breakdown plasma correspond to higher temperature and also electron density; therefore, we infer that there is an electron (or excitation) temperature gradient across the length of the plasma. Temperatures on the order of 100 kK can be found early in the plasma decay [28]. Use of Oks's theory allowed us to enhance significantly the agreement of theory and our experimental results.

Superposition spectra occur due to recombination or due to onset of chemical reactions. Analysis of both atomic and molecular emission spectra following laser-induced optical breakdown utilize molecular diagnostics based on line-strength files. The illustrated spectroscopic data display both the Balmer series hydrogen beta line and molecular C_2 Swan band presence. The molecular excitation temperature is determined using modified Boltzmann plots and fitting of spectra from selected molecular transitions [1].

References

[1] Hornkohl J O, Nemes L and Parigger C G 2009 Spectroscopy, dynamics and molecular theory of carbon plasmas and vapors *Advances in the Understanding of the Most Complex High-Temperature Elemental System* ed L Nemes and S Irle (Singapore: World Scientific) ch 4

[2] Witte M J and Parigger C G 2014 *J. Phys.: Conf. Ser.* **548** 012052

[3] Parigger C G, Woods A C and Rezaee M R 2012 *J. Phys.: Conf. Ser.* **397** 012022

[4] Parigger C G 2013 *Spectrochim. Acta* B **78-79** 4

[5] De Lucia F C, Harmon R S, McNesby K L, Winkel R J and Miziolek A W 2003 *Appl. Opt.* **42** 6148

[6] Parigger C G, Plemmons D H, Hornkohl J O and Lewis J W L 1994 *J. Quant. Spectrosc. Radiat. Transfer* **52** 707

[7] Witte M J, Parigger C G, Bullock N A, Merten J A and Allen S D 2014 *Appl. Spectrosc.* **68** 367

[8] Witte M J and Parigger C G 2013 *Int. Rev. At. Mol. Phys.* **4** 63

[9] Parigger C G, Woods A C, Witte M J, Swafford L D and Surmick D M 2014 *J. Vis. Exp.* **84** 51250

[10] Parigger C G, Hornkohl J O, Keszler A M and Nemes L 2003 *Appl. Opt.* **42** 6192

[11] Nemes L, Keszler A M, Hornkohl J O and Parigger C G 2005 *Appl. Opt.* **44** 3661

[12] Woods A C, Parigger C G and Hornkohl J O 2012 *Opt. Lett.* **37** 5139

[13] Parigger C G, Woods A C, Surmick D M, Gautam G, Witte M J and Hornkohl J O 2015 *Spectrochim. Acta* B **107** 132

[14] Gornushkin I B, Merk S, Demidov A, Panne U, Shabanov S V, Smith B W and Omenetto N 2012 *Spectrochim. Acta* B **76** 203

[15] Parigger C G and Oks E 2010 *Int. Rev. At. Mol. Phys.* **1** 13

[16] Parigger C G 2010 *Int. Rev. At. Mol. Phys.* **1** 129

[17] Parigger C G, Woods A C and Hornkohl J O 2011 *Int. Rev. At. Mol. Phys.* **2** 77

[18] Parigger C G, Dackman M and Hornkohl J O 2008 *Appl. Opt.* **47** G1

[19] Parigger C G, Woods A C and Hornkohl J O 2012 *Appl. Opt.* **51** B1

[20] Aragón C and Aguilera J A 2010 *Spectrochim. Acta* B **65** 395

[21] Sherbini A M E L, Hegazy H and Sherbini T M E L 2006 *Spectrochim. Acta* B **61** 532

[22] Detalle V, Heón R and Sabsabi M 2001 *Spectrochim. Acta* B **56** 1011

[23] Griem H 1974 *Spectral Line Broadening by Plasma* (New York: Academic)

[24] Konjević N 1999 *Phys. Rep.* **316** 339

[25] Chen F F 1995 *Introduction to Plasma Physics* (New York: Springer)

[26] Cristoforetti G, De Giacomo A, Dell'Aglio M, Legnaioli S, Tognoni E, Palleschi V and Omenetto N 2010 *Spectrochim. Acta* B **65** 86

[27] Djurović S, Ćirišan M, Demura A V, Demchenko G V, Nikolić D, Gigosos M A and González M Á 2009 *Phys. Rev.* E **79** 046402

[28] Oks E, Parigger C G and Plemmons D H 2003 *Appl. Opt.* **42** 5992

[29] Schrödinger E 1926 *Ann. Phys.* **385** 437

[30] Schrödinger E 1926 *Phys. Rev.* **28** 1049

[31] Laux C 2003 *Plasma Sources Sci. Technol.* **12** 125

[32] Oks E 2006 *Stark Broadening of Hydrogen and Hydrogenlike Spectral Lines on Plasmas: The Physical Insight* (Oxford: Alpha Science International)

[33] Camacho J J, Díaz L, Santos M and Poyato J M L 2011 *Spectrochim. Acta* B **66** 57

[34] Parigger C G, Hornkohl J O and Nemes L 2007 *Appl. Opt.* **46** 4026

[35] Kim J-H and Lee H-J 2006 *J. Korean Phys. Soc.* **49** 3S184

[36] Aydin Ü, Roth P, Gehlen C D and Noll R 2008 *Spectrochim. Acta* B **63** 1060

[37] Gornushkin I B, Shabanov S V, Merk S, Tognoni E and Panne U 2010 *J. Anal. Atom. Spectrom.* **25** 1643

[38] Cremers D E and Radziemski L J 2006 *Handbook of Laser-Induced Breakdown Spectroscopy* (New York: Wiley)

IOP Publishing

Quantum Mechanics of the Diatomic Molecule (Second Edition)

Christian G Parigger and James O Hornkohl

Chapter 21

Laser plasma carbon Swan bands fitting with current databases

21.1 Introduction

This chapter presents analysis of carbon Swan, C_2, laser-plasma-emission records using line-strength data, C_2-Swan-lsf, and the ExoMol astrophysical database. Line-strength data fitting of 0.25 nm spectral resolution ExoMol-computed spectra for a 6000 K temperature C_2 Swan system, d $^3\Pi_g \longrightarrow$ a $^3\Pi_u$, $\Delta v = 0, \pm 1$, reveals a temperature of 5640 K. The 6% lower temperature is associated primarily with the accuracy of the transition wavelengths in the ExoMol versus C_2-Swan-lsf data. The analysis of experiment data examines spectra that are recorded following laser-induced optical breakdown in carbon monoxide. The laser-plasma data are recorded with 0.35 nm spectral resolution. The temperature inferences are elaborated when using nonlinear fitting with both databases. The results show temperatures in excess of 6000 K for the $\Delta v = -1$ sequence, and for a time delay of 30 μs from laser-plasma initiation. The accuracy of the C_2 Swan bands line-strength data is of the order of 1 pm. These line-strength data are also utilized for analysis of laser-induced fluorescence (LIF) experiments that employ a spectral resolution of 5.5 pm, and a temperature of 2716 K is inferred. Accurate C_2 databases show many applications in laboratory diagnosis and interpretation of astrophysical plasma records.

Signatures of the diatomic carbon molecule, C_2, [1–3] occur in plasma-emission following the generation of laser-induced optical breakdown of carbon-containing materials, liquids, and gases, including carbon monoxide gas [4]. The occurrence of C_2 Swan bands in combustion of hydrocarbons and emissions from white dwarf stars, e.g., ProcyonB [5, 6], are two specific examples. Usually, accurate diatomic line strengths data are preferred in the analysis [7–9] of recorded data. However, recent interest in exo-planet spectroscopy motivates determination of new molecular databases, viz. the ExoMol [10] database. The ExoMol database lists various C_2

doi:10.1088/978-0-7503-6204-7ch21

isotopologues; however, this work focuses on $^{12}C_2$. The molecular transition of interest is the C_2 d $^3\Pi_g \longrightarrow$ a $^3\Pi_u$, $\Delta v = 0, \pm 1$ Swan band system [11].

Spectroscopy [11–17] of laser-plasma reveals clean C_2 Swan bands for several dozens of microseconds from the initial laser-plasma generation using pulse widths of a few nanoseconds. For diatomic carbon spectroscopy, one can utilize the ExoMol or other databases in conjunction with the PGOPHER program [18, 19] for diatomic molecular spectroscopy. The ExoMol $^{12}C_2$ data files for the states and the transitions are converted in this work to line-strength files for the purpose of comparison with previously communicated and extensively tested C_2 [7–9] line-strength data that are conveniently accessed with MATLAB [20] scripts.

21.2 Experiment and analysis overview

The laser-plasma experiment for recording of C_2 Swan bands comprises a standard laser-induced breakdown spectroscopy arrangement [4]. A Continuum YG680S-10 Nd:YAG device is operated at the fundamental IR wavelength of 1064nm, 7.5 ns pulse width, 300 mJ energy per pulse, and a rate of 10 Hz. The laser beam is focused into a cell containing 99.97% purity carbon monoxide with at most 0.02% nitrogen gas impurity, at 22.4 kPa (3.25 psi) above atmospheric pressure. An optical multichannel analyzer manages electric gates and recordings of an intensified, 1024 pixels linear diode array mounted at the exit plane of a MonoSpec 27 Thermo Jarrel-Ash 0.275 monochromator. A 1800 grooves/mm holographic grating has a reciprocal linear dispersion of 2 nm/mm. Wavelength and sensitivity calibrations utilize standard light pen-ray and tungsten light sources, respectively. Typically, averages over 200 individual laser-plasma events are accumulated with a gate width of 1 μ s. The $\Delta v = -1$ sequence of the C_2 Swan system discussed in this work is recorded at a gate delay of 30 μ s. Over and above the 1 μs average, the data represent a line-of-sight average of the expanding plasma.

The analysis of the diatomic molecular spectra utilizes line-strength data, the Boltzmann equilibrium spectral program (BESP), and the Nelder–Mead downhill simplex, nonlinear fitting algorithm [21] in conjunction with the Nelder–Mead temperature (NMT) program for computation and fitting of theory and experiment spectra. The molecular C_2 Swan line strengths 'C_2-Swan-lsf' [8] are established using the Wigner–Witmer diatomic eigenfunctions [22, 23] and standard molecular spectroscopy methods [7]. In turn, the ExoMol states and transition files for C_2 [24, 25] are utilized for the generation of line-strength data that can be used with BESP and NMT. The ExoMol data show Einstein A-coefficients that are converted to line strengths [26–29].

The C_2-Swan-lsf and ExoMol databases show vacuum wave numbers. In both BESP and NMT programs, air wavelengths are computed by including refractive index of air variations with wave number [30]. The details are elaborated in [8, 26], including the preference of considering Gaussian profiles. Extensions to combined Lorentzian and Gaussian profiles, or Voigt profiles, can be implemented using standard spectroscopy approaches [31].

21.3 Results

This section focuses on the diatomic molecular C_2 d $^3\Pi_g \longrightarrow$ a $^3\Pi_u$, $\Delta v = 0, \pm 1$ sequences and progressions. First, recorded spectra of the $\Delta v = -1$ sequence [4] are re-evaluated by fitting a linear, spectroscopically broad background in addition to minimizing the difference of theory and experiment spectra for the C_2-Swan-lsf data. Second, fitting of the recorded data is accomplished with ExoMol data that are transformed from Einstein A-coefficients to line strengths. Third, ExoMOl C_2 data in the range of 440–590 nm are computed and then analyzed with C_2-Swan-lsf line strengths. Finally, comparisons are included of LIF data and C_2-Swan-lsf computed spectra in order to exhibit the accuracy of that theory data set. For these comparisons, a separate LIF-program (not communicated here) is utilized because for LIF the wave numbers for the lower states of the transitions are needed [7].

21.3.1 Analysis of $\Delta v = -1$ Swan spectra with NMT program and C_2-Swan-lsf line strengths

A previous analysis of the $\Delta v = -1$ sequence [4] shows a temperature of 6745K for a spectral resolution of 0.35 nm (11.5 cm^{-1}) when assuming zero background contributions. Over and above clearly developed Swan spectra for a time delay of 30 μs from laser-plasma initiation, there are background contributions from other radiating species. This background radiation is modeled to vary linearly with wavelength. The background contributions are computed simultaneously with fitting the spectra for temperature determination while keeping the same spectral resolution. The NMT script would allow one to also fit the spectral resolution in the fitting of theory with experiment data.

Figure 21.1 illustrates spectra determined from temperature fitting with constant Gaussian line-width, $\Delta\bar{\lambda}$. The simulated spectrum utilizes the C_2-Swan-lsf data in the experimental range 528–565 nm. Analysis of the measured data leads to a C_2 excitation temperature of T = 7360 K.

21.3.2 Analysis of $\Delta v = -1$ Swan spectra with NMT program and ExoMol C_2 line strengths

For the analysis with ExoMol C_2 data, the states and transition files for C_2 [24, 25] are collated in a table that is compatible with the NMT program, including conversion of Einstein A-coefficients to line strengths. Tables 21.1 and 21.2 reveal the number of lines that agree within specified wave number values and the number of transitions, respectively.

Figure 21.2 illustrates spectra determined from temperature and linear background fitting with constant Gaussian line-width, $\Delta\bar{\lambda}$. The results indicate a temperature of T = 5740 K that is 1890 K lower than that obtained with C_2-Swan-lsf fitting, see figure 21.1. The temperature discrepancy is attributed to the spectroscopically different line positions.

The ExoMol C_2 data appear to successfully model in part the apparent differences near 543 nm in figure 21.1 that suggest the presence of so-called 6–7 high

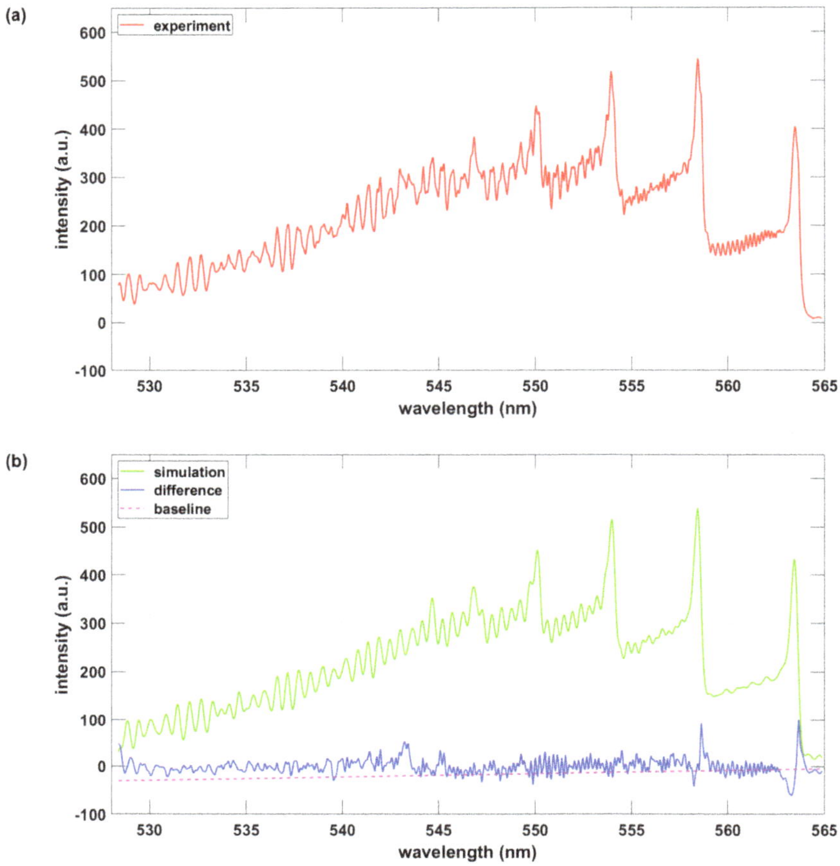

Figure 21.1. **(a)** Experiment. **(b)** NMT fitting with C_2-Swan-lsf data, T = 7360 K, fixed $\Delta\bar{\lambda} = 0.35$ nm. Reproduced from [32]. CC BY 4.0.

Table 21.1. Subset of the C_2 ExoMol data that agree within $\Delta\tilde{\nu}$ of 5032 transitions in the C_2-Swan-lsf data in the range 528.36–564.85 nm (17 704–18 926 cm^{-1}).

Database	$\Delta\tilde{\nu}<0.05$ cm^{-1}	$\Delta\tilde{\nu}<0.2$ cm^{-1}	$\Delta\tilde{\nu}<0.5$ cm^{-1}	$\Delta\tilde{\nu}<2.0$ cm^{-1}	$\Delta\tilde{\nu}<10.0$ cm^{-1}
ExoMol C_2	1147	2215	2980	4073	4789

Table 21.2. Number of transitions in the range 528.36–565.85 nm (17 704–18 926 cm^{-1}).

Database C_2 Swan	528–565 nm	528–565 nm $A_{coeff} > 10^3 s^{-1}$
ExoMol C_2	283 005	37 696
C_2-Swan-lsf	5032	5032

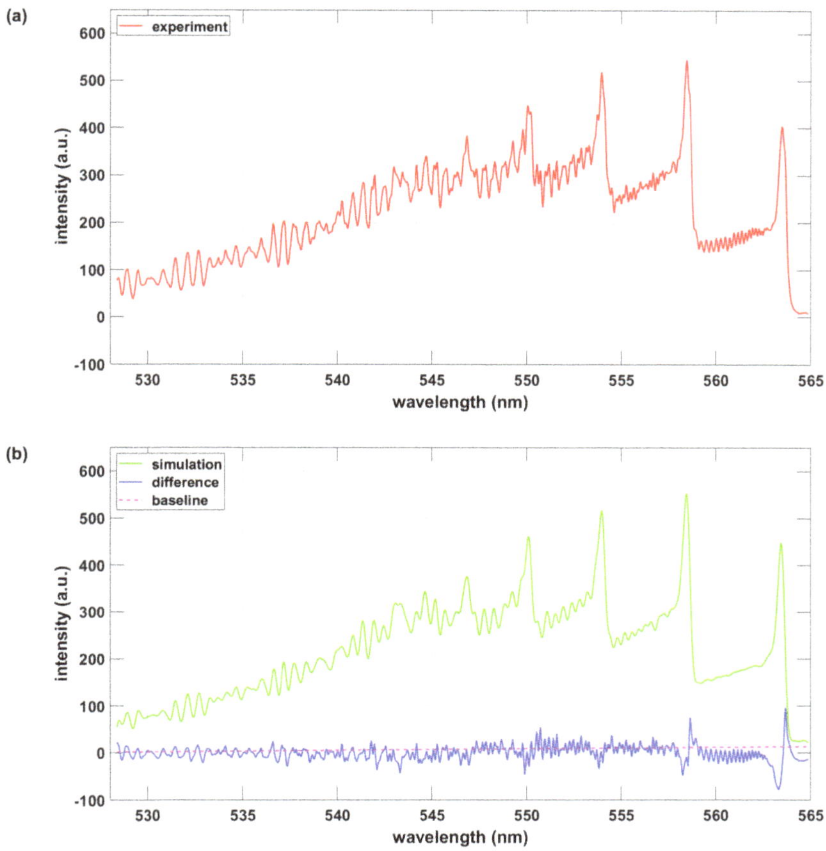

Figure 21.2. **(a)** Experiment. **(b)** NMT fitting with ExoMol C_2 data, T = 5740 K, fixed $\Delta\bar{\lambda} = 0.35$ nm. Reproduced from [32]. CC BY 4.0.

pressure band of C_2, a known intensity anomaly in the C_2 Swan system. However, subtle differences occur for the 2–3, 3–4, and 4–5 bands near 554, 550, and 547 nm, respectively. The 0–1 and 0–2 bands near 564 and 558 nm, respectively, reveal similar differences between experiment and theory spectra.

21.3.3 Swan spectra $\Delta v = 0, \pm 1$: ExoMol C_2 and C_2-Swan-lsf data comparisons

Figure 21.3 illustrates ExoMol C_2 computed, or numerical experiment, data in the wavelength range 440–590 nm that are fitted using the NMT script and C_2-Swan-lsf data. The differences in temperature of 360 K can primarily be associated with the wave numbers that are listed in the ExoMol C_2 database. There may also be differences in the Frank–Condon factors and *r*-centroids, but this is not further evaluated in this work. Figure 21.3(b) exhibits obvious differences near the heads of the Swan bands.

The number of transitions in the ExoMol C_2 database and in the range of 440 –590 are well over one million, or of the order of 100 times more transitions than

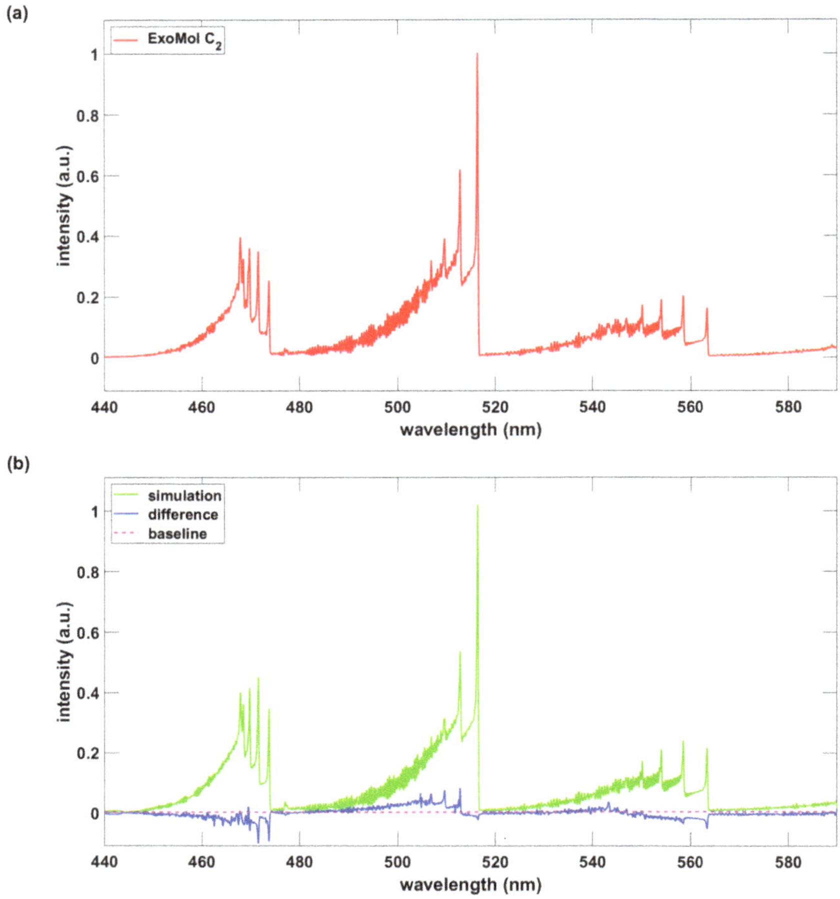

Figure 21.3. **(a)** Numerical experiment data, $T = 6000$ K, $\Delta\bar{\lambda} = 0.25$ nm. **(b)** NMT fitting with C_2-Swan-lsf data, inferred temperature from fixed line-width fitting: T = 5640 K. Reproduced from [32]. CC BY 4.0.

Table 21.3. Number of transitions in the ranges 440–540 nm (16 950–22 725 cm^{-1}).

Database Swan	440–590 nm	440–590 nm $A_{coeff} > 10^3 s^{-1}$
ExoMol C_2	1 251 235	169 566
C_2-Swan	17 689	17 689

those in the C_2-Swan line-strength data. Table 21.3 illustrates comparisons, and it indicates that for Einstein A-coefficients larger than 10^3 s^{-1} there are of the order of 10 times more lines that are included in the ExoMol C_2 database than that for C_2-Swan data.

Table 21.4 displays agreements of lines within the indicated wave number range for otherwise the same identification for upper and lower levels of the transitions. Of

Table 21.4. Subset of the C_2 ExoMol data that agree within $\Delta\tilde{\nu}$ of 17 689 transitions in the C_2-Swan-lsf data in the range 440–590 nm (16 950–22 725 cm^{-1}).

Database	$\Delta\tilde{\nu}$<0.05 cm^{-1}	$\Delta\tilde{\nu}$<0.2 cm^{-1}	$\Delta\tilde{\nu}$<0.5 cm^{-1}	$\Delta\tilde{\nu}$<2.0 cm^{-1}	$\Delta\tilde{\nu}$<10.0 cm^{-1}
ExoMol C_2	3123	6901	9617	14 094	16 998

Table 21.5. Number of transitions in the range 507.723–516.696 nm (19 354–19 696 cm^{-1}).

Database Swan	507.723–516.696 nm	507.723–516.696 nm $A_{coeff} > 10^3 s^{-1}$
ExoMol C_2	77 832	8708
C_2-Swan	1535	1535

Table 21.6. Subset of the C_2 ExoMol data that agree within $\delta\tilde{\nu}$ of 1535 transitions in the C_2-Swan-lsf data in the range 507.723–516.696 nm (19 354–19 696 cm^{-1}).

Database	$\Delta\tilde{\nu}$<0.05 cm^{-1}	$\Delta\tilde{\nu}$<0.2 cm^{-1}	$\Delta\tilde{\nu}$<0.5 cm^{-1}
ExoMol C_2	651	1194	1337

the 169 566 ExoMol C_2 transitions, 16 988 or about 10% are within 10 wave numbers of those listed in the 17 689 C_2-Swan data. However, only 3123 of 169 566 transitions, or about 2%, are in agreement within the accuracy of 0.05 cm^{-1}.

21.3.4 Laser-induced fluorescence and C_2-Swan line strengths

The C_2 line-strength database has been extensively tested [9], including in the analysis of laser-plasma-emission spectra. Tables 21.5 and 21.6 summarize number of transitions in the C_2 ExoMol versus C_2-Swan databases and the number of lines that agree, respectively. The use of accurate line strengths extends to analysis of LIF data [33] of the $\Delta v = 0$ sequence, and comparisons with Doppler-limited dye laser excitation spectra of the $\Delta v = +1$ C_2 Swan band [33, 34]. Figure 21.4 displays recorded and fitted LIF spectra of C_2 in the range of 507.723−516.696 nm. Analysis and fitting of LIF data requires knowledge of lower state wave numbers, and consequently a different script (not communicated here) as discussed in [7]. The laser step size in the experiment amounted to 0.002 cm^{-1} (0.05 pm), and the full width half maximum of the fitted spectrum amounts to 0.22 cm^{-1} (5.5 pm), or about four times larger than the typical resolution of the C_2-Swan-lsf data.

21.4 Discussion

The agreement of the ExoMol C_2 and C_2-Swan-lsf databases line position is marginal when using accuracies of the order of 0.05 cm^{-1}, or of the order of

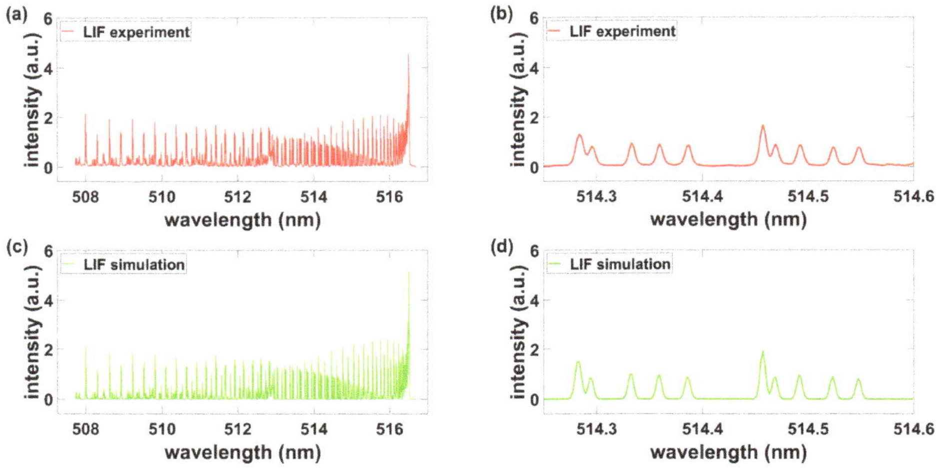

Figure 21.4. **(a)** LIF data, $T = 2716$ K, laser step size: 0.000 05 nm. **(b)** Expanded experiment data region. **(c)** Fitted data, $T = 2716$ K, $\Delta\bar{\lambda} = 0.0055$ nm. **(d)** Expanded fitted data region. Reproduced from [32]. CC BY 4.0.

1 pm. For spectral resolutions of 10 cm^{-1}, or about 0.25 nm, and for the $\Delta v = 0 \pm 1$ sequences and progressions, about 6% lower temperature is inferred when fitting 6000 K, Exomol C_2 theory data with C_2-Swan-lsf data. Consequently, use of the C_2-Swan-lsf database is recommended. For measurements with spectral resolutions of 11.5 cm^{-1}, or an average resolution of 0.35 nm, significant differences also occur; namely, a 25% lower temperature would be predicted when using the Exomol C_2 database. The C_2-Swan-lsf line-strength table is generated by fitting high-resolution Fourier-transform data rather than computation from first principles.

References

[1] Pretty W E 1927 *Proc. Phys. Soc.* **40** 71

[2] Johnson R C 1927 *Phil. Trans. Royal Soc.* A **226** 157

[3] Phillips J G and Davis S P 1968 *The Berkeley Analysis of Molecular Spectra, Vol. 2, I. The Swan System of the C_2 Molecule* (Berkeley, CA: University of California Press)

[4] Parigger C G, Plemmons D H, Hornkohl J O and Lewis J W L 1994 *J. Quant. Spectrosc. Radiat. Transfer* **52** 707

[5] Dufour P, Blouin S, Coutu S, Fortin-Archambault M, Thibeault C, Bergeron P and Fontaine G 2017 *The Montreal White Dwarf Database: A Tool for the Community. In: Astronomical Society of the Pacific (ASP) Conf. Series 509, Proc. 20th European White Dwarf Workshop*; P-E Tremblay, B Gaensicke and T Marsh (San Francisco, CA, USA: Astronomical Society of the Pacific) Available online: http://dev.montrealwhitedwarfdatabase.org

[6] Parigger C G, Helstern C M, Gautam G and Drake D A 2019 *J. Phys.: Conf. Ser.* **1289** 012001

[7] Hornkohl J O, Nemes L and Parigger C G 2009 *Spectroscopy, dynamics and molecular theory of carbon plasmas and vapors. In: Advances in the Understanding of the Most Complex High-Temperature Elemental System* ed L Nemes and S Irle (Singapore: World Scientific) ch 4

[8] Parigger C G 2023 *Foundations* **3** 1

[9] Parigger C G and Hornkohl J O 2020 *Quantum Mechanics of the Diatomic Molecule with Applications* (Bristol: IOP Publishing)
[10] Tennyson J *et al* 2020 *J. Quant. Spectrosc. Radiat. Transf.* **255** 107228
[11] Ochkin V N 2009 *Spectroscopy of Low Temperature Plasma* (Weinheim: Wiley)
[12] Kunze H-J 2009 *Introduction to Plasma Spectroscopy* (Heidelberg: Springer)
[13] Fujimoto T 2004 *Plasma Spectroscopy* (Oxford: Clarendon)
[14] Demtröder W 2014 *Laser Spectroscopy 1: Basic Principles* 5th edn (Heidelberg: Springer)
[15] Demtröder W 2015 *Laser Spectroscopy 2: Experimental Techniques* 5th edn (Heidelberg: Springer)
[16] Miziolek A W, Palleschi V and Schechter I 2006 *Laser Induced Breakdown Spectroscopy* (New York: Cambridge University Press)
[17] Singh J P and Thakur S N (ed) 2020 *Laser-Induced Breakdown Spectroscopy* 2nd edn (New York: Elsevier)
[18] Western C M 2017 *J. Quant. Spectrosc. Radiat. Transfer* **186** 221
[19] McKemmish L K 2021 *WIREs Comput. Mol. Sci.* **11** e1520
[20] *MATLAB Release R2022a Update 5* (MA: Natick)
[21] Nelder J A and Mead R 1965 *Comp. J.* **7** 308
[22] Wigmer E and Witmer E E 1928 *Z. Phys.* **51** 859
[23] Hettema E H 2000 *On the Structure of the Spectra of Two-Atomic Molecules According to Quantum Mechanics. In: Quantum Chemistry: Classic Scientific Papers* (Singapore: World Scientific)
[24] Yurchenko S N, Szabo I, Pyatenko E and Tennyson J J 2018 *Mon. Notices Royal Astron. Soc.* **480** 3397
[25] McKemmish L K, Syme A-M, Borsovszky J, Yurchenko S N, Tennyson J, Furtenbacher T and Császár A G 2020 *Mon. Notices Royal Astron. Soc.* **497** 1081
[26] Parigger C G 2023 *Atoms* **11** 62
[27] Condon E U and Shortley G 1953 *The Theory of Atomic Spectra* (Cambridge: Cambridge University Press)
[28] Hilborn R C 1982 *Am. J. Phys.* **50** 982
[29] Thorne A P 1988 *Spectrophysics* 2nd edition (London: Chapman and Hall)
[30] Ciddor P E 1996 *Appl. Opt.* **35** 1567
[31] Corney A 1977 *Atomic and Laser Spectroscopy* (Oxford: Clarendon)
[32] Parigger C G 2023 *Preprints* **2023** 2023050423
[33] Hornkohl J O, Parigger C G and Lewis J W L 1996 On the use of line strengths in applied diatomic spectroscopy *Technical Digest Series of the Laser Applications to Chemical and Environmental Analysis Conf.* (Washington, DC: Optica Publishing Group)
[34] Suzuki T, Saito S and Hirota E 1985 *J. Molec. Spectrosc.* **113** 399

Chapter 22

Aluminum monoxide, AlO

This chapter on aluminum monoxide (AlO) elaborates aspects for two detailed applications; namely, measurement and accurate analysis of laser ablation from alumina [1], Al_2O_3, and inferences from emission spectroscopy in aluminum combustion. The former experiments were conducted in the Center for Laser Application at the University of Tennessee Space Institute (UTSI), and the latter were conducted in collaborative research at Sandia National Laboratories, Albuquerque, NM. This chapter also discusses work communicated at a recent spectral line shape conference [2].

In-depth studies of AlO spectra from alumina ablation included measurements and analysis with a modified diatomic Boltzmann plot and with the Nelder–Mead fitting algorithm of the entire spectrum [1]. Figure 22.1 displays a typical AlO overview spectrum and the corresponding fitted spectrum.

Measurement of the AlO sequences and progression [3] is accomplished using 266 nm radiation from a frequency quadrupled Nd:YAG laser. At the alumina target, an irradiance of 5 GW cm^{-2} was available at the target surface. The resulting plasma from the laser-surface interaction was imaged onto the slit of a 0.275 m Jarrell-Ash spectrometer. A 350 nm cut-on filter was used to effectively reduce second-order interferences. The AlO spectrum was measured by the use of an intensified 1024 pixel linear diode array (LDA) and an optical multichannel analyzer (OMA) that was synchronized to 50 Hz. The LDA was read synchronous to the laser repetition rate, and four additional readouts were used to reduce dark-count buildup during the laser's off cycle [1]. Extensive analysis, including conditioning the data with a 15-point 6-degree Savitzky-Golay filter, revealed standard deviations of the order of 1% for the inferred temperature. Moreover, the modified diatomic Boltzmann plot (DBP) and the Nelder–Mead temperature (NMT) fitting routines agreed within 3%. The DBP utilizes prominent peaks rather than fitting the entire AlO spectrum, and hence the DBP may be better suited than NMT.

doi:10.1088/978-0-7503-6204-7ch22

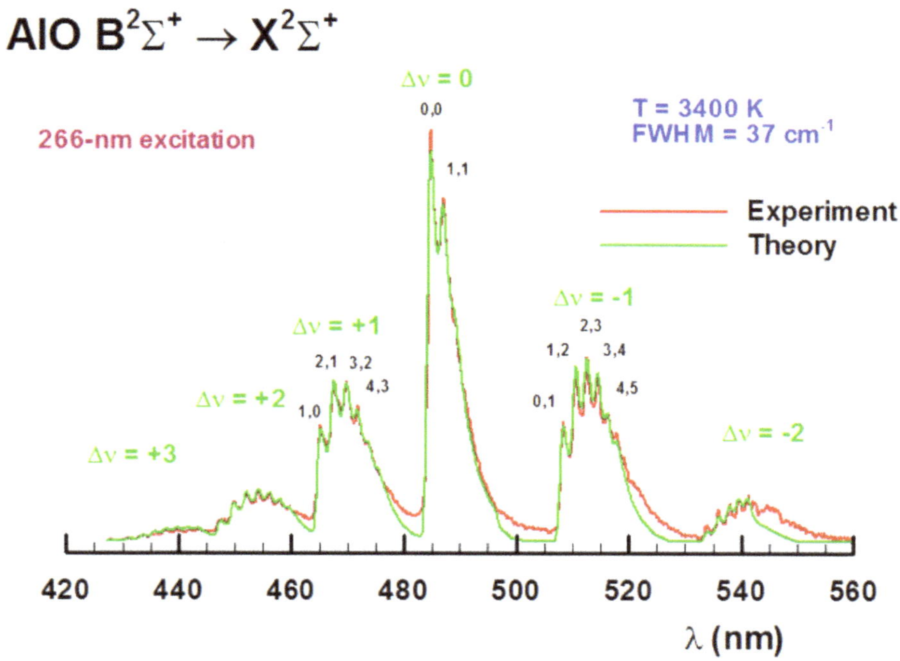

Figure 22.1. Experimental and fitted AlO spectra [1] at a spectral resolution of 0.89 nm (37 cm^{-1}), 20 μs time delay after plasma generation, 10 μs gate width.

In Surmick and Parriger [2], the aim is characterization of the temperature decay of laser-induced plasma near the surface of an aluminum target from laser-induced breakdown spectroscopy (LIBS) measurements of an aluminum alloy sample. Laser-induced plasma are initiated by tightly focusing 1064 nm, nanosecond pulsed Nd: YAG laser radiation. Temperatures are inferred from AlO spectra viewed at systematically varied time delays by comparing experimental spectra to theoretical calculations with a Nelder–Mead algorithm. The temperatures are found to decay from 5173 ± 270 to 3862 ± 46 K from 10 to 100 μs time delays following optical breakdown. The temperature profile along the plasma height is also inferred from spatially resolved spectral measurements and the electron density is inferred from Stark broadened Hβ spectra.

22.1 Laser-induced breakdown spectroscopy

Time-resolved LIBS is a valuable experimental tool for determining key plasma parameters, such as the temperature and electron number density. The laser-induced plasma created following breakdown has many applications, including use in forensics, geophysics, and, of particular interest, combustion diagnostics. For the purposes of studying combustion, aluminum LIBS studies are of significant because aluminum is commonly found in solid rocket propellants and high explosives, where the goal is commonly to infer temperature [4, 5]. Determinations of the electron number density of aluminum laser-induced plasma have also proven to be useful in diagnostic applications, such as with aluminum alloys [6, 7]. In previous time-resolved, aluminum laser ablation studies, the temperature of laser-induced plasma were determined from Boltzmann plots of aluminum atomic emissions and were found to be in the 4000–4250 K range [4, 8]. The plasma temperature may also be determined from the temperature dependence of diatomic molecular spectra, such as AlO. Temperature results from previous studies were found to be 4000 K at 50 μs [9] and 4880 K at 45 μs [10] from AlO emissions. Electron densities were also inferred in reference [9] and were found to be 1023 m − 3. In this work, we seek to use AlO diatomic molecular emissions to characterize the temperature decay of laser-induced plasma near the surface of an aluminum target in laboratory air over a broad range of times following optical breakdown using time-resolved LIBS measurements. Inferences are also made on the temperature distribution along the plasma height and the electron number density.

22.2 Experimental details for AlO measurements

Time-resolved spectral measurements are used to characterize the temperature decay of laser-induced plasma initiated near the surface of an aluminum target. Laser-induced plasma are formed by tightly focusing laser radiation from a Q-switched, Nd:YAG laser operating in its fundamental mode with a 12 ns pulse width and an energy per pulse of 190 mJ. Line of sight spectral measurements are made with a Jobin-Yvon HR640 spectrometer with 0.1 nm spectral resolution coupled to gated, intensified detectors. An intensified, LDA coupled to an OMA (Princeton Instruments 1460) was used to collect AlO spectra from 10 to 100 μs time delays in 5 μs steps with gate widths of 5, 10, and 20 μs for 10–20, 20–100, and 70–100 μs delays, respectively. A gated intensified charge-coupled detector was used to infer the temperature dependence along the plasma height at a time delay of 60 μs with a 6 μs gate width. Time resolution was achieved by synchronizing the laser repetition rate with the measurement rate of the gated detectors with the use of waveform, pulse, and delay generators. References [8, 11] provides a detailed description of the apparatus. Prior to analysis, all spectra were properly calibrated for detector response and background sensitivity.

22.3 Selected results

To infer the temperature as a function of time following optical breakdown, spectra collected with intensified, LDA were fit to theoretical calculations of the AlO $B^2\Sigma^+ \rightarrow X^2\Sigma^+ \Delta v = 0$ transition. Comparisons were made with a Nelder–Mead fitting algorithm, which was utilized for its ability to fit multiple parameters simultaneously, including a variable baseline offset. The Nelder–Mead algorithm is a minimization method in which a geometric simplex is reduced in size until a satisfactory tolerance level is reached [12, 13]. The theoretical spectra are calculated from tables of accurately calculated line strengths, where the line strength is calculated in its factorized form from the electronic transition strength, the Hönl–London factor, and Franck–Condon factor [1, 14, 15]. The baseline offset of all fitted AlO spectra were quadratic in nature and errors in the inferred temperatures were found by considering variations of the spectral resolution and baseline offset. Analysis showed that the error in the inferred temperatures was in the 45–75 K range, with the exception of the inference made at 10 μs with an error of 236 K. This resulted from the presence of Hβ superposition spectra. The results of the temperature analysis are shown in figure 1 with an exponential decay trend. The trend line shows that the plasma temperature decay slows at later times, as was expected. To infer the temperature dependence along the plasma height, spatially resolved spectra with 0.013mm resolution were fit using the Nelder–Mead algorithm. The temperature at the center of the plasma was found to be 5780 ± 236 K at 5.2 mm slit height and decreased to 5396 ± 315 K at 6.5 mm. The temperature was found to increase above the plasma edge to a value of 5820 ± 230 K at a slit height of 9.1 mm. The larger temperatures and errors are likely to result from nonoptimal plasma and detector conditions; however, the temperature profile suggests that aluminum particles are combusting in a plume of above the plasma, given the ignition temperature of aluminum is 2750 K and the inferred temperatures from the linear diode experiment (figures 22.2 and 22.3).

At time delays of 10, 15, and 20 μs following breakdown, superposition of Balmer series Hβ spectra with AlO measurements were observed. The Hβ spectra may be

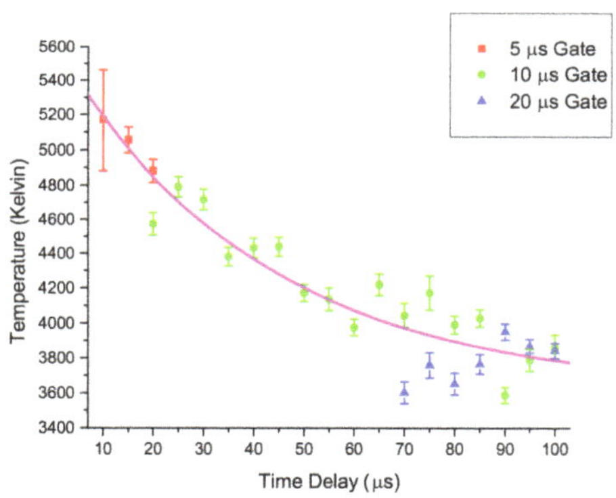

Figure 22.2. Inferred temperatures as a function of time with an exponential decay trend line. Reproduced from [2], © IOP Publishing Ltd. CC BY 3.0.

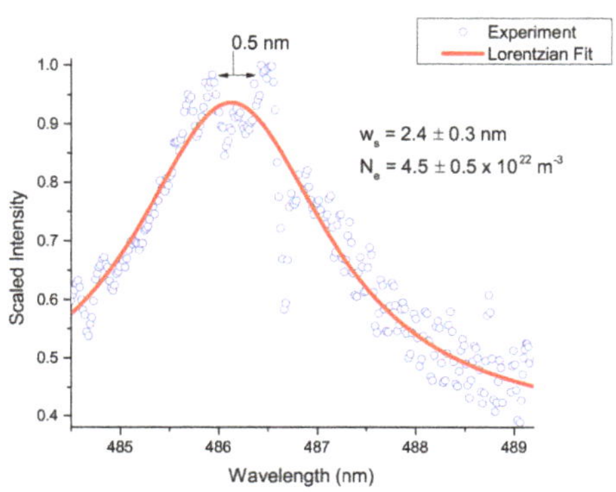

Figure 22.3. Fit of Hβ superposition spectra at a 10 μs delay to a Lorentzian line profile. Reproduced from [2], © IOP Publishing Ltd. CC BY 3.0.

extracted and used to infer the electron number density from the Stark broadened emission. The line width of the spectra is taken to be due to Stark effects and is inferred by fitting a Lorentzian line profile to the Hβ spectra. The Stark widths are used with the empirical theory of Konjević [16] and the convergent theory of Oks [17] to infer the electron number density, *Ne*. To determine the *Ne* with the theory of Oks, published data were fitted to determine the *Ne* versus Hβ width dependence [18]. Figure 2 shows the fitted Hβ spectrum from the 10 μs delay along with its corresponding Stark width and electron number density. The noticeable dip in the residual in figure 2 near 486.5 nm is due to the subtraction of the the fitted AlO spectra and represents the 1–1 vibrational peak from the transition of interest. The 0.5 nm Hβ peak separation is also indicated in this figure. Though not considered in this work, this peak separation may also be used to infer the electron density [18]. The inferred *Ne* are found to be 4.5 ± 0.5, 3.8 ± 0.5, and 3.0 $\pm$ 0.6 $\times$ 1022 $m - 3$. Errors in the inferred *Ne* result from AlO superposition influences.

Accurate theoretical calculations of the diatomic, molecular AlO spectra have been instrumental in fitting to experimental data, particularly for studies of with spectral resolutions of the order of 0.1 nm [14, 15]. The accuracy of such fitting is based, partially, on the ability to calculate the line strength tables with a resolution ranging from one to several orders of magnitude greater than that of the measured spectra. Recent efforts involving laser ablation experiments of an aluminum alloy target have been performed in order to accurately determine the temperature decay of aluminum containing laser-induced plasma using a line strength table with such accuracy. A summary of aluminum laser ablation efforts at UTSI may be found in references [1, 2, 9, 10, 19]. Figure 22.4 illustrates the Δv = 0 sequence at a spectral resolution of 0.09 nm and a corresponding, measured spectrum from two-dimensional slit height versus wavelength data. Representative results of these investigations conducted with a one-dimensional array detector are presented below. Figure 22.5 shows measured and fitted AlO spectra [20], using the AlO line strength table, at two time delays following plasma initiation.

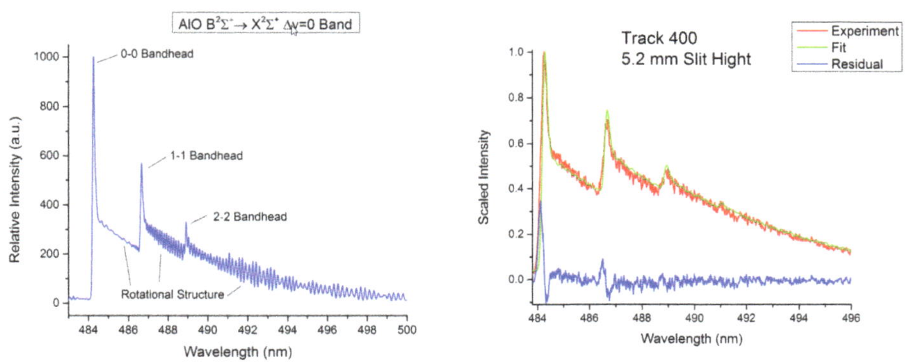

Figure 22.4. Calculated $T = 4$ kK, $\Delta\lambda = 0.09$ nm (left-hand panel), measured and fitted (right-hand panel) AlO emission spectra. Reprinted from [20] with permission of Elsevier.

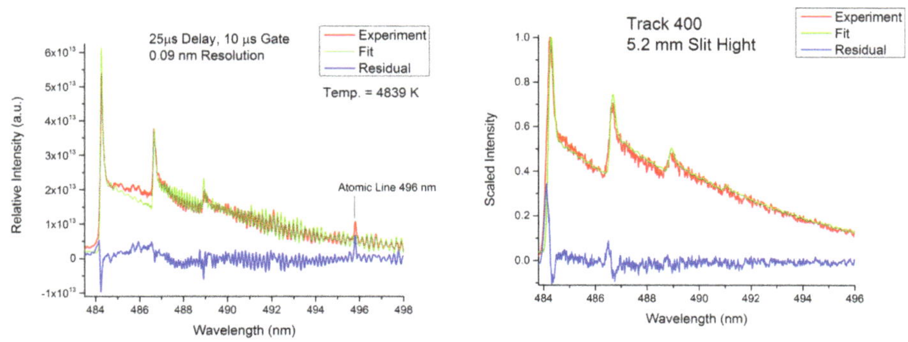

Figure 22.5. Fitted (green line) and measured (red line) AlO $B^2\Sigma^+ \longrightarrow X^2\Sigma^+$, $\Delta v = 0$ spectra collected at $25\,\mu s$ (left-hand panel) and $80\,\mu s$ (right-hand panel) following optically induced breakdown. The inferred temperatures are 4.839 kK and 3.699 kK, respectively. Reprinted from [20] with permission of Elsevier.

The objectives of the AlO work was to characterize the temporal temperature decay of laser-induced plasma near the surface of an aluminum target and was accomplished with time-resolved LIBS measurements. A Nelder–Mead fitting algorithm was employed to infer the temperature from the experimental spectra, and errors were determined from spectral resolution and baseline offset variations. The temperatures were found to decrease as a function of time from 5173 ± 290 to 3862 ± 46 K from 10 to 100 μs time delays. The decay of the temperature was also found to slow at later time delays. Spatially resolved AlO measurements were used to infer the temperature along the height of the plasma and showed that the temperature increases above the plasma edge. The electron density of the plasma was inferred from measurements of Hβ spectra superimposed on AlO measurements at time delays preceding 20 μs and were found to decrease from 4.5 ± 0.5 to $3 \pm 0.5 \times 1016$ cm $-$ 3. Observations of the slowing of the temperature decay, the temperature increase above the plasma, and Hβ superposition spectra suggest that combustion occurs.

References

[1] Dors I G, Parigger C G and Lewis J W L 1998 *Opt. Lett.* **23** 1778

[2] Surmick D M and Parigger C G 2014 *J. Phys.: Conf. Ser.* **548** 012046

[3] Bransden B H and Joachain C J 2003 *Physics of Atoms and Molecules* 2nd edn (New York: Prentice-Hall)

[4] Piehler T N, DeLucia F C, Munson C A, Homan B E, Miziolek A W and McNesby K L 2005 *Appl. Opt.* **44** 3654

[5] De Lucia F C, Harmon R S, McNesby K L, Winkel R J and Miziolek A W 2003 *Appl. Opt.* **42** 6148

[6] Sabsabi M and Cielo P 1995 *Appl. Spectrosc.* **44** 3654

[7] Rai A K, Yueh F and Signh J P 2003 *Appl. Opt.* **42** 2078

[8] Lightstone J M, Carney J R, Boswel C J and Wilikinson J 2007 *AIP Conf. Proc.* **955** 1255

[9] Parigger C G, Hornkohl J O and Nemes L 2007 *Appl. Opt.* **46** 4026

[10] Surmick D M and Parigger C G 2014 *Appl. Spectrosc.* **68** 992

[11] Parigger C G, Woods A C, Witte M J, Swafford L D and Surmick D M 2014 *J. Vis. Exp.* **84** 51250

[12] Nelder J A and Mead R 1965 *Comput. J.* **7** 308

[13] Lagarias J C, Reeds J A, Wright M H and Wright P E 1998 *SIAM J. Optim.* **9** 112

[14] Parigger C G and Hornkohl J O 2011 *Spectrochim. Acta* A **81** 403

[15] Nemes L, Hornkohl J O and Parigger C G 2005 *Appl. Opt.* **44** 3686

[16] Konjević N, Ivković M and Sakan N 2012 *Spectrochim. Acta* B **76** 16

[17] Ispolatov Y and Oks E 1994 *J. Quant. Spectrosc. Radiat. Transfer* **51** 129

[18] Oks E, Parigger C G and Plemmons D H 2003 *Appl. Opt.* **42** 5992

[19] Parigger C G 2006 *AIP Conf. Proc.* **874** 101

[20] Parigger C G, Gautam G, Woods A C, Surmick D M and Hornkohl J O 2014 *Trends Appl. Spectrosc.* **11** 1

IOP Publishing

Quantum Mechanics of the Diatomic Molecule (Second Edition)

Christian G Parigger and James O Hornkohl

Chapter 23

AlO laser-plasma emission spectra analysis with current databases

23.1 Introduction

This chapter elaborates on analysis of aluminum monoxide (AlO) laser-plasma emission records using line-strength data and the ExoMol astrophysical database. A nonlinear fitting program computes comparisons of measured and simulated diatomic molecular spectra. Predicted cyanide spectra of the AlO, $B\ ^2\Sigma^+ \longrightarrow X\ ^2\Sigma^+$, $\Delta v = 0, \pm1, \pm2, +3$ sequences and progressions compare nicely with 1 nm resolution experimental results. The analysis discusses experiment data captured during laser ablation of Al_2O_3 with 266 nm, 6 mJ pulses. The accuracy of the AlO line-strength data is better than 1 pm. This work also presents a comparison of the AlO line strength and of ExoMol data for the $^{27}Al^{16}O$ diatomic molecule. Accurate AlO databases show a volley of applications in laboratory and astrophysical plasma diagnosis.

The diatomic molecule AlO occurs in plasma-emission following laser ablation of aluminum containing samples, including alumina (Al_2O_3) [1] or aluminum containing alloys [2]. Combustion of aluminized propellants also reveals nice AlO flame emission spectra [3]. Usually, accurate diatomic line strengths data are preferred in the analysis [4–6] of recorded data. However, recent interest in exoplanet spectroscopy has motivated the determination of molecular databases, viz. ExoMol [7]. The ExoMol database lists various AlO isotopologues; however, this work focuses on $^{27}Al^{16}O$. The transition of interest is the AlO $B\ ^2\Sigma^+ - X\ ^2\Sigma^+$ band system, which is similar in principle to previously communicated cyanide, CN $B\ ^2\Sigma^+ - X\ ^2\Sigma^+$ band system [8].

Spectroscopy of laser-plasma reveals clean AlO band system for delays of the order of several dozens of microseconds from the initial ablation plasma generation using pulse widths of a few nanoseconds [9–15]. For AlO spectroscopy, one can employ the ExoMol database in conjunction with the PGOPHER program for

doi:10.1088/978-0-7503-6204-7ch23

simulating rotational, vibrational, and electronic spectra [16]. There are, of course, other databases that can be accessed [17] for diatomic molecules, including HITEMP, which, for example, shows hydroxyl (OH) data [18]. The ExoMol AlO data files for the states and the transitions are converted in this work to line-strength files for the purpose of utilizing previously communicated and extensively tested line-strength data that are freely available along with MATLAB [19] scripts for a subset of transitions associated with the AlO B–X band systems [4–6].

23.2 Experimental and analysis details

The data from laser ablation experiments with frequency quadrupled Q-switched Nd:YAG radiation [1] show a range of 430–540 nm, and the published comparisons with line-strength data reveal a temperature of 3432 K at a delay of 20 μs. The measurements use standard laser-induced breakdown spectroscopy equipment. The analysis in that work utilizes AlO-lsf line strengths and the Nelder–Mead downhill simplex, nonlinear fitting algorithm [20]. The analysis communicated in this work is designed such that the same nonlinear method can be used, but the ExoMol database for AlO is recast in a set of transitions with line-strength data that are determined from Einstein A-coefficients.

The computation of diatomic molecular spectra utilizes line-strength data. The Boltzmann equilibrium spectral program (BESP) and the Nelder–Mead temperature (NMT) program allow one to respectively compute an emission spectrum and fit theoretical to experimental spectra. The construction of the communicated molecular AlO line strengths ‘AlO-lsf’ [5] first makes use of Wigner–Witmer eigenfunctions and a diatomic line position fitting program, second computes Frank–Condon factors and r-centroids, and third combines these factors with the rotational factors that usually decouple from the overall molecular line-strength due to the symmetry of diatomic molecules. In turn, the ExoMol states and transition files for AlO [21, 22] are utilized for the generation of line-strength data that can be used with BESP and NMT.

The ExoMol data show Einstein A-coefficients that are converted to line strengths [23–25], S, for electric dipole transitions, using

$$A_{ul} = \frac{16\pi^3}{3g_u h\varepsilon_0 \lambda^3}(e\ a_0)^2 S_{ul}, \quad g_u = 2(2J_u + 1). \tag{23.1}$$

Here, A_{ul} denotes the Einstein A-coefficient for a transition from an upper, u, to a lower, l, state, and h and ε_0 are Planck’s constant and vacuum permittivity, respectively. The elementary charge is e, the Bohr radius is a_0, and the transition strength is S_{ul}. The line strength, S, that is used in the MATLAB scripts is expressed in traditional spectroscopy units (stC^2 cm^2). The wavelength of the transition is λ, g_u is the upper state degeneracy, and J_u the total angular momentum of the upper state. In the establishment of line-strength data, Hund’s case (a) basis functions are preferred in connection with application of the Wigner and Witmer [26, 27] diatomic eigenfunction.

23.3 Results

This section elaborates analysis of recorded AlO spectra of the B $^2\Sigma^+ \longrightarrow$ X $^2\Sigma^+$, $\Delta v = 0, \pm1, \pm2, 3$ sequences and progressions. The use of ExoMol data and computed sets of line-strength data that appear to be in use for extragalactic studies [7] would alleviate computation of specific transitions that are investigated in laser-plasma laboratory experiments. The ExoMol database shows 4 945 580 transitions and 94 862 states including the three lowest electronic states, X $^2\Sigma^+$, A $^2\Pi$, B $^2\Sigma^+$, C $^2\Pi$, D $^2\Sigma^+$, and E $^2\Delta$, e.g., 54 585 A states and 10 781 B states. The 10 781 B states lead to 774 575 B–X transitions.

The AlO-lsf B–X data contain 33 484 transitions. The differences in number of transitions are in part due to the number of rotational states, the cutoffs for Einstein A-coefficients and associated line strengths (see equation (23.1)), or the establishment of sets of computed molecular parameters that fit data from high-resolution, Fourier-transform spectroscopy. The line positions are determined from high-resolution data with a standard deviation comparable to the estimated experimental errors of the high-resolution line positions. The obtained, simulated line position accuracies are typically better than 0.05 cm^{-1}.

In this work, a Gaussian profile models the spectrometer and intensified linear-array detector transfer function. However, a measured system transfer function or a Voigt function can replace the selected Gaussian profile provided that changes are implemented in the MATLAB source scripts for the recently communicated BESP and NMT scripts [5]. Note that the PGOPHER program allows one to accomplish Voigt profile fits.

23.3.1 Analysis with NMT program and ExoMol line strengths

The AlO B–X line positions and Einstein A-coefficients (that are converted to line strengths) are collected in a data file that is compatible with the mentioned NMT-spectral fitting program. Figure 23.1 illustrates spectra determined from temperature fitting with constant Gaussian line-width, $\Delta\lambda$. The simulated spectrum utilizes only AlO B–X transitions in the experimental range of 430–540 nm. Analysis of the measured data with the AlO-lsf data [1] reveals a temperature of $T = 3329$K, and a fitted FWHM of 1 nm (43 cm^{-1}).

23.3.2 ExoMol AlO and AlO-lsf data comparisons

The simulated spectra are composed of quite a few individual rotational-vibrational transitions of the AlO B–X $\Delta v = 0, \pm1, \pm2, +3$ sequences and progressions. Tables 23.1 and 23.2 summarize the number of lines in the data files.

Table 23.3 displays agreements of lines within the indicated wave number range, and otherwise the same identification for upper and lower levels of the transitions.

The differences in accuracy of the line positions can cause systematic errors in analysis of plasma-emission spectra. Visualization of these differences is suggested by (a) generating a 'numerical experiment' spectrum using the AlO B–X data as extracted from the ExoMol database, and then (b) analyzing the synthetic spectrum

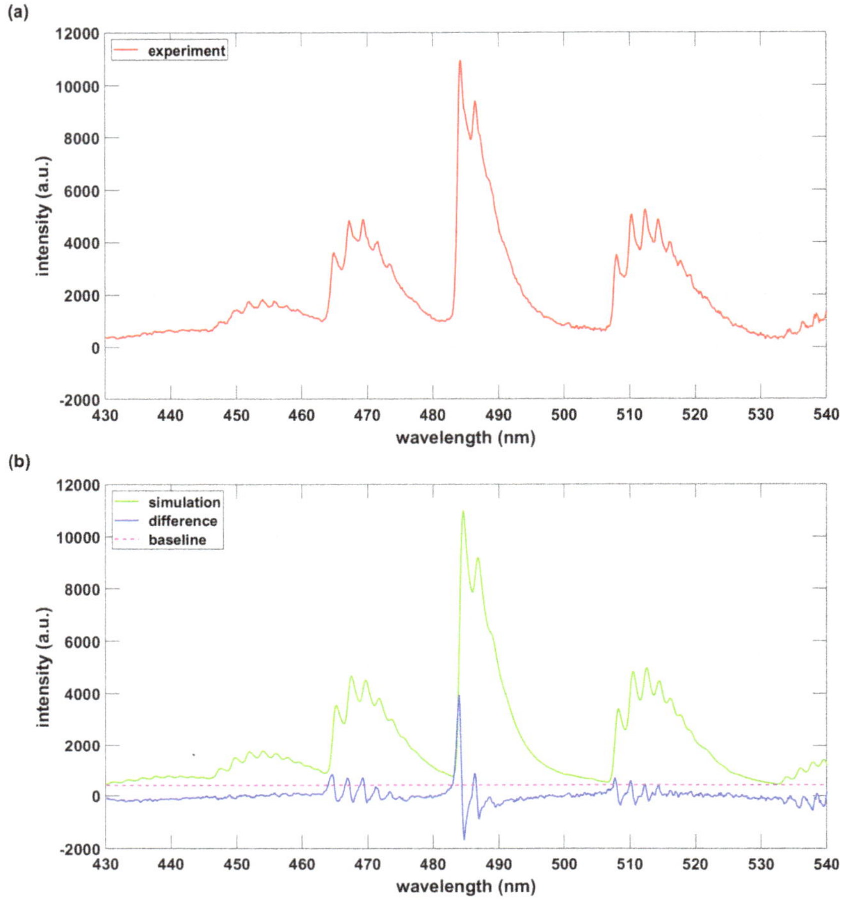

Figure 23.1. **(a)** Experiment. **(b)** NMT fitting with ExoMol B–X data, T = 3380 K, $\Delta\lambda = 1.0$ nm. Reproduced from [28]. CC BY 4.0.

Table 23.1. Number of B–X transitions and those in the experimental range 430–540 nm (18 500 cm^{-1}–23 250 cm^{-1}).

Database	AlO B–X	AlO B–X 430–540 nm
ExoMol	774 575	169 143
AlO-lsf	33 484	29 258

Table 23.2. Number of transitions in the experiment range (see table 23.1) with Einstein A-coefficients, A_{coeff}, larger than 10^3 s^{-1}.

Database	AlO B–X 430 nm to 540 nm $A_{coeff} > 10^3 s^{-1}$
ExoMol	104 260
AlO-lsf	29 258

Table 23.3. Subset AlO B–X lines of the ExoMol data that agree within $\Delta\tilde{\nu}$ of 29 258 AlO B–X transitions in the AlO-lsf data for the experiment range (18 500–23 250 cm^{-1}).

Database	$\Delta\tilde{\nu}<0.05$ cm^{-1}	$\Delta\tilde{\nu}<0.2$ cm^{-1}	$\Delta\tilde{\nu}<1.0$ cm^{-1}	$\Delta\tilde{\nu}<2.0$ cm^{-1}	$\Delta\tilde{\nu}<10.0$ cm^{-1}	$\Delta\tilde{\nu}<20.0$ cm^{-1}
ExoMol	747	3146	10 843	14 425	21 036	22 609

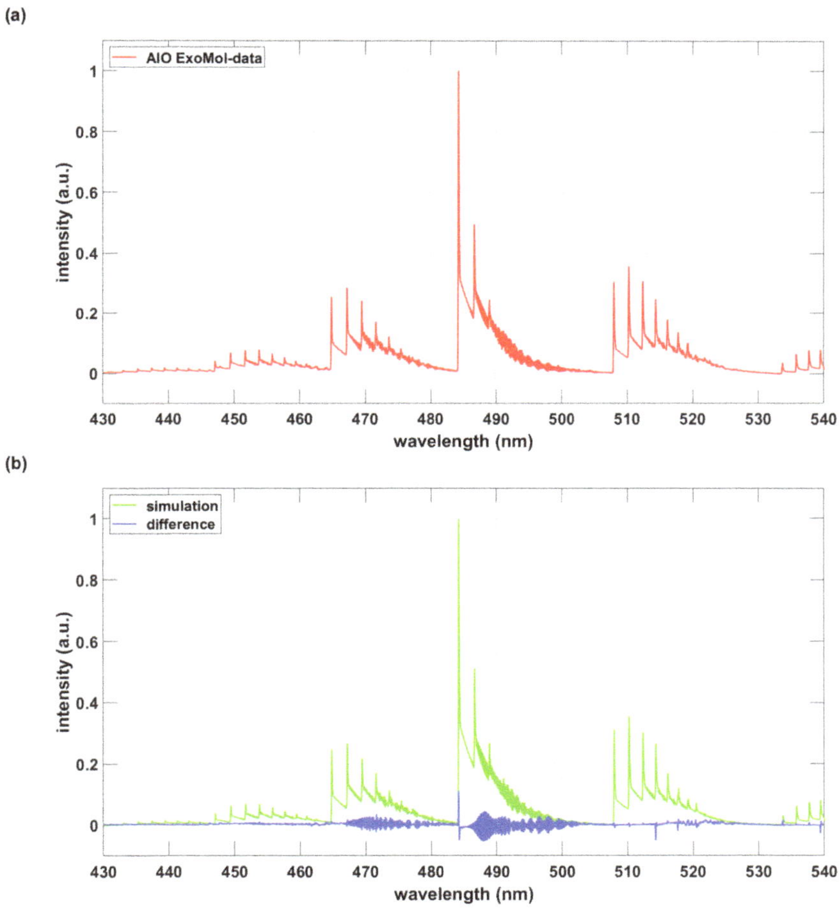

Figure 23.2. **(a)** Numerical experiment data, $T = 3380$ K, $\Delta\bar{\lambda} = 0.1$ nm. **(b)** NMT fitting with AlO-lsf B–X data, inferred temperature from fixed line-width fitting: T = 3200 K. Reproduced from [28]. CC BY 4.0.

with the AlO-lsf database. Figure 23.2 exhibits the ExoMol-database generated and AlO-lsf line-strength data analyzed results.

The obvious undulations in the difference spectrum illustrates the ExoMol inaccuracies indicated in table 23.3. A temperature of $T = 3380$ K and a full width at half maximum, fixed Gaussian line-width, $\Delta\bar{\lambda}$, of 0.1 nm is selected for the 'numerical experiment.' Analysis by only fitting temperature yields $T = 3200$ K, i.e., a temperature that is about 5% lower than specified for the spectrum in figure 23.2(a).

The AlO-lsf line-strength database has been extensively tested [6]. The ExoMol database appears acceptable within $\simeq$ 20 wave numbers, which corresponds to a spectral resolution of $\simeq$ 0.3 nm. High-resolution experimental data may very well be affected by the inaccuracies of the line positions listed in the ExoMol database. The AlO-lsf database accuracy is better than 0.05 cm^{-1}, which corresponds to $\simeq$ 1 pm for the AlO B–X bands.

Further comparisons of the AlO-lsf and AlO-ExoMol databases explore the $\Delta v = 0$ AlO B–X sequence. Figure 23.3 illustrates AlO-ExoMol data computed for a temperature of 3380 K and a spectral resolution of 0.07 nm, and it also shows the NMT-simulated results when only fitting temperature. As expected, there is a difference of approximately 30% between specified (3380 K) and fitted temperature (2460 K). For a spectral resolution of 0.1 nm, the fitted temperature for the $\Delta v = 0$ AlO B–X sequence equals 2920 K; or, in other words, the temperature difference

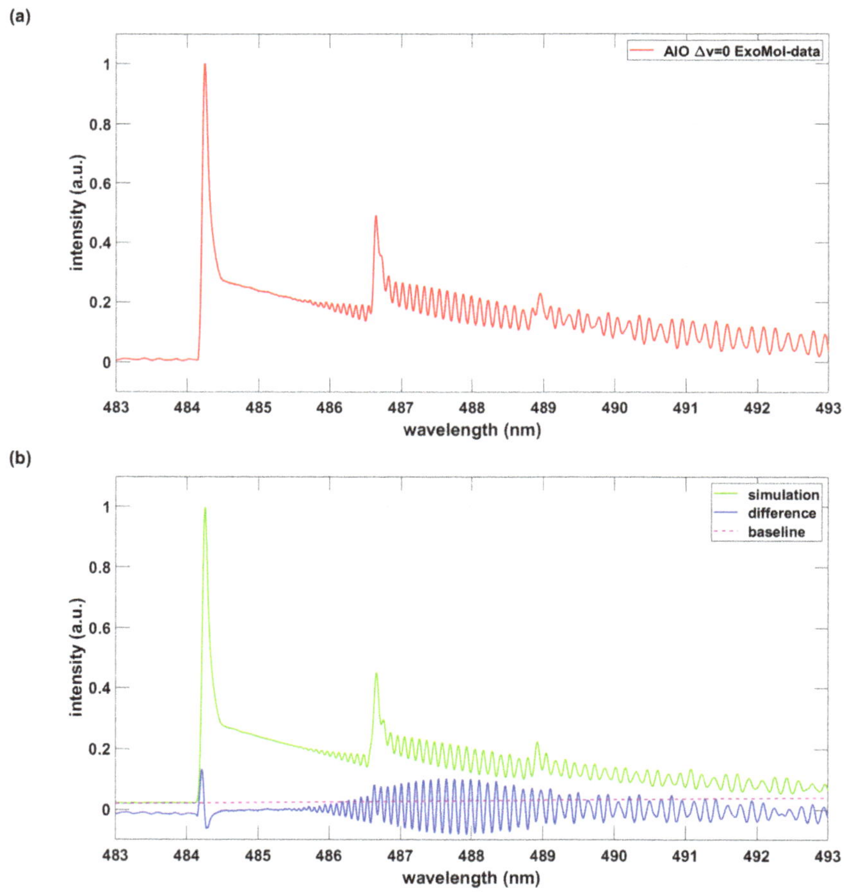

Figure 23.3. **(a)** Numerical experiment data, $T = 3380$ K, $\Delta\bar{\lambda} = 0.07$ nm. **(b)** NMT fitting with AlO-lsf B–X data, inferred temperature from fixed line-width fitting: T = 2460 K. Reproduced from [28]. CC BY 4.0.

Table 23.4. Number of transitions in the experiment range 483–493 nm with Einstein A-coefficients, A_{coeff}, larger than 10^3 s^{-1}.

Database	AlO B–X 483–493 nm $A_{coeff} > 10^3 s^{-1}$
ExoMol	10 159
AlO-lsf	2818

Table 23.5. AlO B–X lines of the ExoMol data that agree within $\delta\tilde{\nu}$ of 2818 transitions in the AlO-lsf data for the range 483–493 nm (20 284 cm^{-1}–20 704 cm^{-1}).

Database	$\Delta\tilde{\nu}$<0.05 cm^{-1}	$\Delta\tilde{\nu}$<0.3 cm^{-1}	$\Delta\tilde{\nu}$<3.0 cm^{-1}
ExoMol	96	506	1517

decreases is approximately 14% lower than specified. Tables 23.4 and 23.5 summarize comparisons of the transition lines with Einstein A-coefficients larger than 10^3 s^{-1}. There are about five times more lines in the ExoMol database for the 10 nm spectral window. Among the 2818 AlO-lsf lines, only 96 ExoMol lines agree within better than 0.05 cm^{-1}, and 1517 ExoMol lines show wave numbers within 3 cm^{-1} (about 0.07 nm) of those of the AlO-lsf data.

The AlO-lsf line-strength database has been extensively tested [5]. The ExoMol database appears acceptable within $\simeq$ 20 wave numbers, i.e., average spectral resolution of $\simeq$ 0.3 nm. Analysis of higher than 0.3 nm resolution data, viz. spectral resolution of 0.07 nm, is affected by inaccuracies of the line positions listed in the ExoMol database.

23.4 Discussion

The AlO $B\ ^2\Sigma^+ - X\ ^2\Sigma^+$, $\Delta v = 0, \pm 1, \pm 2, +3$ sequences and progressions reveal many vibrational and rotational transitions that are usually not individually resolved in the study of laser-induced plasma emissions in the spectral range of 430–540 nm. Analysis of the 1 nm spectral resolution experimental emission spectrum with ExoMol line strengths and the NMT program shows AlO excitation temperature of $\simeq$ 3380 K, which is consistent with previous analysis with AlO-lsf line strengths.

The agreement of the ExoMol AlO B–X and AlO-lsf line position is marginal when using accuracies of the order of 0.05 cm^{-1}, or of the order of 1 pm. However, for measurements with spectral resolutions of 43 cm^{-1}, or of the order of 1 nm, almost exactly identical results are inferred from fitting of measured ablation spectra. A significant advantage of the AlO-lsf database is its accuracy in predicting line position compared to the ExoMol database. The AlO-lsf line-strength table is generated by fitting high-resolution Fourier-transform data rather than computation from first principles. Analysis of predicted ExoMol spectra for different AlO isotopologues may be of interest in future research.

References

[1] Dors I G, Parigger C G and Lewis J W L 1998 *Opt. Lett.* **23** 1778

[2] Surmick D M and Parigger C G 2014 *Appl. Spectrosc.* **68** 992

[3] Parigger C G, Woods A C, Surmick D M, Donaldson A B and Height J L 2014 *Appl. Spectrosc.* **68** 362

[4] Parigger C G and Hornkohl J O 2011 *Spectrochim. Acta* A **81** 404

[5] Parigger C G 2023 *Foundations* **3** 1

[6] Parigger C G and Hornkohl J O 2020 *Quantum Mechanics of the Diatomic Molecule with Applications* (Bristol: IOP Publishing)

[7] Tennyson J *et al* 2020 *J. Quant. Spectrosc. Radiat. Transf.* **255** 107228

[8] Parigger C G 2022 *Int. Rev. At. Mol. Phys.* **14** 7

[9] Kunze H-J 2009 *Introduction to Plasma Spectroscopy* (Heidelberg: Springer)

[10] Fujimoto T 2004 *Plasma Spectroscopy* (Oxford: Clarendon)

[11] Ochkin V N 2009 *Spectroscopy of Low Temperature Plasma* (Weinheim: Wiley)

[12] Demtröder W 2014 *Laser Spectroscopy 1: Basic Principles* 5th edn (Heidelberg: Springer)

[13] Demtröder W 2015 *Laser Spectroscopy 2: Experimental Techniques* 5th edn (Heidelberg: Springer)

[14] Miziolek A W, Palleschi V and Schechter I (ed) 2006 *Laser-Induced Breakdown Spectroscopy (LIBS)—Fundamentals and Applications* (New York: Cambridge University Press)

[15] Singh J P and Thakur S N (ed) 2007 *Laser-Induced Breakdown Spectroscopy* (Amsterdam: Elsevier Science)

[16] Western C M 2017 *J. Quant. Spectrosc. Radiat. Transfer* **186** 221

[17] McKemmish L K 2021 *WIREs Comput. Mol. Sci.* **11** e1520

[18] Rothman L S, Gordon I E, Barber R J, Dothe H, Gamache R R, Goldman A, Perevalov V I, Tashkun S A and Tennyson J 2010 *J. Quant. Spectrosc. Radiat. Transf.* **111** 2139

[19] *MATLAB Release R2022a Update 5* (Natick, MA: The MathWorks, Inc.)

[20] Nelder J A and Mead R 1965 *Comp. J.* **7** 308

[21] Patrascu A T, Yurchenko S N and Tennyson J 2015 *Mon. Notices Royal Astron. Soc.* **449** 3613

[22] Bowesman C A, Shuai M and Yurchenko S N 2021 *Mon. Notices Royal Astron. Soc.* **508** 3181

[23] Condon E U and Shortley G 1953 *The Theory of Atomic Spectra* (Cambridge: Cambridge University Press)

[24] Hilborn R C 1982 *Am. J. Phys.* **50** 982 http://arxiv.org/ftp/physics/papers/0202/0202029.pdf

[25] Thorne A P 1988 *Spectrophysics* 2nd edition (London: Chapman and Hall)

[26] Wigner E and Witmer E E 1928 *Z. Phys.* **51** 859
Hettema H 2000 *Quantum Chemistry: Classic Scientific Papers* (Singapore: World Scientific) p 287

[27] Hettema E H 2000 *On the Structure of the Spectra of Two-Atomic Molecules According to Quantum Mechanics. In: Quantum Chemistry: Classic Scientific Papers* (Singapore: World Scientific)

[28] Parigger C G 2023 *Preprints* **2023** 2023050423

IOP Publishing

Quantum Mechanics of the Diatomic Molecule (Second Edition)

Christian G Parigger and James O Hornkohl

Chapter 24

Hydroxyl, OH

Ultraviolet hydroxyl (OH) radiation measurements and studies have been employed successfully to characterize atmospheric phenomena and combustion processes [1–5]. The use of the hydroxyl-radical in combustion diagnostics has been reported frequently, and OH is a prime molecule for the study of thermal and chemical phenomena [6–18]. The preferred techniques comprise laser-induced fluorescence, two-dimensional imaging possibly with single pulse excitation, and degenerate four wave mixing for accurate measurement of temperature at various stages of a particular chemical process. This chapter discusses various aspects of OH measurements [19, 20] in nanosecond laser-induced breakdown spectroscopy.

The hydroxyl-radical is also a potential monitor of plasma temperature. Emission spectroscopy is primarily employed to determine the spectroscopic temperature of OH that is formed by recombination. In the characterization of laser-induced air plasma particularly the $A^2\Sigma \leftrightarrow X^2\Pi$ UV system of OH is of interest. Applications of laser-induced breakdown plasma include the initiation of combustion, or laser-ignition. Prior to the use of OH as an indicator and monitor of laser-ignition, knowledge is required of the spectral signatures of the laser-induced breakdown plasma that can ignite combustible mixtures. Ignition delays in laser-induced combustion can be in the order of some 10 μs, yet air-plasma recombination spectra indicate a temperature of 3 kK and a number density of 3×10^{16} cm^{-3} at a time delay of 200 μs [20]. OH is clearly detected in time-resolved spectral studies of laser-induced breakdown micro-plasma in standard ambient temperature and pressure (SATP) laboratory air in the wavelength range of 305–322 nm.

Laser-induced breakdown spectroscopy (LIBS) is utilized to characterize optical breakdown in SATP laboratory air. Optical breakdown is accomplished by focusing 1064 nm laser radiation of 3.5 ns pulse width; intensities of typically 10 TW/cm^2 resulted in the focal volume. A Coherent Infinity 40-100 Nd:YAG laser was operated at a frequency of 10 or 100 Hz. The spectra were resolved with a 0.64 m Jobin-Yvon spectrometer and measured by the use of an intensified array detector

doi:10.1088/978-0-7503-6204-7ch24

and a multichannel analyzer. The gate-widths were 10 μs for data recorded between 20 and 50 μs after optical breakdown, 20 μs for data recorded at 60 and 80 μs delays, and 50 μs for data recorded at delays of 100 and 200 μs. Wavelength calibrations were performed with standard light sources and the sensitivity correction was accomplished with a deuterium lamp. The detector's data were corrected for dark-noise contribution. Figures 24.1 and 24.2 show time-resolved recombination emission spectra that were measured at time delays of 60 and 100 μs, respectively, after the optical breakdown event in air. The spectral resolution amounted to 0.32 nm (32 cm $^{-1}$).

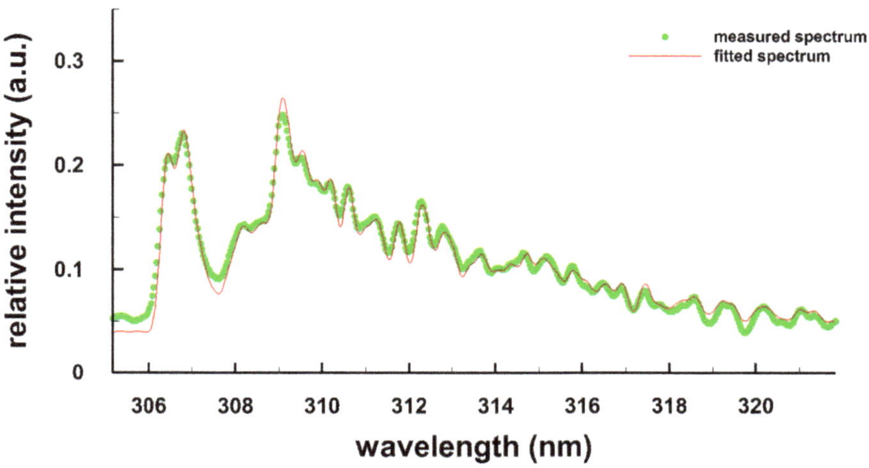

Figure 24.1. Measured and fitted air breakdown spectra, 60 μs time delay, $T = 4$ kK. Adapted with permission from [20] © The Optical Society.

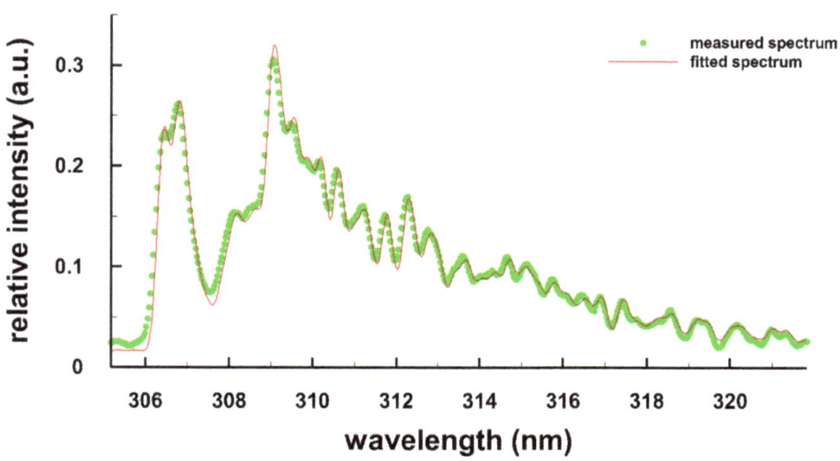

Figure 24.2. Measured and fitted air breakdown spectra, 100 μs time delay, T = 3.1 kK. Adapted with permission from [20] © The Optical Society.

Predominantly, the N_2 second positive system contributes to the spectral emissions at a delay time of some 10 μs, OH emissions can be recognized at delay times of 20 μs. The band structure of the N_2 second positive system is greatly reduced at a delay time of 40 μs, and subsequent emissions in the plasma decay are primarily from the $A^2\Sigma \leftrightarrow X\ ^2\Pi$ UV system of OH in the wavelength region of 305–322 nm, but significant background radiation occurs at delay times of 60 μs (figure 24.1).

The spectral fitting of the emissions recorded at a delay time of 60 μs was further investigated. Figure 24.1 reflects results from the application of previously discussed algorithms [21–23] for the computation of diatomic line-strengths [24–26] that are constructed from OH spectral databases [27–30]. The data reduction included a systematic variation of the background contributions which were computed by the use of the NEQAIR code. The calculated background was subtracted from the measured spectrum; subsequently, the Nelder–Mead [31, 32] least-square fitting algorithm was employed to determine the best-fit temperature. The background subtraction was accomplished by scaling experimental and synthetic spectra such that the integrated emissions are identical. For time delays of more than 150 μs, or for temperatures of approximately 3 kK, the background contribution is in the order of 1% or less of the integrated emission. Consequently, the temperature is exclusively found from the OH spectral distribution. The background radiation contains information of species number densities; however, for increased delay times or lower temperatures, the error bars increase for the species number densities. Although NEQAIR wavelength positions are inadequate for precise fitting, the integrated emissions in the region of 305–322 nm are accurately predicted. Furthermore, the assumption of thermodynamic equilibrium in the OH spectra produced by optical breakdown of air is reasonable for time delays of the order of 100 μ s.

References

[1] Levin D A, Laux C O and Kruger C H 1999 *J. Quant. Spectrosc. Radiat. Transfer* **61** 377

[2] Crosley D R and Jeffries J B 1996 Temperature measurements by laser-induced fluorescence of the hydroxyl radical *Temperature: Its Measurement and Control in Science and Industry* ed L J F Schooley vol 6 (New York: American Institute of Physics)

[3] Battles B E and Hanson R K 1995 *J. Quant. Spectrosc. Radiat. Transfer* **54** 521

[4] Levin D A, Laux C O and Laux C H 1995 A general model for the spectral calculation of oh radiation in the ultraviolet *AIAA paper 95-1990 presented at the 26th AIAA Plasmadynamics and Lasers Conf.(June 19-22, 1995) (San Diego, CA)*

[5] Laoux C O 1993 Optical diagnostics and radiative emission of air plasmas *PhD Dissertation* Department of Mechanical Engineering. Stanford University, Stanford, CA

[6] Rakestraw D J, Farrow R L and Dreier T 1990 *Opt. Lett.* **15** 709

[7] Kohse-Höinghaus K, Meier U and Attal-Tretout B 1990 *Appl. Opt.* **29** 1560

[8] Attal-Tretout B, Schmidt S C, Crete E, Dumas P and Taran J P 1990 *J. Quant. Spectrosc. Radiat. Transfer* **43** 351

[9] Desgroux P and Cottereau M J 1991 *Appl. Opt.* **30** 90

[10] Ewart P and Kaczmarek M 1991 *Appl. Opt.* **30** 3996–9

[11] Cignoli F, Benecchi S and Zizak G 1992 *Opt. Lett.* **17** 229

[12] Nyholm K, Maier R, Aminoff C G and Kaivola M 1993 *Appl. Opt.* **32** 919
[13] Nyholm K, Fritzon R and Alden M 1993 *Opt. Lett.* **18** 1672
[14] Seitzman J M, Hanson R K, Debarber P A and Hess C F 1994 *Appl. Opt.* **33** 4000
[15] Rahn L A and Brown M S 1994 *Opt. Lett.* **19** 1249
[16] Allen M G, McManus K R, Sonnenfroh D M and Paul P H 1995 *Appl. Opt.* **34** 6287
[17] Palmer J L and Hanson R K 1996 *Appl. Opt.* **35** 485
[18] Neuber A A, Janicka J and Hassel E P 1996 *Appl. Opt.* **35** 4033
[19] Parigger C G, Lewis J W L, Plemmons D P and Hornkohl J O 1996 Nitric oxide optical breakdown spectra and analysis by the use of the program NEQAIR *Laser Appl. Chem. Bio. Env. Anal.* **3** 85
[20] Parigger C G, Guan G and Hornkohl J O 2003 *Appl. Opt.* **42** 5986
[21] Tellinghuisen J 1972 *J. Mol. Spectrosc.* **44** 194
[22] Tellinghuisen J 1974 *J. Comput. Phys.* **6** 221
[23] Zare R N, Schmeltekopf A L, Harrop W J and Albritton D L 1973 *J. Mol. Spectrosc.* **46** 37
[24] Tatum J B 1967 The interpretation of intensities in diatomic molecular spectra *Astrophys. J. Suppl.* **XIV** 21
[25] Hornkohl J O, Parigger C G and Lewis J W L 1991 *J. Quant. Spectrosc. Radiat. Transfer* **46** 405
[26] Hornkohl J O and Parigger C G 1996 *Am. J. Phys.* **64** 623
[27] Dieke G H and Crosswhite H M 1962 *J. Quant. Spectrosc. Radiat. Transfer* **2** 97
[28] Coxon J A 1980 *Can. J. Phys.* **58** 933
[29] Coxon J A and Foster S C 1981 *Can. J. Phys.* **60** 41
[30] Coxon J A, Sappey A D and Copeland R A 1991 *J. Mol. Spectrosc.* **145** 41
[31] Nelder J A and Mead R 1965 *Comput. J.* **7** 308
[32] Salieri F, Quarteroni A and Sacco R 2000 *Numerical Mathematics* (New York: Springer)

IOP Publishing

Quantum Mechanics of the Diatomic Molecule (Second Edition)

Christian G Parigger and James O Hornkohl

Chapter 25

Hydroxyl laser-plasma emission spectra analysis with current databases

This chapter communicates application of the NMT.m script for the fitting of recorded experimental data. Furthermore, it gives a summary of how line-strength data are established.

25.1 Summary for computation of line-strength data

25.1.1 Wigner–Witmer diatomic eigenfunction

The Hund's case (a) basis functions (explicitly, see equation (25.4) below) are derived from the Wigner and Witmer [1, 2] diatomic eigenfunction,

$$\langle \rho, \zeta, \chi, \mathbf{r}_2, \dots, \mathbf{r}_N, r, \theta, \phi \,|n\mathrm{v}JM\rangle = \sum_{\Omega=-J}^{J} \langle \rho, \zeta, \mathbf{r}'_2, \dots, \mathbf{r}'_N, r \,|n\mathrm{v}\rangle\, D^{J*}_{M\Omega}(\phi, \theta, \chi). \quad (25.1)$$

The coordinates include the distance, ρ, of one electron (the electron arbitrarily labeled 1, but it could be any one of the electrons) from the internuclear vector, $\mathbf{r}(r, \theta, \phi)$, the distance, ζ, of that electron above or below the plane perpendicular to $\mathbf{r}$ and passing through the center of mass of the two nuclei (the coordinate origin), the angle, χ, for rotation of that electron about the internuclear vector $\mathbf{r}$, and the remaining electronic coordinates $\mathbf{r}_2, \dots, \mathbf{r}_N$ in the fixed and $\mathbf{r}'_2, \dots, \mathbf{r}'_N$ in the rotating coordinate system. The vibrational quantum number v has been extracted from the quantum number collection n that represents all required quantum numbers except J, M, Ω, and v. The Wigner–Witmer diatomic eigenfunction is well-suited for the description of the diatomic molecule. The exact separation of the Euler angles represents a clear advantage over the Born–Oppenheimer approximation for the diatomic molecule in which the angle of electronic rotation, χ, is unnecessarily separated from the angles describing nuclear rotation, θ and ϕ. Equation (25.1) can be derived by writing the general equation for coordinate (passive) rotations α, β, and γ of the eigenfunction, replacing two generic coordinate vectors with the diatomic vectors $\mathbf{r}(r, \theta, \phi)$ and $\mathbf{r}'(\rho, \zeta, \chi)$, and equating the angles of coordinate rotation to the angles of physical rotation ϕ, θ, and ϕ. The general equation for coordinate rotation holds in isotropic space, and therefore the quantum numbers J, M, and Ω in the Wigner–Witmer eigenfunction include all electronic and nuclear spins.

The rotation matrix element $D^J_{M\Omega}(\phi, \theta, \chi)$ and its complex conjugate $D^{J*}_{M\Omega}(\phi, \theta, \chi)$ do not fully possess the mathematical properties of quantum mechanical angular momentum. It is well-known that a sum of Wigner D-functions is required to build an angular momentum state. The equation

$$J'_{\pm}\, D^{J*}_{M\Omega}(\phi, \theta, \chi) = \sqrt{J(J+1) - \Omega(\Omega \mp 1)}\; D^{J*}_{M,\,\Omega\mp 1}(\phi, \theta, \chi) \tag{25.2}$$

is a mathematical result readily obtained from equation (25.1) and

$$J'_{\pm}\,|J\Omega\rangle = \sqrt{J(J+1) - \Omega(\Omega \pm 1)}\; |J, \Omega \pm 1\rangle, \tag{25.3}$$

in which the prime on the operator $J'_{\pm}$ indicates that it is written in the rotated coordinate system where the appropriate magnetic quantum number is Ω.

The Hund's case (a) basis function based upon the Wigner–Witmer diatomic eigenfunction, with the spin ket $|S\Sigma\rangle$, equals

$$\begin{aligned} |a\rangle &= \langle \rho, \zeta, \chi, \mathbf{r}'_2, \ldots, \mathbf{r}'_N, r, \theta, \phi\, |nvJM\Omega L\Lambda S\Sigma\rangle \\ &= \sqrt{\frac{2J+1}{8\pi^2}}\,\langle \rho, \zeta, \mathbf{r}'_2, \ldots, \mathbf{r}'_N, r\,|nv\rangle\; |S\Sigma\rangle\; D^{J*}_{M\Omega}(\phi, \theta, \chi). \end{aligned} \tag{25.4}$$

As noted above, a sum of $|a\rangle$ basis functions is required to build an eigenstate of angular momentum.

The prediction of diatomic spectra and establishment of the CNv-lsf database involves (a) determination of accurate wave numbers for the transition and rotational line strengths, viz. Hönl–London values (see details below in section 25.1.2); (b) vibrational transition strengths, viz. Frank–Condon factors from eigenfunctions for the diatomic potential; and (c) expansion of the electronic transition moments employing r-centroids. The product of these three factors yields the line strength, or alternatively the Einstein A-coefficient using equation (19.2).

25.1.2 Diatomic line position fitting algorithm

A traditional task of a diatomic spectroscopist is the computation of a set of molecular parameters from experimentally measured vacuum wave numbers $\tilde{\nu}_{\text{exp}}$ that are associated with J' and J, and in turn infer from that set of molecular parameters the $\tilde{\nu}_{\text{exp}}$ with a standard deviation comparable to the estimated experimental error. In practice, an experimental line list frequently shows gaps, viz. spectral lines are missing. Following a successful fitting process, one can use the molecular parameters to predict all lines.

Trial values of upper- and lower-state molecular parameters, typically taken from previous works by others for the band system in question, are used to compute upper

H' and lower H Hamiltonian matrices in the case (a) basis given by equation (25.4) for specific values of J' and J. The upper and lower Hamiltonians are numerically diagonalized,

$$T' = \tilde{U}' \, H' \, U', \tag{25.5}$$

$$T = \tilde{U} \, H \, U, \tag{25.6}$$

giving the upper T' and lower T term values. The vacuum wave numbers $\tilde{\nu}$, labeled using i and j for the dimensions (levels) of the upper and lower Hamiltonians,

$$\tilde{\nu}_{ij} = T'_i - T_j, \tag{25.7}$$

are determined, and the rotational strength,

$$S_{ij}(J', J) = (2J + 1) \left| \sum_n \sum_m \tilde{U}'_{in} \langle J\Omega; q, \Omega' - \Omega \, | J'\Omega' \rangle \; U_{mj} \; \delta(\Sigma'_n \Sigma_m) \right|^2, \tag{25.8}$$

is evaluated. The degree of the tensor operator, q, responsible for the transitions amounts to $q = 1$ for electric dipole transitions. For nonzero rotational factors, $S(J', J)$, the vacuum wavenumber $\tilde{\nu}_{ij}$ is added to a table of computed line positions to be compared with the experimental list $\tilde{\nu}_{\text{exp}}$ versus J' and J. The Clebsch–Gordan coefficient, $\langle J\Omega; q, \Omega' - \Omega \, | J'\Omega' \rangle$, is the same one appearing in the pure case (a) formulae for $S(J', J)$. For specific values of J' and J, one constructs tables for $\tilde{\nu}_{\text{exp}}$ and computed $\tilde{\nu}_{ij}$. The errors $\Delta\tilde{\nu}_{ij}$,

$$\Delta\tilde{\nu}_{ij} = \tilde{\nu}_{ij} - \tilde{\nu}_{\text{exp}}, \tag{25.9}$$

are computed where each $\tilde{\nu}_{ij}$ is the one that most closely equals one of the $\tilde{\nu}_{\text{exp}}$. Once values of $\tilde{\nu}_{ij}$ and $\tilde{\nu}_{\text{exp}}$ are matched, each is marked unavailable until a new list of $\tilde{\nu}_{ij}$ is computed. The indicated computations are performed for all values of J' and J in the experimental line list, and corrections to the trial values of the molecular parameters are subsequently determined from the resulting $\Delta\tilde{\nu}_{ij}$. The entire process is iterated until the parameter corrections become negligibly small. As this fitting process successfully concludes, one obtains a set of molecular parameters that predict the measured line positions $\tilde{\nu}_{\text{exp}}$ with standard deviations that essentially equal the experimental estimates for the accuracy of the $\tilde{\nu}_{\text{exp}}$. Of course, once sets of transition lines and strengths along with appropriate designations are determined for a particular diatomic molecule, one can employ these sets (that so-to-speak resemble a digital 'fingerprint' of the diatomic molecule) in the analysis of recorded spectra.

25.2 Hydroxyl analysis example

Figure 25.1 shows NMT output in graphical form when using the data in the file OH-LSF.txt. The data file OH100micros.dat is included in the supplement of [3]. The fitting program also incorporates a slight, overall wavelength offset of WLoffset = −0.05nm for the data file OH100micros.dat. Measurement of laser-plasma emissions shows a background from other species as discussed in [4].

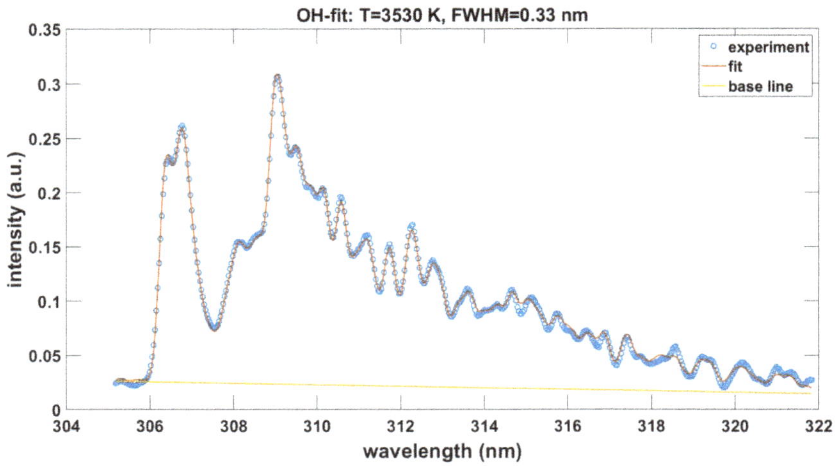

Figure 25.1. Measured and with OH-LSF.txt fitted OH emission spectra, $\Delta v = 0$, $\delta\lambda = 0.33$ nm, T = 3.53 kK. Reproduced from [3]. CC BY 4.0.

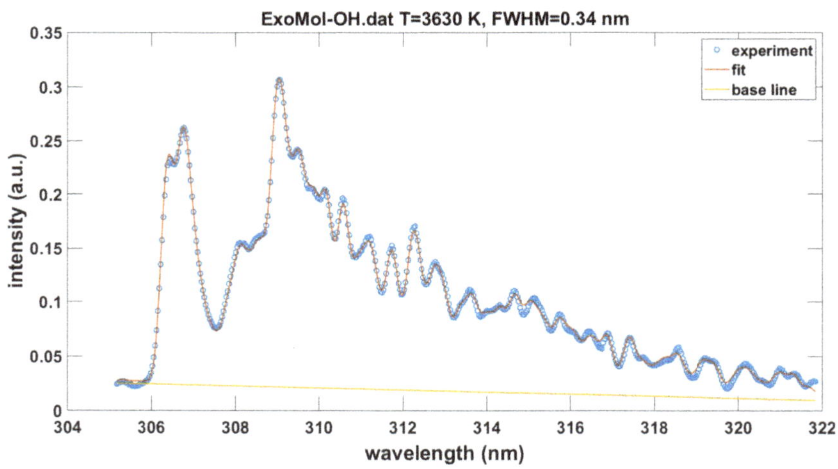

Figure 25.2. Measured and with ExoMol-OH.dat fitted OH emission spectra, $\Delta v = 0$, $\delta\lambda = 0.34$ nm, T = 3.63 kK. Reproduced from [3]. CC BY 4.0.

Fitting with the ExoMol [5] $^{16}O^{1}H$ database [6–8] requires preparation of provided transition and state files to be consistent with the NMT.m input portion of the program. Figure 25.2 displays the results. The temperatures differ by 0.1 kK, spectral resolution by 0.01 nm, and there is a slightly different linear background.

25.3 Analysis comparisons

The HITEMP [9] database file 13_HITEMP2020.par for OH (57019 lines) predicts a spectrum that is practically identical to the one from ExoMol (54 276 transitions, 1878 states) in the wavelength range of 305.17–321.83 nm for the experimental data OH100micros.dat. There are other OH databases that can be applied in the analysis

Table 25.1. Comparison of ExoMol-OH.dat and OH-LSF.txt line strengths in the wavelength range from 305.17 to 321.83 nm in the experimental data OH100micros.dat.

Data file	Transition lines	Equal lines	Vibrational levels
ExoMol.dat	856	512	0,1,2,3,4
OH-LSF.txt	528	512	0,1

of the OH emission spectra, e.g., see LIFBASE [10] with associated OH transition probabilities [11]. LIFBASE shows data for molecules of interest in this work, namely OH (A–X), NO(A–X,B–X,C–X,D–X), CN(B–X), and N_2^+(B–X). Table 25.1 shows comparisons for the wavelength range of the communicated OH UV (A–X) date file OH100micros.dat. There are 328 extra lines in the ExoMol.dat file, with most lines showing Einstein A coefficients that are larger than $1 \times 10^3\ \mathrm{s}^{-1}$ and higher vibrational levels than those for OH-LSF.txt. Subsequent to correction of an overall term-value offset in the ExoMol-OH data, T_{offset}, of $T_{offset} = 1809.4876$, the 512 transitions are labeled 'equal' for transition wave numbers that differ by less than $0.5\mathrm{cm}^{-1}$. It is noteworthy that 497 out of the 512 lines agree within $0.1\ \mathrm{cm}^{-1}$.

The ExoMol database shows Einstein A coefficients that are converted to line strengths, S_{ul}, for electric dipole transitions [12], using (MKS units)

$$A_{ul} = \frac{16\pi^3}{3g_u h\varepsilon_0 \lambda^3}(ea_0)^2 S_{ul}, \qquad g_u = 2(2J_u + 1), \tag{25.10}$$

where A_{ul} denotes the Einstein A-coefficient for a transition from an upper, u, to a lower, l, state, and h and ε_0 are Planck's constant and vacuum permittivity, respectively. The elementary charge is e, the Bohr radius is a_0, and S_{ul} is the transition strength. The line strength, S, that is used in the MATLAB scripts is expressed in traditional cgs units ($\mathrm{stC}^2\ \mathrm{cm}^2$, see table 15.5). The wavelength of the transition is λ, g_u is the upper state degeneracy, and J_u the total angular momentum of the upper state.

For the 512 lines with practically equal transition wave numbers, the ratios of ExoMol-OH.dat and OH-LSF.txt strengths show the mean value 1.093 with a standard deviation of 0.071. This line-strength variation may have several causes, including differences in Hönl–London terms, Frank–Condon factors, and/or r-centroids. The fitted temperature indicates a 2.8% increase from 3.53 kK to 3.63 kK, despite a mean 9.3% difference in line strengths. The NMT.m fitting script requires only relative intensities in the inference of temperature from measured spectra.

References

[1] Wigner E and Witmer E E 1928 *Z. Phys.* **51** 859
Hettema H 2000 *Quantum Chemistry: Classic Scientific Papers* (Singapore: World Scientific) p 287

[2] Hettema E H 2000 *On the Structure of the Spectra of Two-Atomic Molecules According to Quantum Mechanics. In: Quantum Chemistry: Classic Scientific Papers* (Singapore: World Scientific)

[3] Parigger C G 2023 *Foundations* **3** 1

[4] Parigger C G, Guan G and Hornkohl J O 2003 *Appl. Opt.* **42** 5986

[5] Tennyson J *et al* 2020 *J. Quant. Spectrosc. Radiat. Transf* **255** 107228

[6] Brooke J S A, Bernath P F, Western C M, Sneden C, Afşar M M, Li G and Gordon I E 2016 *J. Quant. Spectrosc. Radiat. Transf* **138** 142

[7] Yousefi M, Bernath P F, Hodges J and Masseron T 2018 *J. Quant. Spectrosc. Radiat. Transf.* **217** 416

[8] Bernath P F 2020 *J. Quant. Spectrosc. Radiat. Transf.* **240** 106687

[9] Rothman L S, Gordon I E, Barber R J, Dothe H, Gamache R R, Goldman A, Perevalov V I, Tashkun S A and Tennyson J 2010 *J. Quant. Spectrosc. Radiat. Transf.* **111** 2139

[10] Luque J and Crosley D R 2021 *LIFBASE: Database and Spectral Simulation for Diatomic Molecules* (Menlo Park, CA: SRI International)

[11] Luque J and Crosley D R 1998 *J. Chem. Phys.* **109** 439

[12] Tatum J *Stellar Atmospheres* Open Education Resource LibreTexts Project: LibreTexts Physics, shared under CC BY-NC 4.0 licence, University of Victoria, BC, Canada

IOP Publishing

Quantum Mechanics of the Diatomic Molecule (Second Edition)

Christian G Parigger and James O Hornkohl

Chapter 26

OH laser-induced breakdown spectroscopy and shadowgraphy

26.1 Introduction

This chapter communicates measurement and analysis of diatomic molecular hydroxyl (OH) spectra after generation of laser-induced plasma. A relative laboratory-air humidity of the order of 25% causes the occurrence of OH recombination radiation that is recorded with optical-emission spectroscopy. A Q-switched, 150 mJ, 6 ns pulsed Nd:YAG laser beam at the fundamental wavelength of 1064nm is focused in air with f/5 optics. Formation of OH is clearly discernible at time delays of several dozen microseconds after plasma initiation. Optical emissions are dispersed by a 0.64 m Czerny–Turner spectrometer and an intensified charge-coupled device records the data along the wavelength and slit dimensions.

The OH molecule is well-known to occur in combustion of hydrocarbons or hydrogen-oxygen reactions. Laser spectroscopy [1–4] allows one to ascertain plasma composition. In this work, laser-induced breakdown [5–8] induces plasma in standard ambient temperature (SATP) and pressure laboratory air. Recombination radiation of OH occurs due to moisture in SATP air; however, OH emission signals show spectroscopic interference from the N2 second positive system that usually occurs in nitrogen discharges [9]. Recorded spectra in air breakdown dwarf signal strengths that can be measured in combustion processes that utilize oxygen as an oxidizer, e.g., combustion of hydrocarbons [10, 11]. Laser-plasma in air, and for the wavelength range of 300–320 nm, reveals significant contributions from the N2 second positive system of nitrogen at time delays of typically 20 μs after optical breakdown [12, 13], and OH emission signals are clearly recognizable in the time delay range of 50–110 μs [14]. Spatiotemporal information is obtained by utilizing the slit dimension for the spatial resolution. The temporal resolution is obtained from a systematic set of time delays that are selected for an intensified charge-coupled device mounted at the exit plane of a spectrometer. It is

doi:10.1088/978-0-7503-6204-7ch26

also of interest to possibly associate measured spectra with shadowgraphs [15]. The shadowgraphs are recorded by employing a digital camera, thereby allowing one to associate spatial connections with the time-resolved, recorded spectra. The average concentration of OH molecules is inferred from equilibrium species distributions computations that employ freely available code for chemical equilibrium with applications [16].

26.2 Experiment results

The experimental arrangement consists of a set of components typical for time-resolved, laser-induced optical-emission spectroscopy [5–7, 12–15], or nanosecond laser-induced breakdown spectroscopy, see also chapter 16.

Figure 26.1 illustrates selected shadowgraphs for time delays of 54.25 μs and 104.25 μ s. The laser-induced shock wave is no longer present and well-developed vortices and fluid flows occur towards the incoming laser beam, as seen in figure 26.1 [14]. These fluid dynamic effects would impact time-resolved data collected along a narrow slice along the horizontal direction.

The spectroscopy experiments encompassed recording emission spectra along the slit height and in the spectral range of 302–321.3 nm, and for time delays of 10–110 μs in 10 μs steps [14]. Figure 26.2 illustrates recorded data for time delays of 30 μ s, 50 μ s, 70 μ s, and 100 μ s, individually scaled from minimum to maximum using rainbow or pseudo-color. The displayed data in figure 26.2 are consistent with previously recorded spectra using an ultraviolet enhanced, intensified linear diode array [13].

The intensified detector has the capability of recording 1024 spatially resolved data; however, four vertical pixels are grouped together for increased sensitivity of the recorded 256 spectra. The figures illustrate reasonable integrated signals from 100 consecutive laser-plasma events that are dispersed in approximately 100 spectra in the central region of the detector. Figure 26.2(a) displays OH and N_2 second positive band edges near 306 nm and 315 nm, respectively. Figures 26.2(c) and (d)

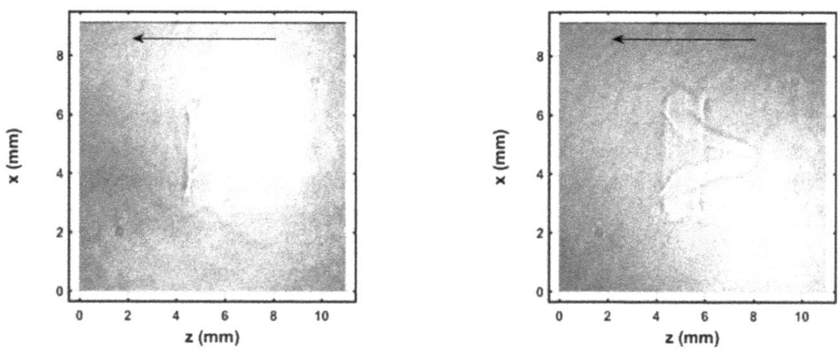

(a) (b)

Figure 26.1. Recorded shadowgraphs of expanding laser-induced plasma initiated with 170 mJ, 6 ns, 1064nm focused beam incident from right. Backlight 5 ns, 532 nm, time delay (a) 54.25 μs and (b) 104.25 μs. Reproduced from [17]. © IOP Publishing Ltd. CC BY 3.0.

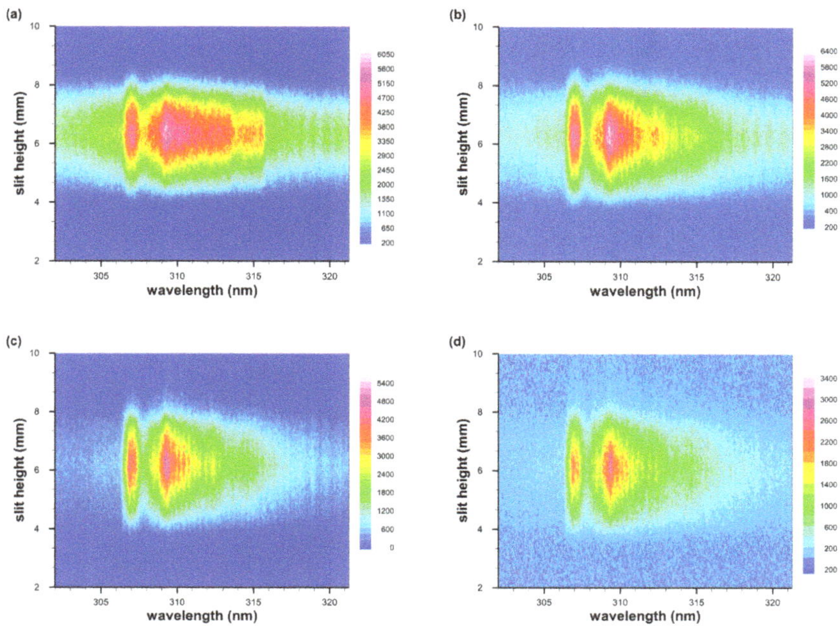

Figure 26.2. Recorded spectra of slit height versus wavelength. Gate width: 10 μ s, time delays (a) 30 μ s, (b) 50 μ s, (c) 70 μ s, and (d) 100 μs. Reproduced from [17]. © IOP Publishing Ltd. CC BY 3.0.

depict primarily OH emission spectra. Experimental averaging over 100 consecutive laser-plasma events enhances the signal-to-noise ratio by one order of magnitude. Conversely, as one collapses the 1024 vertical pixels to a single super-pixel, one mimics a linear diode array capable of recording single-shot OH spectra in laboratory air. Inspection of the spectra reveals that there is a slight curvature that would cause decreased resolution when averaging the central spectra. Figure 26.3 displays vertically averaged data of figure 26.2 showing typical liner-diode array type spectra.

In combustion investigation of, for example, hydrocarbon laser ignition, OH signals are significantly larger than those obtained in laboratory-air breakdown. However, planar laser-induced fluorescence (or planar LIF) is usually applied in laser-initiated combustion that allows one to correlate shadowgraphs with fluid physics expansion of the kernel [10, 11]. The recorded optical-emission data of OH are difficult to connect with the measured shadowgraphs [12]. However, spectra of CN that are recorded within the first microsecond after initiation of laser plasma in the laboratory can be associated with the expanding shock wave [18, 19]. An increase in electron density is inferred near the shock wave from analysis of a carbon atomic line superposed with the CN emission spectrum, and Abel inversion techniques allows one to associate increased CN density near the expanding shock wave.

However, integration of the measured spectra (see figure 26.2) along the spectral window of 302–321.3 nm for each slit-height position is encouraging for comparison

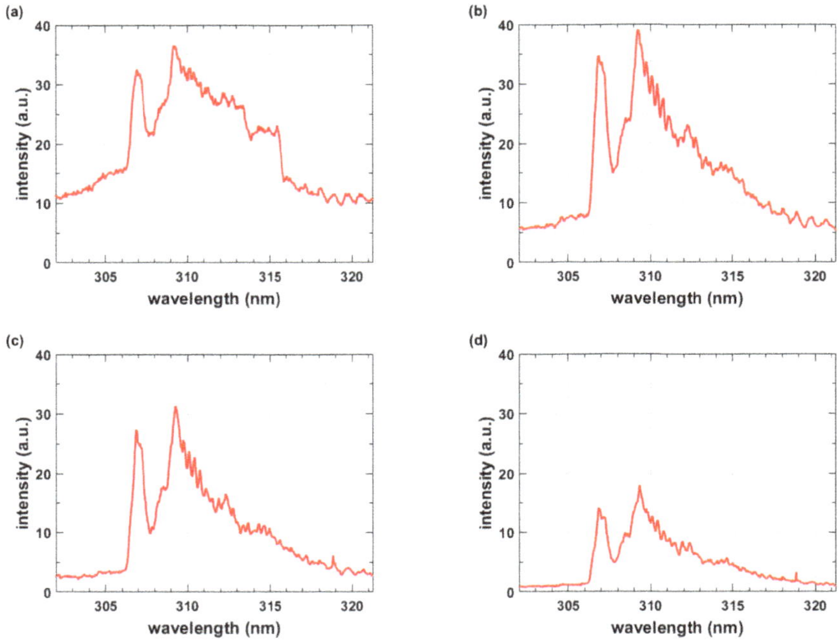

Figure 26.3. Vertical average between the slit heights 4.2 mm and 8.2 mm of recorded data. Time delays (a) 30 μ s, (b) 50 μ s, (c) 70 μ s, and (d) 100 μs. Reproduced from [17]. © IOP Publishing Ltd. CC BY 3.0.

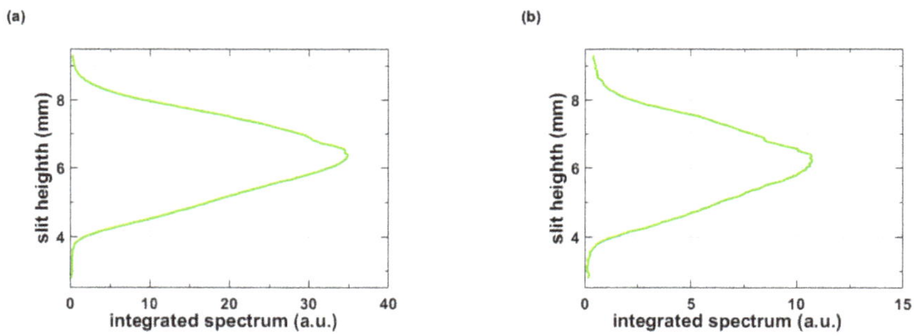

Figure 26.4. Integrated spectra for each slit height. Time delays (a) 50 μs and (b) 100 μ s. The maxima near 6 mm appear to indicate OH maxima near the central vertical structure in the corresponding shadowgraphs of figure 26.1. Reproduced from [17]. © IOP Publishing Ltd. CC BY 3.0.

of the results from shadowgraphs and spectroscopy. Figure 26.4 displays values of the integrated spectra for the slit height. The maximum of the integrated spectra near 6 mm appears to correspond to the central vertical structure in the shadowgraphs, see figure 26.1. There also appears to be slight variations near 7 mm that would correlate with fluid dynamics towards the laser. In figure 26.1, the slit height or z-direction is parallel to the laser propagation.

26.3 Summary

Spatiotemporal measurements of OH emission spectra in principle allow one to establish a correlation of shadowgraphs and laser spectroscopy results. However, laser spectroscopy results are based on averaging 100 individual events, but shadowgraphs are from single optical breakdown events. The fluid dynamic variations in the 50–100 μs range pose difficulties in establishing a connection analogous to the one obtained for CN for time delays of the order of 1 μ s. The OH band edge is noticeable for time delays of 10 μs from the laser-plasma generation, but with strong N2 second positive contributions. However, for time delays in the range of 50–100 μ s, interference from N_2 is significantly diminished in the 302–321.3 nm wavelength range for $\Delta v = 0$ transitions of the A–X band of OH.

References

[1] Kunze H-J 2009 *Introduction to Plasma Spectroscopy* (Heidelberg: Springer)

[2] Fujimoto T 2004 *Plasma Spectroscopy* (Oxford: Clarendon)

[3] Ochkin V N 2009 *Spectroscopy of Low Temperature Plasma* (Weinheim: Wiley)

[4] Boulous M I, Fauchais P and Pfender E 1994 *Thermal Plasmas–Fundamentals and Applications* (New York: Plenum)

[5] Cremers D E and Radziemski L J 2006 *Handbook of Laser-Induced Breakdown Spectroscopy* (New York: Wiley)

[6] Miziolek A W, Palleschi V and Schechter I 2006 *Laser Induced Breakdown Spectroscopy* (New York: Cambridge University Press)

[7] Singh J P and Thakur S N (ed) 2020 *Laser-Induced Breakdown Spectroscopy* 2nd edn (New York: Elsevier)

[8] De Giacomo A and Hermann J 2017 *J. Phys. D: Appl. Phys.* **50** 183002

[9] Fatima H, Ullah M U, Ahmad S, Imran M, Sajjad S, Hussain S and Qayyum A 2021 *Springer Nature Appl. Sci.* **3** 646

[10] Chen Y-L and Lewis J W L L 2001 *Opt. Express* **9** 360

[11] Qin W, Chen Y-L and Lewis J W L 2005 *International Flame Research Foundation (IFRF) Combust. J.* 200508

[12] Parigger C G, Jordan H B S, Surmick D M and Splinter R 2020 *Molecules* **25** 988

[13] Parigger C G, Guan G and Hornkohl J O 2003 *Appl. Opt.* **42** 5986

[14] Parigger C G 2022 *Int. Rev. At. Mol. Phys.* **13** 15

[15] Rai A K, Pati J K, Parigger C G and Rai A K 2019 *Atoms* **7** 72

[16] McBride B J and Gordon S 2005 *Computer Program for Calculating and Fitting Thermodynamic Functions, NASA RP-1271* **1271**

[17] Parigger C G and Helstern C M 2023 *J. Phys.: Conf. Ser.* **2439** 012004

[18] Parigger C G, Helstern C M, Jordan B S, Surmick D M and Splinter R 2020 *Molecules* **25** 615

[19] Parigger C G 2020 *Spectrochim. Acta* B **179** 106122

IOP Publishing

Quantum Mechanics of the Diatomic Molecule (Second Edition)

Christian G Parigger and James O Hornkohl

Chapter 27

Titanium Monoxide, TiO

Various research efforts regarding laser-induced breakdown spectroscopy (LIBS) of titanium normally focus on the atomic and ionic Ti spectral transition lines. However, after a characteristic time subsequent to laser ablation, these lines are no longer discernable. During this temporal window, the diatomic molecular transition lines of titanium monoxide (TiO) are prominent in the laser-induced plasma (LIP) emissions. This chapter discusses recent TiO research [1], which was also presented at a recent line shape conference [2].

The molecule TiO has long been studied in the contexts of stellar emissions, allowing for some of the molecular transition bands to be accurately computed from theory. In this research, optical emission spectroscopy (OES) of LIP generated by laser ablation of titanium is performed in order to infer temperature as a function of time subsequent to plasma formation. The emission spectra of the resulting ablation plume is imaged as a function of height above the sample surface. Temperatures are inferred over time delays following plasma formation ranging from 20–200 μs. Computed TiO $A^3\Phi - X^3\Delta$, $\Delta v = 0$ transition lines are fitted to spectral measurements in order to infer temperature. At $t_{\text{delay}} = 20$–80 μs, the observed plume contains two luminescent regions each with a distinctly different temperature. As the plume evolves in time, the two regions combine and an overall temperature increase is observed.

27.1 Introduction

LIBS, as an experimental technique, has grown in popularity, in large part due to the simplicity of the required equipment. The traditional LIBS apparatus consists of a single laser, serving as an excitation source, and a spectrometer equipped with a detector for gathering spectral emissions [3]. The success and growth of future LIBS applications require a deeper understanding of the fundamental processes occurring as a plasma is observed at various time delays, given various experimental conditions [4]. To this extent, much of the ongoing fundamental research utilizing

doi:10.1088/978-0-7503-6204-7ch27

LIBS seeks to observe and analyze specific atomic and molecular species present in the LIP [5–9]. By closely examining the properties of spectra from specific atoms and molecules at various regions of the plasma as it evolves in time, a description of the characteristic processes going on throughout the plasma can be obtained. Of particular interest is the time domain for which molecular recombination overcomes electron collisions as the dominant mechanism concerning plasma emissions.

As LIP evolves in time, the characteristic emitting species dominating the spectra will also evolve. An initial LIBS spectrum, typically within the first tens of nanoseconds, is dominated by a spectral continuum contribution. Mechanisms such as inverse Bremsstrahlung and radiative recombination involving free electrons are responsible for this spectral continuum. As the plasma expands and cools, ionic spectral line transitions appear in the plasma emissions, superposed to the continuum contribution. The more highly ionized species will become prominent first, followed by the lesser ionized species and ultimately the neutral atomic transition lines [10–12]. Three-body recombination occurring with plasma expansion causes a rapid decrease of the electron density in the plasma [13]. As a consequence of the decreasing electron density in the plasma, each emission line will become progressively more narrow during the plasma expansion [14].

After a characteristic time, depending on electron density, the elements in the plasma and the plasma expansion, emissions from radiative molecular transitions will become discernible as the prominence of atomic spectral lines observed in the plasma decreases [15]. Of particular interest for this research is the diatomic molecular radiative transitions of the plasma. Typically, this phenomenon is attributed to the recombination of atoms to form molecules. However, the detection of diatomic transitions of certain species at far earlier stages of plasma evolution when compared to other diatomic transitions in similar conditions leads to the conclusion that not all diatomic transitions are a product of recombination. In these scenarios, it is hypothesized that the molecule was present prior to plasma formation. However, many diatomic molecules observed in LIP are unstable in typical laboratory environments. For such molecules, LIBS offers a suitable method for observation. The spectral transitions of such molecules can be collected typically for tens of microseconds. Often times, these emissions are quite luminous, from the ultraviolet to the infrared.

For the purposes of analyzing the laser ablation plasma above the Ti surface at time delays on the order of tens to hundreds of microseconds, TiO spectral transitions are utilized to infer temperature. This required the computation of new, accurate line-strength tables. Diatomic quantum theory is utilized in computing line-strengths of spectral transitions [6, 16]. The specifics concerning the computation of the line-strengths used in this research are presented elsewhere [1, 17]. Given a few parameters, such as resolution and temperature, the spectral line-strengths can be used generate a synthetic spectrum representative of the radiant transitions of a diatomic molecule. In order to infer micro-plasma parameters, the experimentally collected spectra believed to contain the diatomic spectral signatures are fitted with the computed spectra for the molecule [1, 17, 18].

Previous studies observed the TiO $B^3\Pi \rightarrow X^3\Delta$ and $A^3\Phi \rightarrow X^3\Delta$, $\Delta v = 0$ transition bands in the ablation plasma [1, 19]. The collected spectra were then fitted with computed spectra in order to infer temperature as a function of delay time. Temperature inferences, resulting from the use of an intensified linear diode array arrangement to gather time-resolved spectra, typically contained a local minimum occurring between a 40 μs and 60 μs delay time. At 60 μs, when the expanding plasma is thought to be cooling, this unexpected result indicated an increase in temperature. In order to further investigate the situation, the current research observes the TiO $A^3\Phi - X^3\Delta$, $\Delta v = 0$ transition band at time delays ranging from 20–200 μs. Utilizing a gated two-dimensional intensified charge-coupled detector (ICCD) detector, the gathered emissions are time-resolved and imaged as a function of height above an ablation surface. The gathered spectra are then fitted with computed spectra representing the TiO $A^3\Phi - X^3\Delta$ band to infer temperature.

27.2 Experiment

A Ti 6-4 sample is ablated by radiation from a Nd:YAG laser with 160 mJ per pulse and a nominal pulse width of 13 ns. The laser radiation is supplied at 10 Hz and focused to approximately a 1 mm spot size. The beam propagates vertically downward onto the flat Ti surface such that the beam path is parallel to the slit aperture of a Jobin-Yvon spectrometer. The resulting ablation plume emissions are imaged onto the slit in a 1:1 manner. The spectrometer is equipped with an Andor iStar ICCD camera, providing both temporal and spatial resolution. In order to enhance the signal-to-noise ratio, the 1024×1024 pixels provided by the detector are consolidated into 32 spectra (1024 pixels across) gathered along the height of the plume for each image. Additionally, 200 accumulations are used to produce each image. At $t_{\text{delay}} = 20$–$200\ \mu s$, spectra in the range of 705–715 nm are collected every 20 μs utilizing a 2 μs gate width. In this spectral region, at these time delays, TiO transition lines are collected. The gathered spectra are then fitted with computed spectra from the TiO $A^3\Phi - X^3\Delta$, $\Delta v = 0$ transition band in order to infer temperature.

27.3 Results

Figure 27.1 provides an example of an experimentally obtained spectrum fit with a computed spectrum for the TiO γ transition. The collected spectrum of figure 27.1 represents a single horizontal slice of a larger image. Figure 27.2 is an example of such an image. The plot just below the image in figure 27.2 represents the vertical sum of the pixels in the image. Such a summation is likened to using a linear diode array detector. While spectra computed from theory are used to infer the TiO transition temperature along the height of the image, the spectrum obtained by integrating the height of the detector can be similarly analyzed. The result is typically a temperature corresponding to the inferred temperature at the hottest region of the image.

By studying figures 27.2–27.6, some very interesting phenomena become apparent. The temperature inferred from the TiO molecular transitions increases with

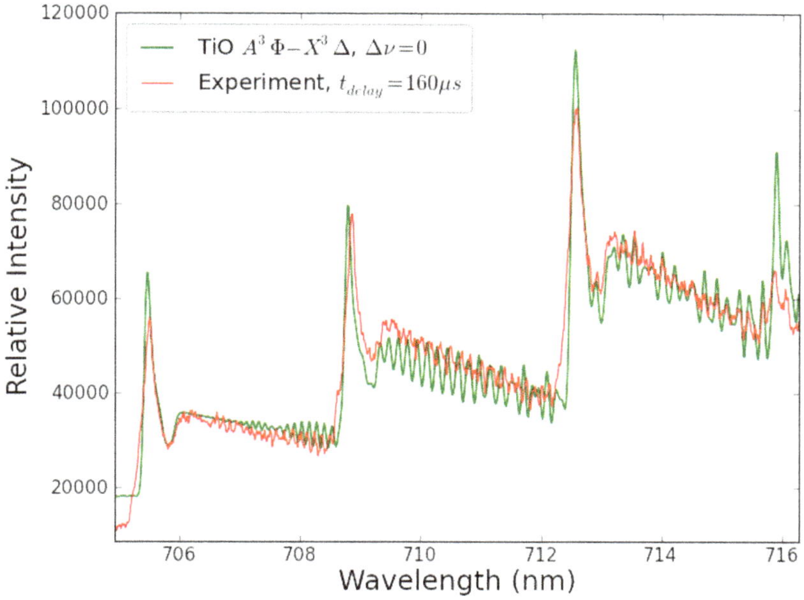

Figure 27.1. This spectrum represents a horizontal slice of one the gathered images fit with computed spectra corresponding to $T = 3400$ K. The measured spectrum was collected at $t_{\text{delay}} = 160\ \mu s$ [2].

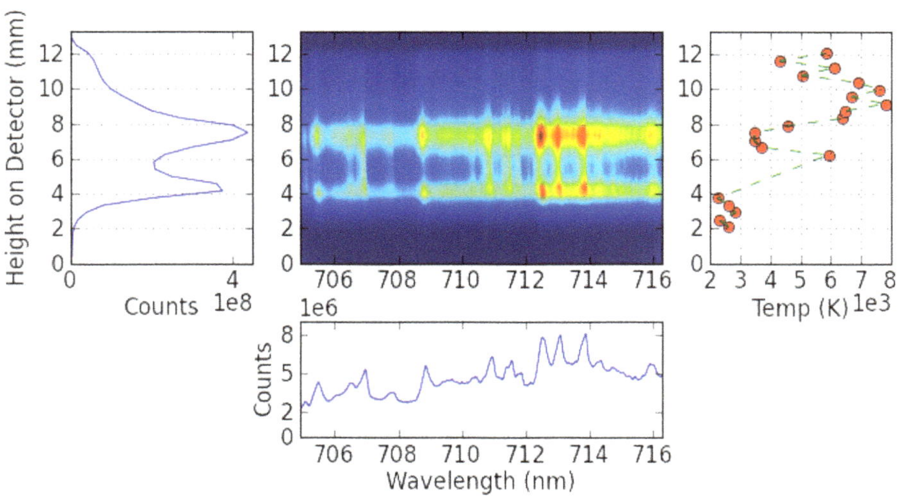

Figure 27.2. Recorded spectra of laser ablation plasma above a titanium surface recorded at $t_{\text{delay}} = 40\ \mu$s with a 2 μs gate width [2].

height along the ablation plume. This phenomenon is consistent throughout $t_{\text{delay}} =$ 20–200 μs. More interestingly, it is clear that between $t_{\text{delay}} =$ 20–60 μs there are two distinct regions of the plasma. One region rests several millimeters above the sample surface, while the other appears to rise from the surface before the two ultimately combine. After the two regions combine ($\sim$100 μs), there is a noticeable bulge in uppermost part of the intensity profile.

When the temperature inferences made from emissions of the two regions are compared, it is apparent that the inferred temperature of the TiO emissions from the region closer to the sample surface is distinctly lower than that of the region further above the surface. This is most clearly evident in figures 27.2 and 27.3. In particular, when examining the inferred temperatures of figure 27.2, it is important to keep in mind that we are struggling to make temperature inferences resulting from fitting computed TiO spectra to the collected spectra on account of signal-to-noise issues. Notice at the earlier time delays, the characteristic structure of the TiO $A^3\Phi - X^3\Delta$ band is hardly discernable. Much of the spectral signatures observed in this time domain are atomic line shapes. When inspecting the inferred temperature, more confidence is given to those values inferred from the brighter regions of the plume. These regions are made brighter by the luminescent TiO transitions. This conveniently allows for succinct comparison between the two brightest regions of figure 27.2 because the TiO spectral signatures in the dimmer regions are less discernable from the other spectral contributions.

Examining the plots to the left-hand side of the images in figures 27.2–27.6 provides insight into the luminosity along the height of the plasma. The motion of the lower peak towards the upper peak can be used as a measure of the kinetics of the luminous lower region of the plasma. These plots for figures 27.2–27.4 also illustrate an increase in luminosity of the lower region as it approaches the upper. At later time delays, t_{delay}= 100–200 μs, a hump is observed at the top luminous region of the ablation plume. The height of the plasma corresponding to the slight dip below the hump is determined to be hotter than the regions above or below the dip. While the hump appears to be a consequence of a snowplow effect at the top of the plasma, two-dimensional spatial imaging of the plume would possibly provide an

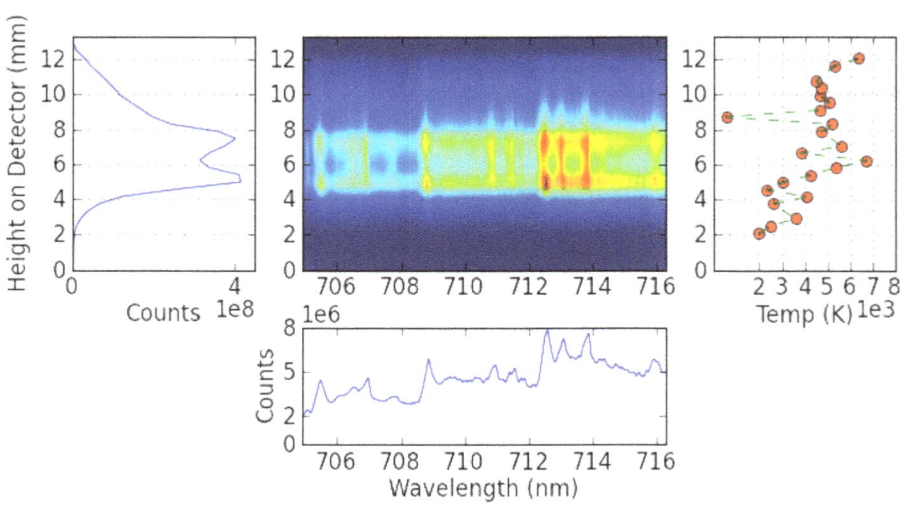

Figure 27.3. Recorded spectra of laser ablation plasma above a titanium surface recorded at $t_{\text{delay}} = 60\ \mu$s with a 2 μs gate width [2].

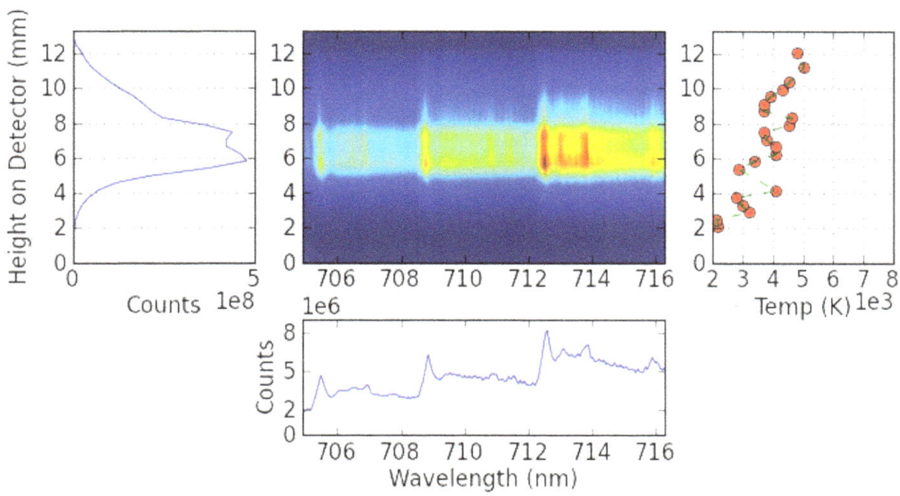

Figure 27.4. Recorded spectra of laser ablation plasma above a titanium surface recorded at $t_{\text{delay}} = 80\ \mu\text{s}$ with a 2 μs gate width [2].

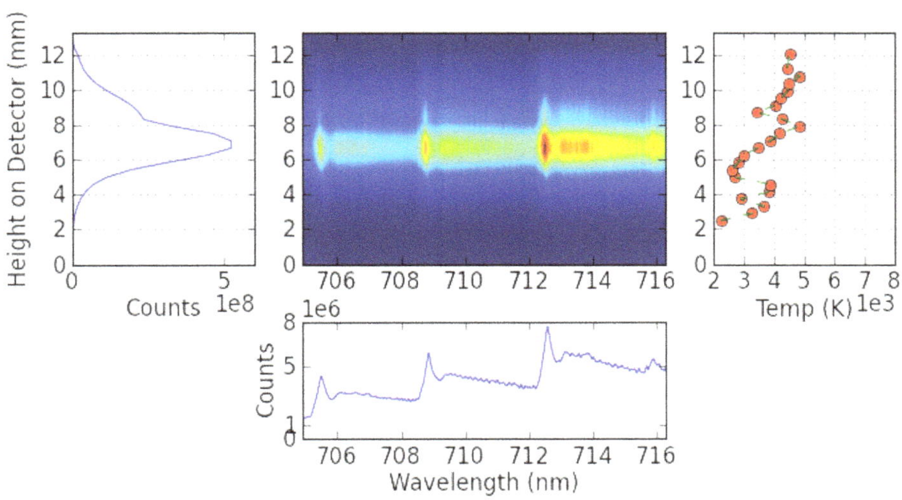

Figure 27.5. Recorded spectra of laser ablation plasma above a titanium surface recorded at $t_{\text{delay}} = 100\ \mu\text{s}$ with a 2 μs gate width [2].

explanation. The current apparatus observes primarily along the axial distance perpendicular to the surface; however, imaging in the radial direction would likely provide a crown-shaped profile, as in figure 8 of Aguilera *et al* [20].

Comparison of the inferred temperature from the upper region of figure 27.2 shows remarkable agreement with the results obtained previously at this same time delay ($t_{\text{delay}} = 40\ \mu s$) using this same window (2 μs) [19]. If one makes the same comparison with the upper region of figure 27.3 and Woods *et al* [19] at $t_{\text{delay}} = 60\ \mu s$, there is some agreement for the inferred temperatures here as well. When the spectra

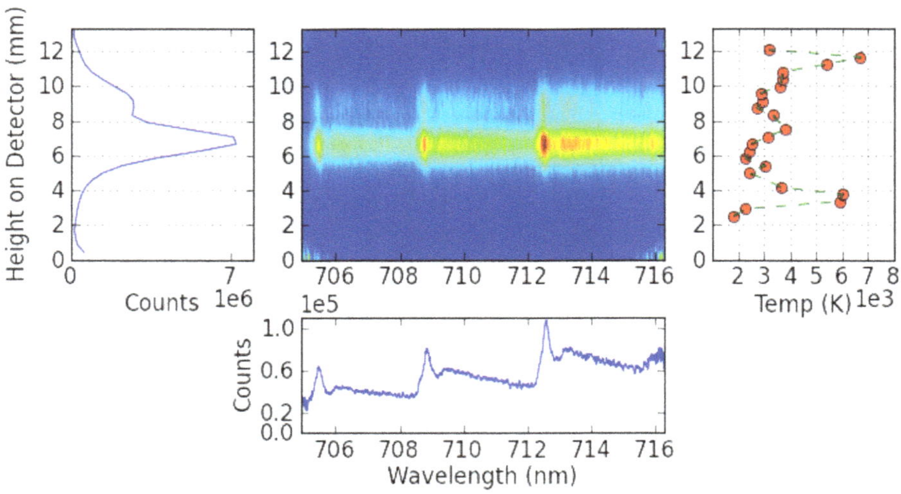

Figure 27.6. Recorded spectra of laser ablation plasma above a titanium surface recorded at $t_{\text{delay}} = 200$ μs with a 2 μs gate width [2].

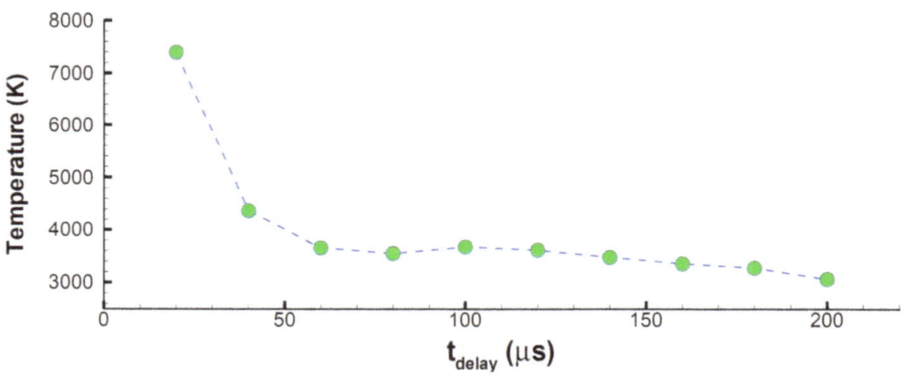

Figure 27.7. Temperature as function of time delay inferred from the analysis of the spectra resulting from integrating along the height of the detector [2].

obtained by integrating over the height of the images are analyzed, the inferred temperatures typically correspond to a temperature slightly elevated from that of the upper luminescent region. Figure 27.7 illustrates temperature versus time delay.

Examining figure 27.7, a local temperature increase of $\approx$ 100 K occurs at $t_{\text{delay}} = 100$ μs. Previous investigations concerning both the TiO $B^3\Pi \rightarrow X^3\Delta$ and $A^3\Phi - X^3\Delta$ systems have also found an unexpected temperature increase at later time delays [1, 19]. Inspection of the inferred temperatures along the height of the image over $t_{\text{delay}} =$ 40–100 μs indicate that the lower region undergoes drastic heating, as the cooler TiO molecules of the lower luminescent region interact with the warmer TiO molecules of the upper region. This interaction results in an overall increase in TiO temperature in the plasma.

While more investigations are required to fully characterize this bright region rising from the surface, some hypotheses can be formed. By the ability to fit computed TiO spectra to this emitting region, it is determined that the TiO molecule is the dominant emitter. In the upper region, the TiO transitions are likely the result of recombination. This is consistent with the delay time in which the transitions become discernible given the excitation source and the evolution of the temperature and electron density, as determined by the measurements made within the first hundreds of nanoseconds. The drastic temperature difference at $t_{\text{delay}} = 40\ \mu s$ between the upper and lower luminescent regions remains largely unexplained. Since the cooler plasma temperatures are typically found to be in the lower regions, perhaps a sudden flurry of recombination occurs near the sample surface, overtaking electron impacts as the driving mechanism for emissions. This trend then rises along the plasma until it reaches the upper region. However, because the temperatures inferred from the TiO transitions in the lower region are starkly less than those of the upper region and gradually increase as the regions interact, an alternate theory would maintain that the mechanism causing the radiative transitions of the diatomic molecule is different between the two regions. During laser ablation at this intensity, the sample surface will become molten for a characteristic duration [21]. Considering that the lower region is much cooler and appears to heat as it leaves the surface, the TiO emissions in this region may possibly be emanating from bulk titanium being ejected from the sample surface.

References

[1] Woods A C, Parigger C G and Hornkohl J O 2012 *Opt. Lett.* **37** 5139
[2] Woods A C and Parigger C G 2014 *J. Phys.: Conf. Ser.* **548** 012037
[3] Parigger C G, Woods A C, Witte M J, Swafford L D and Surmick D M 2014 *J. Vis. Exp.* **84** 51250
[4] Amoruso S, Bruzzese R, Spinelli N and Velotta R 1997 *J. Phys.* B **36** 3227
[5] Parigger C G and Oks E 2010 *Int. Rev. At. Mol. Phys.* **1** 13
[6] Hornkohl J O, Nemes L and Parigger C G 2009 *Spectroscopy, dynamics and molecular theory of carbon plasmas and vapors. In: Advances in the Understanding of the Most Complex High-Temperature Elemental System* ed L Nemes and S Irle (Singapore: World Scientific) Chapter 4
[7] Witte M J, Parigger C G, Bullock N A, Merten J A and Allen S D 2014 *Appl. Spectrosc.* **68** 367
[8] Parigger C G, Swafford L D, Woods A C, Surmick D M and Witte M J 2014 *Spectrochim. Acta* B **99** 28
[9] Parigger C G, Woods A C, Surmick D M, Swafford L D and Witte M J 2014 *Spectrochim. Acta* B **99** 15
[10] Albert O, Roger S, Glinec Y, Loulergue J C, Etchepare J, Boulmer-Leborgne C, Perriére J and Millon E 2003 *Appl. Phys.* A **76** 319
[11] De Giacomo A 2003 *Spectrochim. Acta* B **58** 71
[12] Capitelli M 2004 *Spectrochim. Acta* B **59** 271
[13] Casavola A, Colonna G and Capitelli M 2003 *Appl. Surf. Sci.* **208** 85
[14] Griem H 1974 *Spectral Line Broadening by Plasma* (New York: Academic)

[15] De Giacomo A, Dell'Aglio M, Santagata A and Teghil R 2005 *Spectrochim. Acta* B **60** 935
[16] Parigger C G and Hornkohl J O 2011 *Spectrochim. Acta* A **81** 403
[17] Parigger C G, Woods A C, Keszler A, Nemes L, Hornkohl J O and Conf A I P 2012 *AIP Conf. Proc.* **1464** 628
[18] Hermann J, Perrone A and Dutouquet C 2001 *J. Phys.* B **34** 153
[19] Woods A C, Parigger C G and Hornkohl J O 2012 *Int. Rev. At. Mol. Phys.* **3** 103
[20] Aguilera J A and Aragón C 1861 *Spectrochim. Acta* A **59** 204
[21] Bogaerts A and Chen Z 2005 *Spectrochim. Acta* A **60** 1280

Christian G Parigger and James O Hornkohl

Chapter 28

Nitric Oxide, NO

Nitric oxide (NO) spectra are studied with laser-induced breakdown spectroscopy (LIBS) in laboratory air and in a 1:1 mole ratio N_2:O_2 gaseous mixture. This chapter summarizes LIBS of NO that was recently presented at a line shape conference [1]. The temporally-resolved excitation temperatures are determined by fitting recorded spectra with predicted diatomic spectra [1] that are constructed from diatomic line strengths. For time delays of 25 and 50 μs, the fitted temperatures are 6800 and 6100 K, respectively. Analysis of experiments in the 1:1 N_2:O_2 mixture is also accomplished with the nonequilibrium air radiation (NEQAIR) program [2].

Optical breakdown plasmas are generated by focused Nd:YAG ir laser radiation of 1–100 TW/cm^2 intensity. Typically, rich emission spectra result. The decaying laser-induced plasma can be characterized by the application of time-resolved spectroscopy techniques. Primarily, recombination spectra occur subsequent to optical breakdown [3–5]. It is of interest to obtain temperature and species density information from such spectra.

In the near-ultraviolet wavelength region of 205–300 nm, the emission spectra are comprised of overlapped electronic transitions of primarily NO and N_2, and to a lesser degree of O_2 and atomic species. The spectral analysis of these superimposed spectra of such multiple species is most challenging, and the computer code for NEQAIR is applied in the spectral analysis of the transient micro-plasma.

The program NEQAIR [6, 7] is widely used to predict and interpret optical spectra from air plasmas. The majority of NEQAIR applications concern atmospheric radiation, but the program is also applied in simulation facilities and has other applications, such as the prediction of laser sustained plasmas. In this paper, we apply the nonequilibrium air radiation code, in the form of NEQAIR8 [8, 9], to our knowledge for the first time, in the analysis of time-dependent optical breakdown spectra.

The analysis of the recorded experimental spectra from air breakdown is based on the use of calculated number densities of a dry air plasma composition at a pressure

doi:10.1088/978-0-7503-6204-7ch28

of 100 kPa [10]. With the assumption of local thermodynamic equilibrium, a one-parameter least-square fitting procedure is applied to find the spectroscopic temperature.

One of the many applications of LIBS is its use in diagnostics, such as combustion studies. Quantitative analysis efforts include investigations of both atomic and molecular spectra [11]. In such works, diatomic species of interest were CN, N2, and NH [12, 13]. Of particular interest to combustion are quantitative calculations and analysis of NO spectra for studies of nitrogen containing flames and exhaust plumes of rockets. The near-ultraviolet spectra of the NO γ-band in the 140–340nm range are considered [14].

In order to quantitatively study the emission of diatomic molecules, diatomic line strengths for NO are utilized. Detailed aspects of NO computations are deferred to appendix D. The use of line strengths is effective for the calculation of diatomic, molecular spectra due to its symmetric nature [15]. In this chapter, the line-strength approach is briefly reviewed. The formal definition of the line strength is the total sum of degenerate states that produce the same spectral line

$$S_{ul} = \sum_u \sum_l \left| \langle u | T_k^{(q)} | l \rangle \right|^2 \tag{28.1}$$

where u and l represent the upper and lower states, respectively, and $T_k^{(q)}$ is the kth component of the irreducible tensor operator of rank q that gives rise to the spectral transition. This operator is the electric dipole operator and the line strength may be expressed as

$$S_{ul} = S(n'v', nv)S(J', J) \tag{28.2}$$

where $S(n'v', nv)$ is the electronic-vibration transition strength and $S(J', J)$ is the Hönl–London line-strength factor. Expressing the line strength as in equation (28.2) allows for a direct analytical calculation of the line strength, and in turn the intensity of a diatomic emission spectra, which may be expressed as

$$I_{ul} = \frac{16\pi^3 c(a_0 e)^2 C_{abs} N_0}{3\varepsilon_0 Q} C_\nu \tilde{\nu}_{ul}^4 S_{ul} e^{-hF_u/k_B T}, \tag{28.3}$$

where a_0 is the Bohr radius, e is the electron charge, c is the speed of light, Q is the partition function, N_0 is the total population of the species, C_{abs} and C_ν are the absolute and relative calibration factors, h is Planck's constant, F_u is the upper term value, k_B is Boltzmann's constant, and T is the temperature. A simple, yet rigorous selection rule results from the expression in equation (28.2). Given the relative ease in calculating the Hönl–London line strength and the ability for the Hönl–London line strength to be zero, allowed transitions are those with a nonvanishing line strength. In this study, we apply this method to infer the emission temperature of the NO emission spectra from measurements made in the nominal 230–240 nm range for detailed studies and in the range of 210–300 nm for overview investigations.

28.1 Experimental details

The measurement of NO breakdown spectra, laser-induced plasma is generated in standard ambient temperature and pressure (SATP) laboratory air. Optical breakdown is initiated by focusing 1064nm, 12 nanosecond pulsed laser radiation from a Q-switch, Nd:YAG laser operating in its fundamental mode with an average energy per pulse of 190 mJ. Light from the plasma emission is dispersed by a 0.64 meter Jobin-Yvon Czerny–Turner spectrometer equipped with 3600 grooves/mm grating. Spectra are recorded with an intensified, linear diode array coupled to an optical multichannel analyzer. Time synchronization for the laser pulse firing rate and the measurement rate of the linear diode array is achieved with the use of waveform, pulse, and delay generators. Exact experimental details and a video [16] summarize the timing apparatus used in this experiment and in other, similar laser-induced plasma studies involving atomic and diatomic spectra. Prior to analysis, all spectra are properly calibrated for the detector background and sensitivity response.

28.2 Results

The NO spectra are fitted to theoretical calculations tabulated from the line-strength method. Comparisons between experiment and theory are made using a Nelder–Mead algorithm, which is used for its ability to use fit multiple parameters simultaneously, including a variable baseline offset. The Nelder–Mead minimizes parameters by iteratively reducing the size of a geometric simplex where the number of vertices in the simplex is one more than the number of parameters to be fitted [17, 18]. The parameters considered while fitting are the temperature, spectral resolution, and offset. The spectral resolution parameter is held constant for each fit and is determined from the physical limitations of the experimental apparatus. Given the wavelength region and the high groove density of the grating used, the spectral resolution of the experimental spectra was 0.056 nm, which also accounts for slit width effects. The constant offset parameter was allowed to vary. NO spectra were collected and analyzed for 25, 50, and 75 μs time delays for air breakdown experiments and the inferred temperatures were found to be 6.8, 6.1, and 5.8 kK for 25, 50, and 75 μs time delays, respectively. The fitting results for the 25, 50, and 75 μs data are depicted in figures 28.1–28.3. As can be seen in the fits shown in figures 28.1 and 28.2, the diatomic line-strength method has the capability to calculate and fit spectra that include many rotational lines.

28.3 Comparison with overview spectra

The recorded spectra and inferred temperatures versus time delay in the spectral 230–240 nm agree with with previous overview spectra recorded in NO LIBS experiments in SATP air. Furthermore, the NO LIBS spectra are analogous to those measured in plasma torches [19]. In these overview experiments, laser-induced optical breakdown was accomplished by the focusing of the Nd:YAG 1064nm radiation of 3.5 ns pulse width. A Coherent Infinity 40-100 Nd:YAG laser was operated at a frequency of 10 or 100 Hz. Optical breakdown was generated in SATP air and in laboratory cell that

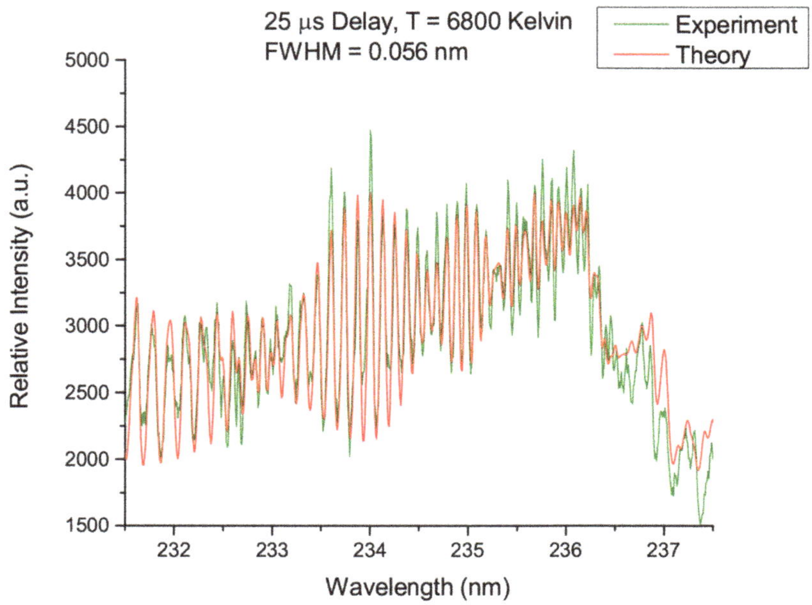

Figure 28.1. Fit to NO laser-induced plasma spectra at 25 μs time delay, $T = 6.8$ kK. Reproduced from [1]. © IOP Publishing Ltd. CC BY 3.0.

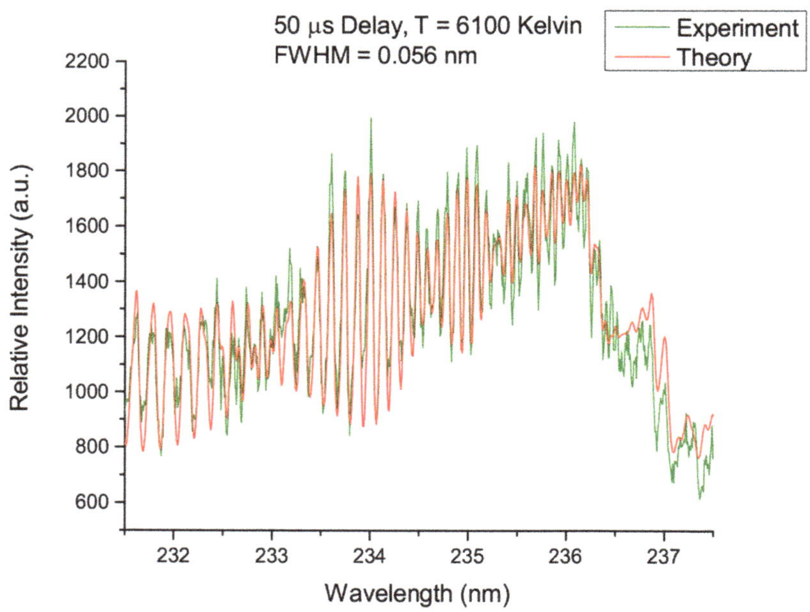

Figure 28.2. Fit to NO laser-induced plasma spectra at 50 μs time delay, $T = 6.1$ kK. Reproduced from [1]. © IOP Publishing Ltd. CC BY 3.0.

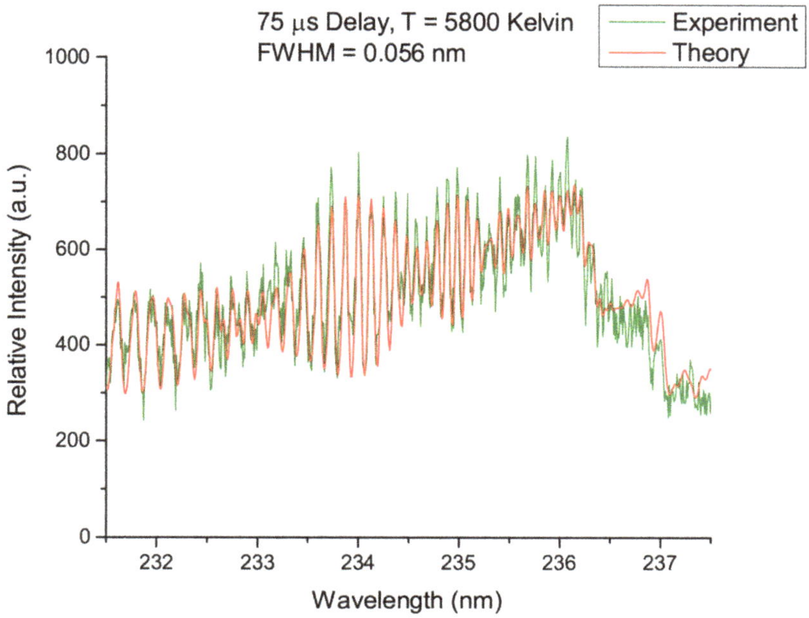

Figure 28.3. Fits to NO laser-induced plasma spectra at 75 μs time delay, $T = 5.8$ kK. Reproduced from [1].

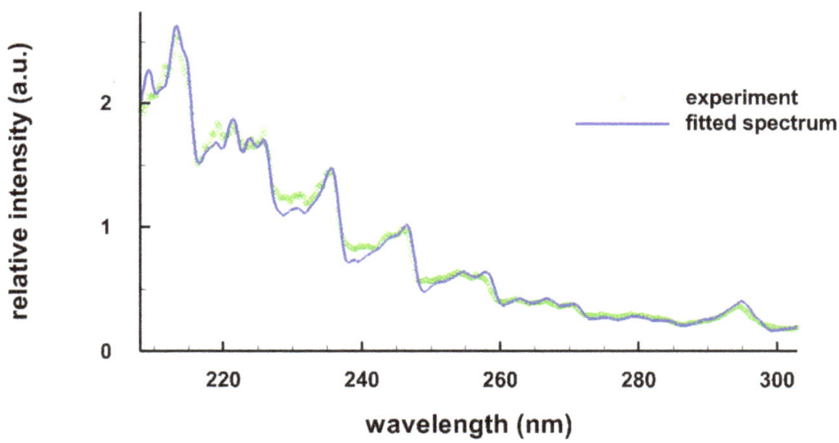

Figure 28.4. NO optical breakdown spectrum, 25 μs time delay, T= 6.7 kK [2].

contained a 1:1 mole ratio nitrogen-oxygen mixture spectrograph. The spectra were resolved with a 0.275 m Jarrel-Ash. Wavelength calibrations were performed with standard light sources, the sensitivity correction was accomplished with a deuterium lamp. The data were detector's dark-noise background subtracted.

Figures 28.4 and 28.5 show the experimental spectra that were recorded at a time delay of 25 μs and 50 μ s, respectively, from the optical breakdown event in air, they also include the fitted low-resolution NO spectra.

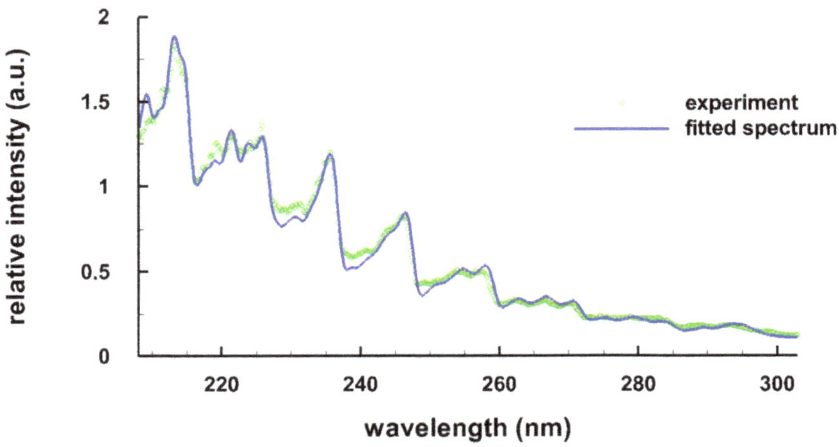

Figure 28.5. NO optical breakdown spectrum, 50 μs time delay, T= 6 kK [2].

Table 28.1. Neutral species number densities [cc^{-1}] for T = 6000 K and T = 6700 K.

T[K]	N	O	Ar	N_2	O_2	NO
6000	2.3×10^{17}	3.9×10^{17}	8.5×10^{15}	6.6×10^{17}	3.3×10^{14}	1.0×10^{16}
6700	4.6×10^{17}	3.1×10^{17}	6.8×10^{15}	3.6×10^{17}	7.8×10^{13}	4.5×10^{15}

Dry air is considered to be a mixture of nitrogen, oxygen, and argon at room temperature, with all other species neglected. Starting from one mole of air at room temperature, the number densities of the species N, O, Ar, N^+, O^+, Ar^+, N_2, N_2^+, O_2, O_2^+, NO, and NO^+, and the free electrons e, can be computed as function of temperature with the chemical equilibrium and applications (CEA) program. The data from CEA serve as input for computation of synthetic spectra with the NEQAIR program, i.e. a Boltzmann distribution with equal translational, rotational, vibrational, and electron temperatures is assumed for the analysis presented here.

Tables 28.1 and 28.2 summarize the temperature and number densities that were utilized as input for the computer code NEQAIR8.

The spectra from the experiments that work were analyzed with the use of the NEQAIR program and the CEA code. Temperatures, T, of 6700 and 6000 K are determined [2, 6, 12] for 25 and 50 μs time delays, respectively. It is important to note, however, that the approximate 1 nm spectral resolution is significantly lower than that presented in figures 28.1 and 28.2. The line-strength method for calculating diatomic spectra provides a viable option for quantifying diatomic spectra. Furthermore, the method is useful for combustion studies that employ laser-induced

Table 28.2. Ionized species and electron densities [cc^{-1}] for $T = 6000$ K and $T = 6700$ K.

T[K]	N^+	O^+	Ar^+	N_2^+	O_2^+	NO^+	e
6000	0	0	0	0	0	2.7×10^{14}	2.7×10^{14}
6700	4.7×10^{13}	4.8×10^{13}	0	0	0	4.4×10^{14}	5.4×10^{14}

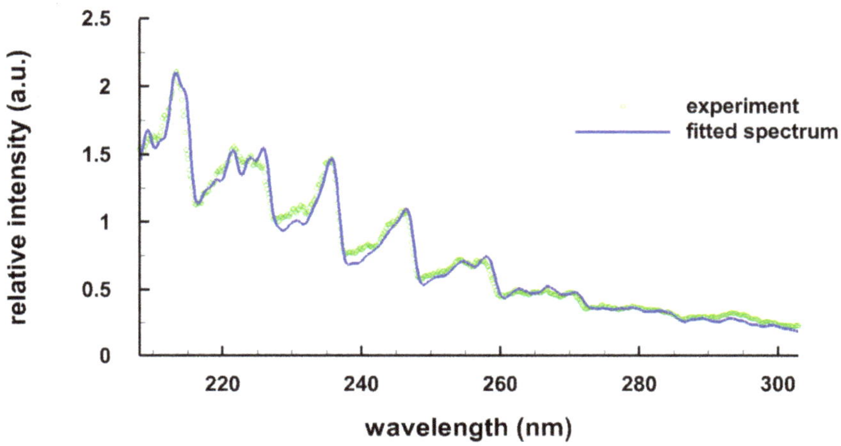

Figure 28.6. N_2:O_2 optical breakdown spectrum, 20 μs time delay.

florescence as a diagnostic method with spectra that often have comb-like structures. In such spectra, the highly resolved nature of a spectra calculated from the line strength is applicable [20]. This is found to be particularly useful in studies of combustion with high temperature flames [21].

The results indicate support for the program NEQAIR for applications in combustion [22–24] and plasma diagnostics. The 'background' contributions appear to be predicted well for selected diatomic transitions, and the synthetic NEQAIR spectra are particularly useful for the analysis of spectral studies with low spectral resolutions. Of course, nonequilibrium computations are of interest, e.g., for different rotational and vibrational temperatures [25], but here we restricted our analysis to equilibrium distributions.

Further measurements investigated NO spectra from optical breakdown in 1:1 mole ratio mixture of $N2$ and $O2$ gas at a pressure slightly above 1 atm, but otherwise keeping the same experimental arrangement as for the SATP air studies. Figures 28.6 and 28.7 show the experimental spectra that were recorded at a time delay of 20 μs and 50 μ s, respectively, from the optical breakdown events in the mixture. Only NO bands are included in the fitting, and O_2^+ spectral data would need to be included for a time delay of 50 μ s.

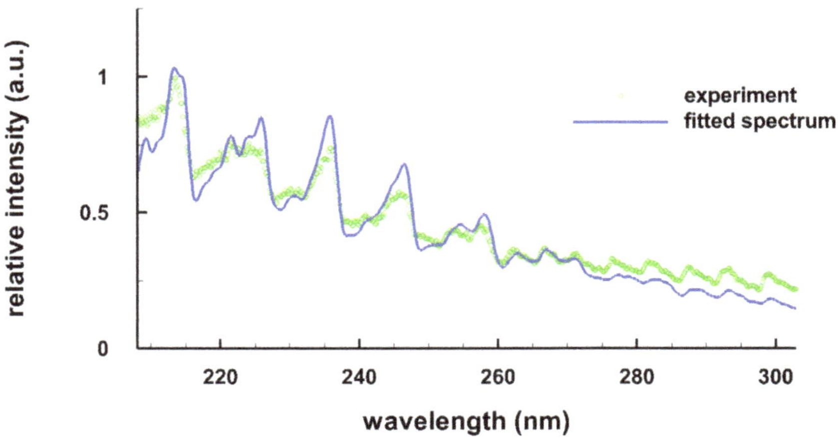

Figure 28.7. N_2:O_2 optical breakdown spectrum, 50 μs time delay. The spectra in the range of 210–260 nm are primarily due to NO, between 260 and 300 nm due to O_2^+.

References

[1] Hornkohl J O, Fleischmann J P, Surmick D M, Witte M J, Swafford L D, Woods A C and Parigger C G 2014 *J. Phys.: Conf. Ser.* **548** 012040

[2] Parigger C G, Lewis J W L, Plemmons D P and Hornkohl J O 1996 Nitric oxide optical breakdown spectra and analysis by the use of the program NEQAIR *Laser Appl. Chem. Bio. Env. Anal.* **3** 85

[3] Hornkohl J O, Parigger C G and Lewis J W L 1991 *J. Quant. Spectrosc. Radiat. Transfer* **46** 405

[4] Parigger C G, Plemmons D H, Hornkohl J O and Lewis J W L 1994 *J. Quant. Spectrosc. Radiat. Transfer* **52** 707

[5] Parigger C G, Plemmons D H, Hornkohl J O and Lewis J W L 1995 *Appl. Opt.* **34** 3331

[6] Park C 1995 *Nonequilibrium Air Radiation (NEQAIR) Program: User's Manual* (Moffet Field, CA: NASA TM 86 707)

[7] Laoux C O 1993 Optical diagnostics and radiative emission of air plasmas *PhD Dissertation* Department of Mechanical Engineering, Stanford University, Stanford, CA

[8] Whiting E E 1995 Private communication

[9] Whiting E E, Park C, Liu Y, Arnold J O and Paterson J A 1996 *NEQAIR96, Nonequilibrium Radiative Transport and Spectra Program: User's Manual* (Moffet Field, CA 94 035: NASA TM 1389)

[10] Boulous M I, Fauchais P and Pfender E 1994 *Thermal Plasmas–Fundamentals and Applications* (New York: Plenum)

[11] Drakes J A, Pruitt D W, Howard R P and Hornkohl J O 1997 *J. Quant. Spectrosc. Radiat. Transfer* **57** 23

[12] Lewis J W L, Parigger C G, Hornkohl J O and Guan G 1999 Laser-induced optical breakdown plasma spectra and analyses with the program NEQAIR *37th Aerospace Sciences Meeting and Exhibit*

[13] Plemmons D H, Parigger C G, Lewis J W L and Hornkohl J O 1998 *Appl. Opt.* **37** 2493

[14] Ochkin V N 2009 *Spectroscopy of Low Temperature Plasma* (Weinheim: Wiley)

[15] Nemes L, Hornkohl J O and Parigger C G 2005 *Appl. Opt.* **44** 3686

[16] Parigger C G, Woods A C, Witte M J, Swafford L D and Surmick D M 2014 *J. Vis. Exp.* **84** 51250
[17] Nelder J A and Mead R 1965 *Comput. J.* **7** 308
[18] Lagarias J C, Reeds J A, Wright M H and Wright P E 1998 *SIAM J. Optim.* **9** 112
[19] Levin D A, Laux C O and Kruger C H 1999 *J. Quant. Spectrosc. Radiat. Transfer* **61** 377
[20] Hammack S D, Carter C D, Gord J R and Lee T 2012 *Appl. Opt.* **51** 8817
[21] Schultz C, Sick V, Heinze J and Stricker W 1997 *Appl. Opt.* **36** 3227
[22] Battles B E and Hanson R K 1995 *J. Quant. Spectrosc. Radiat. Transfer* **54** 521
[23] Harrington J E, Noble A R, Smith G P, Jeffries J B and Crosley D R 1995 Evidence for a new no production mechanism in flames paper # WSS/CI 95
[24] Rumminger M, Heberle N H, Dibble R W, Smith G P, Jeffries J B and Crosley D R 1995 Gas temperature above a porous radiant burner: comparison of measurements and model predictions paper # WSS/CI 95
[25] Labracherie L, Billiotte M and Houas L 1995 *J. Quant. Spectrosc. Radiat. Transfer* **54** 573

IOP Publishing

Quantum Mechanics of the Diatomic Molecule (Second Edition)

Christian G Parigger and James O Hornkohl

Chapter 29

Radial electron density measurements in laser plasma from Abel-inverted hydrogen Balmer beta line profiles

29.1 Introduction

Time-resolved emission spectroscopy is applied to obtain radial electron density values in laser-induced plasma. Hydrogen beta line profiles are recorded following optical breakdown in ultra-high-pure (UHP) hydrogen gas. Asymmetric Abel inversion techniques are utilized in the analysis of data collected for the time delay of 400 ns. The averaged, line-of-sight electron densities are found to be in the range of 1 to 2.5 $\times 10^{17}$ cm^{-3}. The electron densities indicate a factor of 2 variation across the laser-induced plasma.

In various applications of laser-induced plasma, measurements of emission spectra are designed to determine sample composition [1]. Evaluations of plasma characteristics such as electron temperature, T_e, or electron density, N_e, frequently rely on line-of-sight diagnostics. The interest in generating astrophysical conditions in the laboratory for the study of, for example, white dwarf [2, 3] photospheric conditions at $T_e \sim$ 1eV, utilizing hydrogen beta line profiles for measurement of $N_e \sim 1 \times 10^{17}$ cm^{-3} is also noteworthy. For well-established plasma sources and in the study of laser-induced plasma, inversion algorithms can be implemented to fully describe the thermodynamic state and/or species distribution. General approaches make use of the so-called Radon transform and Radon inversion [4–9], especially for situations that require projections along several directions to assemble a comprehensive map of the source. Some of the Radon inverse transform implementations [7–9] are dedicated to medical diagnosis, or are in use for the characterization of nuclear radiation sources, and have been applied successfully in the characterization of laser-induced plasma [10].

In this work, radiation from an Nd:YAG Q-switched device is focused with a single lens to an irradiance of sufficient magnitude for optical breakdown to occur.

doi:10.1088/978-0-7503-6204-7ch29

The spatial distribution of the electron density is of interest, in conjunction with possible simplifications of the inverse algorithm due to symmetries of the experimental arrangement. In an idealized focusing arrangement that describes a clean laser beam and a reasonable low ratio of beam diameter and of focal length, the usual presumptions include a symmetric peak-irradiance distribution together with a diffraction-limited, point-like initiation of plasma. For that reason, the one-parameter Radon transform, or Abel transform, is considered. Strictly following the definition of the Abel transform, lateral symmetry is required for the inversion of the line-of-sight integrated signals. Both self-absorption and Abel inversion have been previously discussed in the context of laser-induced breakdown spectroscopy [11]. However, slight asymmetries are acceptable along the line-of-sight integration. New algorithms [12, 13] and symmetric and asymmetric Abel inversions [14, 15] can be adapted and utilized in the analysis of the Balmer series hydrogen beta data that were recorded using time-resolved laser spectroscopy.

In the set of experiments performed in this work, time-resolved line profiles of the Balmer series hydrogen beta transition, H_β, are recorded for the purpose of determining the variation of the electron density along the line of sight. The laser-induced plasma is generated by guiding the laser beam parallel to the spectrometer slit. For sufficient irradiance, a plasma is generated that extends along the slit. With 1.05:1 plasma imaging and a slit width of 50 μm, a slice of 52.5 μm width of the plasma is dispersed by a holographic grating and recorded with an intensified, two-dimensional array detector comprised of 1024 $\times$1024 pixels. The horizontal dimension of this detector is used to record the spectra, and the vertical dimension is used to spatially resolve the plasma emissions along the slit height. This geometry is selected for investigations of the line-of-sight contributions from the center slice with Abel inverse transforms The spatially resolved spectra along the slit height are utilized to find, first, the radial distribution of the spectra within the imaged slice by applying the Abel inversion of the data for each dispersed wavelength and, second, the electron density distribution from fitting the spatially resolved spectra to known hydrogen beta profiles [16–18].

Shadowgraph and Schlieren measurements of laser-induced plasma in standard ambient temperature and pressure laboratory air show close to spherically symmetric plasma cores for time delays of 100–10 000 ns [19, 20]. The plasma cores reveal internal structure over and above plasma dynamic separation of the shock wave [19, 20]. The laser-induced hydrogen plasma for a time delay of 400 ns also deviates from perfect spherical symmetry. In this work, the contributions along the line of sight for the narrow center slice are modeled to be circularly symmetric, while allowing slight deviation from this symmetry.

The Abel transform in two dimensions describes the integral transform associated with line-of-sight experiments for circularly symmetric plasma. Suppose a lateral distribution of a physical state can only be measured by means of an average and/or line of sight [12] with an integral equation

$$I(z) = 2\int_z^R f(r)\frac{r}{\sqrt{r^2 - z^2}}dr, \tag{29.1}$$

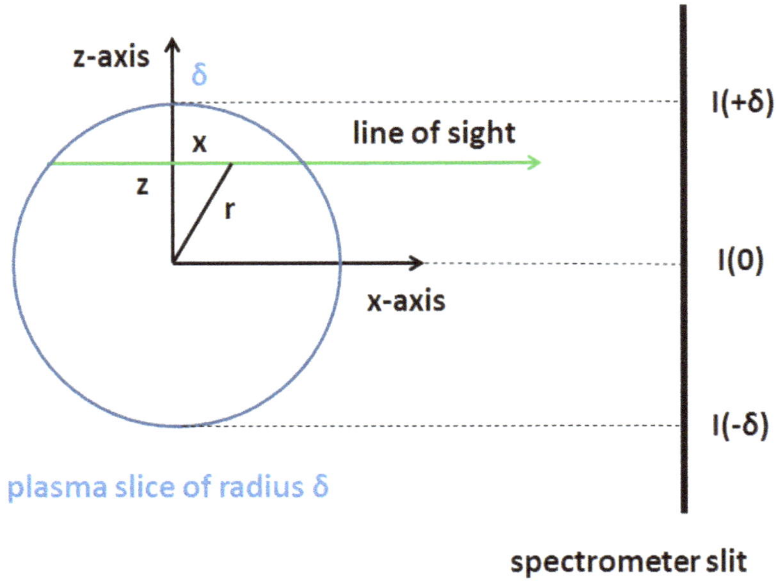

Figure 29.1. Geometry for reconstruction of the line-of-sight measurements with Abel inverse transforms of equation (29.1). Reproduced with permission from [21].

where $I(z)$ and $f(r)$ denote the measured lateral (along the slit height, z) and radial intensity distributions, respectively. The upper integration limit, R, is usually much larger than the circularly symmetric object dimension, δ, or in other words $R \gg \delta$. The factor of 2 indicates top/bottom symmetry. Figure 29.1 elucidates the geometry details for the reconstruction of the recorded time-resolved two-dimensional emission data. The vertically recorded data of the ICCD images correspond to the slit height positions, and the horizontally recorded data correspond to the wavelengths of the spectra.

Equation (29.1) can be inverted using an analytical approach with derivatives, known as the Abel inverse transform. Differentiation, however, is usually difficult to accomplish for spectra. For that reason, a development into a complete set of orthogonal polynomials is designed jointly with a minimization procedure to find the development coefficients. Advantages of this method include [12] computation of the Abel inverse transform without the need for differentiation and without the need for pre-treating the recorded data, leading to inverted intensity values, $I(r)$, with errors of only a few percent for selected test cases [12]. The requirement for circular symmetry can be softened in that top and bottom sections can show differences [15]. One would define a symmetric contribution and also correct for the top to bottom asymmetry. This is accomplished by introducing a symmetrized profile, $I_0(z)$, following the identical analysis procedure discussed by Blades, [15]

$$I_0(z) = [I(z) + I(-z)]/2, \tag{29.2}$$

with the corresponding asymmetric portions

$$G(\pm\ z) = I(\pm\ z)/I_0(z). \tag{29.3}$$

Subsequent to finding the Abel inverse transform of $I_0(z)$ to find $I_0(r)$, the asymmetric profiles can be computed from

$$I(r, \pm\ z) = G(\pm\ z)I_0(r). \tag{29.4}$$

This approach has been applied especially for extracting subtle differences that are of interest in rigorous analysis.

With a two-dimensional array detector coupled to a Czerny–Turner spectrometer, lateral data can be recorded without the need of moving the spectrometer. The plasma core deviates slightly from spherical symmetry for the investigated time delays, and therefore measurements at different angles or for different slices would be required for complete reconstruction of the three-dimensional plasma distribution at a specific time delay. In this work, results are presented for reconstruction from line-of-sight measurements for the near circularly symmetric center slice of width corresponding to the slit width for close to 1:1 imaging.

For the computation of the Abel inverse transform, one can employ the projection-slice theorem that can be discussed with the Fourier–Hankel–Abel transform cycle, or one can rely on a complete series of functions [12] mentioned above. For the analysis of the accumulated spectra and without further application of digital filters, selected Chebyshev series composed of first kind Chebyshev polynomials yield the variation of the emissivity with wavelength, subsequently the electron density can be determined from the laterally resolved hydrogen beta lines. The specific orthogonal, complete series minimizes the error excursions, [22] as anticipated from the use of Chebyshev polynomials. The applied method is based on the development of the Abel inverse transform in terms of a series that includes precalculated transforms. In the process of least-square fitting the recorded data to these transforms, one specifies the number of terms to be included in the series. The number of terms that are included in the series solution affects the accuracy and noise. The more terms that are considered, the higher the noise for the wavelength-dependent emissivity. Most importantly, the particular series solution approach that shows discrepancies in the recovered function of the order of a few percent has been formalized in a Matlab® script [13] that alleviates the implementation of the Abel inverse transform.

29.2 Experimental details

For the experimental studies, laser-induced optical breakdown was initiated in UHP hydrogen gas of 99.999% purity contained in a cell at a pressure of 1.08×10^5 Pa (810 Torr). This particular hydrogen pressure was also selected in a previously communicated detailed study of hydrogen beta line profiles following laser-induced breakdown [23]. Prior to filling the cell with hydrogen, evacuation to pressures of the order of 10^{-4} Pa (10^{-6} Torr) was accomplished to obtain a vacuum that is typically in use for preparing cells for discharges. A Q-switched Nd:YAG laser (DCR-2A(10) PS; Quanta Ray) was operated at the fundamental wavelength of 1064nm and was focused with an f-number of 10 (f/10) using a 125 mm focal length lens through the Quartz windows, with the laser beam entering the cell from the top or parallel to the slit.

For the imaging of the plasma, a single lens was positioned for 1.05:1 imaging with an effective focusing speed of 5.6 to match the f-number of the 0.64 m Jobin–Yvon Czerny–Turner spectrometer. A 1200 groves/mm holographic grating was used in conjunction with an intensified charge-coupled device (ICCD, Andor iStar) that resulted in a spectral resolution of 0.1 nm. The ICCD shows a pixel resolution of 1024 ×1024, with the horizontal 1024 pixels corresponding to the wavelength. With 8 pixel binning, the vertical 128 regions correspond to the location in the plasma column. The camera software was utilized to record signals from 50 consecutive events, to background correct the spectra, and to invert the vertical dimension such that top of the image is identical with the top of the plasma (toward the focusing lens of the Nd:YAG beam). The wavelength and relative intensity calibrations were accomplished with standard light sources and with a calibrated deuterium/tungsten lamp.

The region of interest spanned 5.5 mm for the plasma initiated with 120 mJ, 13 ns pulses, and for a time delay of 400 ns. This particular time delay from optical breakdown results in practically identical results for line-of-sight averaged electron densities [23] when using the Griem and the Oks theory. Moreover, the Balmer series hydrogen beta line can be nicely captured by the spectrometer/detector arrangement. In terms of the 8 pixel binning, 50 spectra of 0.11mm height were recorded in this study of the hydrogen beta line profiles. The arrangement is similar to a recently reported experimental investigation of laser-induced plasma in air [24].

Due to residual moisture in standard ambient temperature and pressure laboratory air, hydrogen can be relatively easily detected but for time delays of the order of 10 × longer from optical breakdown. The presence of oxygen and nitrogen in laboratory air causes significantly more free electrons during the optical breakdown process, and consequently the hydrogen beta emissions can be easily demarcated for time delays of typically 5 μs. In air, nitrogen lines on the red side of the H_β line can affect the appearance of the free-electron background. In turn, use of UHP hydrogen allows one to study line profiles in detail.

It was also established previously that self-absorption of the hydrogen beta line is insignificant [24] for electron densities of the order of 1×10^{17} cm^{-3}. The self-absorption assessment included the use of an additional lens and a mirror at the opposite side of the spectrometer, and by retro-reflection through the plasma, the effect of self-absorption could be quantified. Of course, self-absorption would not only cause a reduction of the measured peak-heights but it would also modify and/or distort the line shapes. The differences of the full width at half maximum (FWHM) and of the hydrogen peak separations are well within the estimated error bars associated with electron density determination from the H_β line. These error bars amount to typically 5–10% for electron densities on the order of $0.5–3 \times 10^{17}$ cm^{-3}. For comparison, self-absorption effects of the hydrogen alpha line [25, 26] would come into play for electron densities, N_e, of the order 20×10^{17} cm^{-3}—for such N_e values, the Balmer series hydrogen beta line cannot be seen in emission. In recent work on symmetric hydrogen Balmer series lines [27], and especially for hydrogen beta [28], both FWHM and peak separation can serve as means to determine the electron density.

29.3 Results

The determination of the electron densities from the background, wavelength, and detector sensitivity-calibrated data is categorized into two distinct approaches: (i) find N_e directly from the recorded line-of-sight set of data that are resolved vertically; and (ii) apply the Abel inverse transform first to obtain the radial, spectrally resolved emission spectra, and then infer the radial N_e variation.

29.3.1 Spatially resolved line-of-sight spectra

Figure 29.2 illustrates the recorded hydrogen beta line profiles for the time delay of 400 ns from the generation of optical breakdown in UHP hydrogen gas. This figure also indicates the fitted, asymmetric hydrogen lines together with the inferred electron densities at a slit height of 1.96 mm. The overall fits to the recorded spectra include the background contributions that are determined simultaneously with the best-fit, least-square hydrogen beta profiles.

The values for the electron density, FWHM, peak separation, and ratio of FWHM are systematically determined by engaging a full-profile fit of the

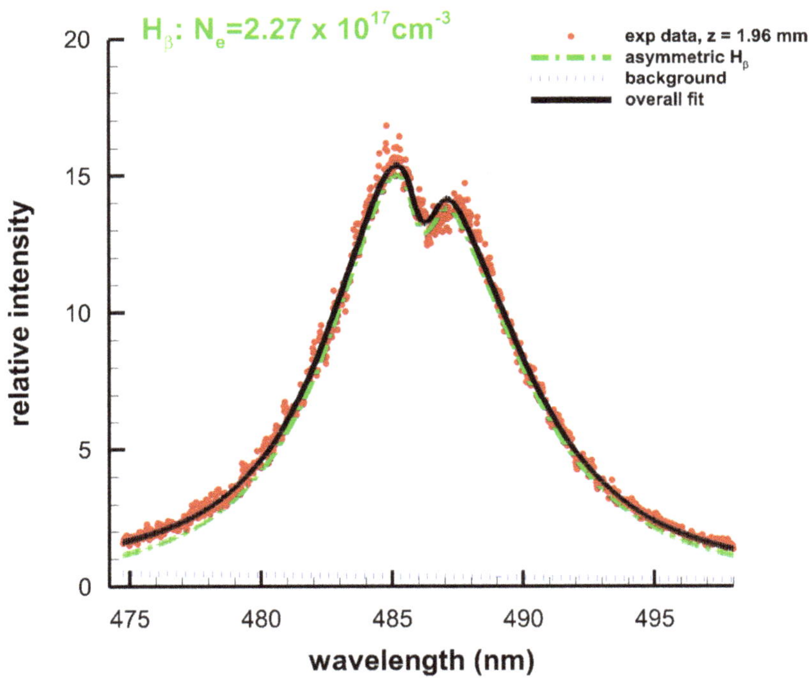

Figure 29.2. Line-of-sight measured and fitted hydrogen beta line profiles for a time delay of τ = 400 ns. Reproduced with permission from [21].

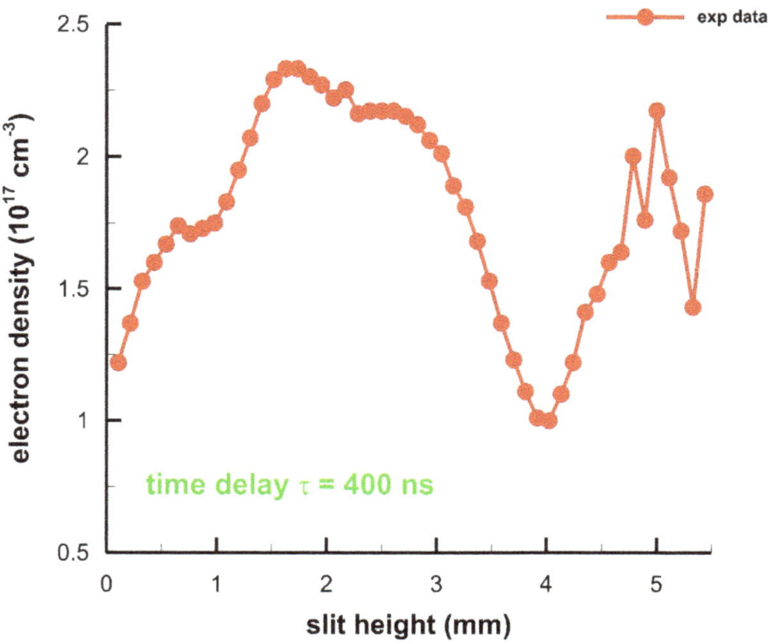

Figure 29.3. Electron densities versus slit height, z, inferred from full-profile fitting of the line-of-sight measured H_β line for a time delay of $\tau = 400$ ns. Reproduced with permission from [21].

least-square kind, including allowance and fitting for a constant background. This constant background shows a very slight increase toward the blue side of the recorded wavelength range. Figure 29.3 illustrates the inferred electron density *versus* slit height. Table 29.1 lists the obtained results for N_e.

The electron densities in figure 29.3 are indicated with reference to the slit height that is measured from bottom to top. The propagation direction of the laser beam that induces optical breakdown is from top to bottom, or in the figure from right-hand to left-hand. The initial peak is associated with the emanating shock wave, followed by a region of diminished N_e and then a slightly increasing but relatively constant electron density prior to decreasing values of N_e for the laser plasma in the forward direction. The electron densities displayed in figure 29.3 are obtained from the line-of-sight data. The values of the electron densities were deduced from full-profile fitting using communicated line profiles [16]. The listed FWHM is determined from the fitted, asymmetric profiles by taking the mean of the two hydrogen beta peaks I_{red} and I_{blue},

$$I_{\text{mean}} = (I_{\text{red}} + I_{\text{blue}})/2, \tag{29.5}$$

and then finding the FWHM from $I_{\text{mean}}/2$.

A recently published formula for the determination of the electron density from the FWHM is applied to provide an alternate value for N_e that includes estimated temperature values, with N_e (cm^{-3}), and the FWHM, $\Delta\lambda$ (nm),

Table 29.1. Slit height, z, H_β electron density, and FWHM from full-profile fitting.

z (mm)	N_e ($10^{17}cm^{-3}$)	FWHM (nm)	N_e^{FWHM} ($10^{17}cm^{-3}$)	r_{N_e}
0.11	1.22	5.10	1.19	1.02
0.22	1.37	5.47	1.32	1.04
0.33	1.53	5.93	1.48	1.04
0.44	1.60	6.16	1.56	1.03
0.54	1.67	6.39	1.64	1.02
0.65	1.74	6.63	1.73	1.01
0.76	1.71	6.53	1.69	1.01
0.87	1.73	6.60	1.71	1.01
0.98	1.75	6.67	1.74	1.01
1.09	1.83	6.95	1.84	0.99
1.20	1.95	7.37	2.00	0.97
1.31	2.07	7.79	2.17	0.96
1.41	2.20	8.26	2.35	0.94
1.52	2.29	8.58	2.48	0.92
1.63	2.33	8.72	2.54	0.92
1.74	2.33	8.72	2.54	0.92
1.85	2.30	8.62	2.50	0.92
1.96	2.27	8.51	2.45	0.93
2.07	2.22	8.33	2.38	0.93
2.18	2.25	8.44	2.42	0.93
2.28	2.16	8.11	2.29	0.94
2.39	2.17	8.15	2.31	0.94
2.50	2.17	8.15	2.31	0.94
2.61	2.17	8.15	2.31	0.94
2.72	2.15	8.08	2.28	0.94
2.83	2.12	7.97	2.24	0.95
2.94	2.06	7.76	2.15	0.96
3.05	2.01	7.58	2.08	0.96
3.16	1.89	7.16	1.92	0.98
3.26	1.81	6.88	1.82	1.00
3.37	1.68	6.43	1.65	1.02
3.48	1.53	5.93	1.48	1.04
3.59	1.37	5.47	1.32	1.04
3.70	1.23	5.12	1.20	1.03
3.81	1.11	4.86	1.12	1.00
3.92	1.01	4.67	1.05	0.96
4.03	1.00	4.66	1.05	0.95
4.13	1.10	4.84	1.11	0.99
4.24	1.22	5.10	1.19	1.02
4.35	1.41	5.58	1.35	1.04
4.46	1.48	5.78	1.42	1.04
4.57	1.60	6.16	1.56	1.03

4.68	1.64	6.29	1.60	1.02
4.79	2.00	7.55	2.07	0.97
4.90	1.76	6.70	1.75	1.00
5.00	2.17	8.15	2.31	0.94
5.11	1.92	7.26	1.96	0.98
5.22	1.72	6.56	1.70	1.01
5.33	1.43	5.64	1.37	1.04
5.44	1.86	7.05	1.88	0.99

Included are results from the electron density, N_e^{FWHM}, obtained from equation (29.6), and the ratio $r_{N_e} = N_e/N_e^{\text{FWHM}}$ [21].

$$\Delta\lambda(H_\beta) = 4.50\left(\frac{N_e}{10^{17}}\right)^{0.71\pm0.03}. \tag{29.6}$$

This formula was derived from tabulated values for the electron densities [23, 29]. The electron density results that were computed using equation (29.6), N_e^{FWHM}, are also indicated in the table. One can see that the electron densities obtained from equation (29.6) agree with the full-profile fits from theory profiles in the temperature range of 10 000–20 000 K within 8%, or in practical terms within the accuracy of extracting the FWHM from moderately noisy hydrogen beta data. The average of the electron density inferred from the FWHM amounts to 1.83×10^{17} cm^{-3}, which is in within 5% of the results from a previous study of line-of-sight averaged spectra for the selected pressure and time delay [23]. Figure 29.4 illustrates the recorded image data corresponding to the results in table 29.1 using pseudo-colored contours for the slit height magnitudes of 0.11–5.44 mm and for the spectral range of 474.4–497.8 nm. The inferred electron densities for the 4.0–5.4 mm region are included in the table; however, due to the signal-to-noise ratio in this region, the N_e values show slightly larger error margins than for the 0.11–4.0 mm region. The color palette on the right-hand side indicates the range of intensities. The double-peak spectra of the hydrogen beta line are clearly discernible at slit heights near 1.5 mm.

The best-fit profiles of the line-of-sight data clearly indicate visually excellent results for the time delay of 400 ns, and equally would show excellent results for time delays of 650 ns. The Abel inverse transforms of the 400 ns data show less noise than the 650 ns data, in part due to the typically a factor of 1.5 wider spectral profiles for the recorded data at a time delay of 400 ns. Therefore, the Abel inverse transform of the data for a delay time of 400 ns are further elaborated in detail. For a time delay of 150 ns, the electron density is determined from partial hydrogen beta profiles. An electron density of 7.5×10^{17} cm^{-3} corresponds to a FWHM of 18.8 nm as calculated using equation (29.6); therefore, a wavelength range of 1.28 times this width is covered by the 24 nm spectral window of the spectrometer-detector system for the experiments reported here.

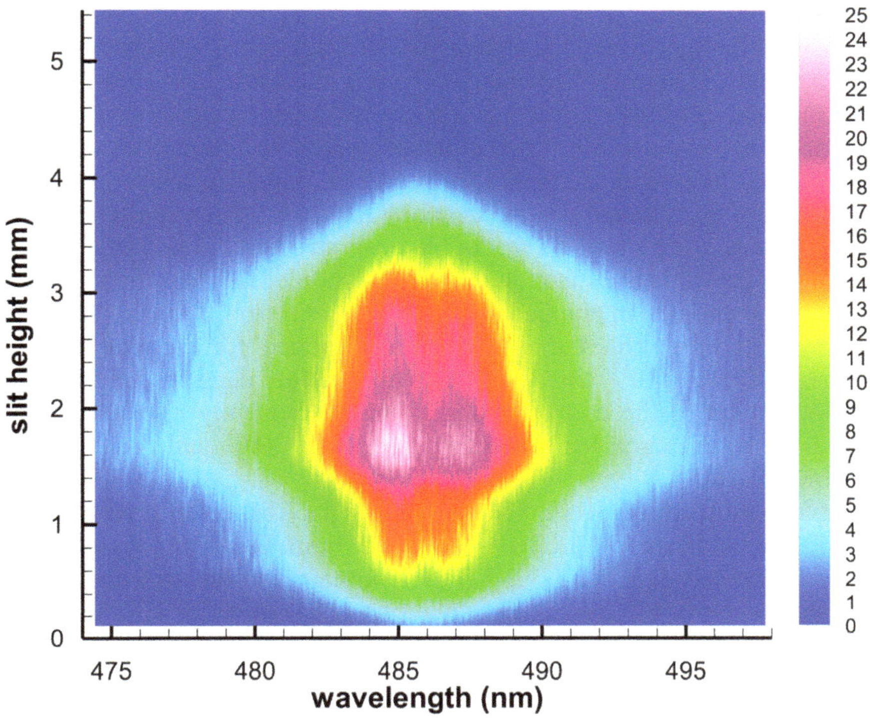

Figure 29.4. Contour image of the recorded hydrogen beta line for a time delay of $\tau = 400$ ns. Reproduced with permission from [21].

29.3.2 Abel-inverted spectra

The radial electron density variation is of interest, for instance, in the evaluation of whether an equilibrium distribution is indeed reasonable. The determination of electron density of course supposes optically thin scenarios. In previous communications, the hydrogen beta line profile was discussed to be optically thin for time delays of the order of 5 μs in air breakdown and for densities of the order of 1×10^{17} cm^{-3}. The 400 ns time delay hydrogen beta data were recorded from optical breakdown in UHP hydrogen gas, the corresponding time delays are of the order of 1/10 of the time delays used for air breakdown. However, self-absorption is not considered to be of significance when evaluating the error bars associated with electron density determination. A method for evaluating whether self-absorption occurs has been recently communicated by means of comparing the electron densities obtained from the peak separation with the ones obtained from the FWHM. In table 29.1, a detailed list of FWHM values is included, yet it is emphasized that the H_β profiles were fitted using full profiles; subsequently, the widths were extracted from these fitted profiles.

The Abel inversion is accomplished by utilizing a polynomial expansion [12, 13] combined with a recently made available scripts (see appendix K) for the software package Matlab®. The spectra that were recorded vertically were utilized for the Abel inversion. For each wavelength, values along the slit-axis are utilized to derive the radial profile, or the wavelength emissivity is resolved radially. In the application of the Abel inversion, slight asymmetries [15] for the bottom and top half are also included. One of the advantages of using this particular implementation is that special filtering of the recorded data is not required. However, it is noted that a reasonable signal-to-noise ratio of the dataset would be required to obtain visual agreement with the typical asymmetric appearance of hydrogen beta line profiles in laser-induced optical breakdown spectroscopy for electron densities larger than 0.1 to 0.5 $\times 10^{17}$ cm^{-3}. Of course, the double-peak structure is also of fundamental nature, and in principle is a function of the resolution of the instrument. For Stark-broadened line profiles in applied LIBS studies, electron densities from double peaked hydrogen beta lines are typically determined for electron densities in the range of 0.5–5 $\times 10^{17}$ cm^{-3}. For higher electron densities in the range of 0.1–100 $\times 10^{17}$ cm^{-3}, the hydrogen alpha line is frequently also used, although care needs to be exercised for values higher than 20 $\times 10^{17}$ cm^{-3} due to H_α self-absorption [25, 26].

Figure 29.5 illustrates the results from applying the Abel inversion algorithm. The symmetric electron density profile is obtained from the symmetrized measured lateral data $I_0(z)$ (see equation (29.2)). The asymmetric profile is reconstructed with the asymmetry portions $G(\pm z)$ (see equations (29.3) and (29.4)) and a close to symmetric profile is obtained. In figure 29.5, negative values for r are used to conveniently display the differences of symmetric and of asymmetric results. The radial position $r = 0$ denotes the center of the plasma slice and it corresponds to the

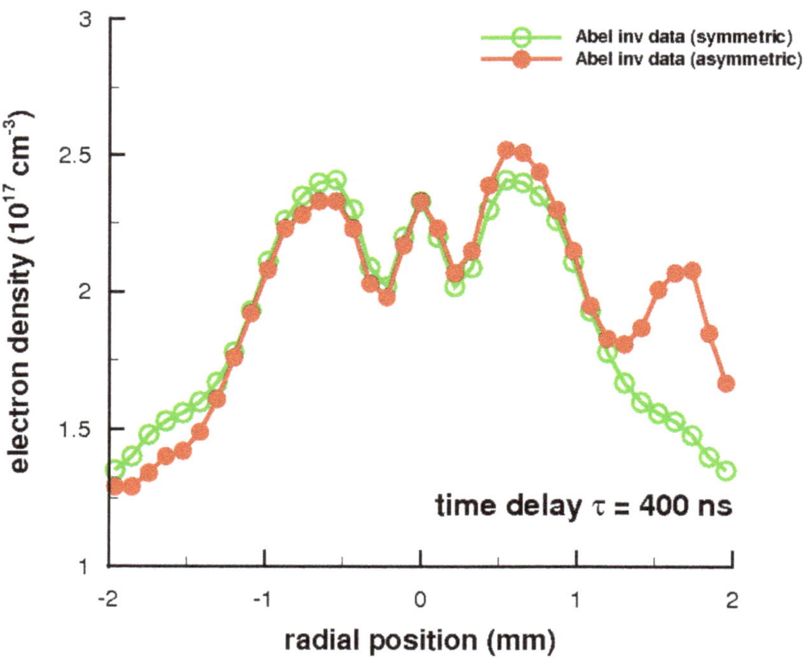

Figure 29.5. Electron densities versus radial position, inferred from Abel inversion of the recorded H_β line profiles for a time delay of $\tau = 400$ ns. Reproduced with permission from [21].

lateral slit height $z = 2.18$ mm. Abel inversions are computed for the lateral slit range $z_{low} = 0.22$ mm to $z_{high} = 4.14$ mm, corresponding to the circularly symmetric plasma slice of radius $\delta = 1.96$ mm for the 400 ns time delay data (see figure 29.4).

At the center, there is a peak in electron density, flanked by maxima, and including a clear peak near $r = 1.6$ mm. The maxima near the $r = \pm 0.54$ mm positions are associated with the breakdown kernel or the very center (viz. epicenter) of the generation of laser-induced breakdown. In view of the circular symmetry, the peaks at $r = \pm 0.54$ mm indicate a ring structure for electron density in the plasma slice. The extra peak near $r = 1.6$ mm denotes an increased electron density, possibly due to expanding shock wave phenomena and/or due to the specific focusing characteristics of the Nd:YAG beam shape that was used. There is an ever so slight increase in asymmetric electron density near $r = -1.6$ mm adjacent to the obvious increase near the $r = 1.6$ mm position. The beam profile resembles a so-called doughnut mode as a result of the specific optical arrangement of the Quanta Ray Nd:YAG laser device that further shows a beam profile with a preferred intensity distribution on one side, in part leading to the asymmetry in the electron density. Other reasons for the occurrence of an asymmetry can be the variations of optical breakdown processes during the capture of an average of 50 consecutive laser-induced plasma events during the experiments. For close to perfect Nd:YAG laser beams, such as the ones used for high-power particle-imaging-velocimetry systems, where close to Gaussian TEM_{00} mode structures are available in conjunction with low M^2 values, the peaks near the positions $r = \pm 1.6$ mm would be expected to be similar in height. In other words, in future experiments that utilize close to TEM_{00} Nd:YAG laser radiation, the measured electron density profiles would be expected to be close to symmetric with respect to the center position. For the reconstruction of the three-dimensional plasma distribution, measurement of more lateral slices or measurements at different angles would be needed because of the slightly asymmetric energy deposition in the generation of laser-induced plasma. The asymmetric energy deposition causes fluid dynamics phenomena, including the development of vortices that have been measured for tens to hundreds of microseconds after air breakdown.

The Abel inverse transforms for spectral emissions are further discussed. Usually, data needs to be filtered in a reasonable manner prior to engaging Abel inversion algorithms. However, such filtering is not applied, yet an average of 50 consecutive laser-plasma events were collected to improve the signal-to-noise ratio. For example, the 400 ns time delay spectra for the radial positions of $r = -0.54$ mm and $r = +0.54$ mm are easily recognizable as Balmer series hydrogen beta lines. Figures 29.6 and 29.7 illustrate the Abel-inverted emission and fitted hydrogen beta line profiles.

The electron density is ~10% lower for $r = -0.54$ mm than for $r = +0.54$ mm, indicating a circular asymmetry of ~10%. Near the edges of the plasma, smaller signals will cause more noise for the Abel inverse transform. Equally, the choice of extra terms in the Chebyshev series would increase the precision, but would also increase unnecessary details or noise that would pose difficulties in the determination

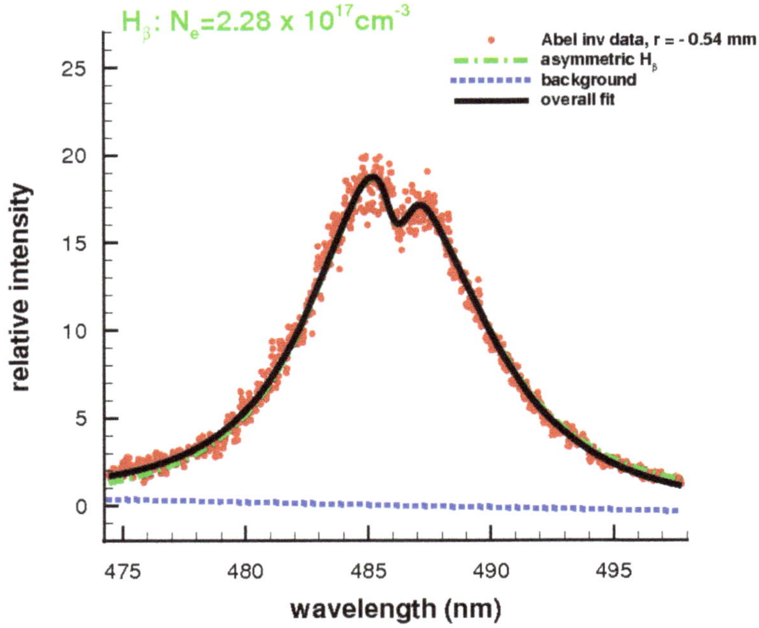

Figure 29.6. The electron density of 2.28×10^{17} cm^{-3} at the position $r = -0.54$ mm is inferred from the Abel-inverted data for the time delay of $\tau = 400$ ns. Reproduced with permission from [21].

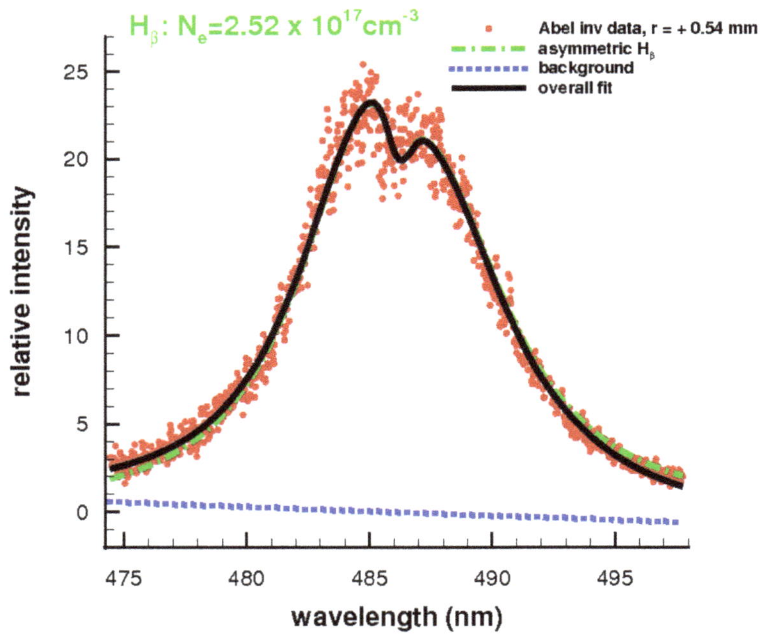

Figure 29.7. The electron density of 2.52×10^{17} cm^{-3} at the position $r = +0.54$ mm is inferred from the Abel-inverted data for the time delay of $\tau = 400$ ns. Reproduced with permission from [21].

of the electron density values. For that reason, the number of terms, N, selected for this Abel inversion analysis amounted to N = 10.

The Abel inversion is also applied for the time delays of 150 ns and 650 ns. Figure 29.8 illustrates the results. For the early 150 ns time delay data, the profiles are nearly symmetric in the center region, except for higher electron density values near $r = 1.62$ mm that are likely due to the specific beam shape of the Nd:YAG radiation. However, there is also a slight peak near $r = -1.62$ mm, which would further indicate phenomena associated with the plasma expansion following laser-induced breakdown.

The center portion reveals a diminished electron density with peaks on either side of center, which would indicate near circularly symmetric electron densities of the plasma kernel inside the volume bounded by the expanding shock that can be recorded with shadowgraph and/or Schlieren imaging. In view of figure 29.5, the electron density near line center reduces with increasing time delay, while sustaining maxima near the $r = \pm 0.54$ mm positions. This is also seen in the expanded electron density values (use the right-hand vertical axis) obtained from the 650 ns data although there are only slight indications of local maxima. One may infer from the indicated results for the electron densities that a standing electron wave pattern is created. Averages from 50 consecutive laser breakdown events and slight variations from event-to-event can easily wash out this pattern for longer time delays.

Furthermore, for the higher electron density of nearly 10×10^{17} cm^{-3} (N_e-axis on the left-hand side) increased errors occur in the N_e determination. The distribution of

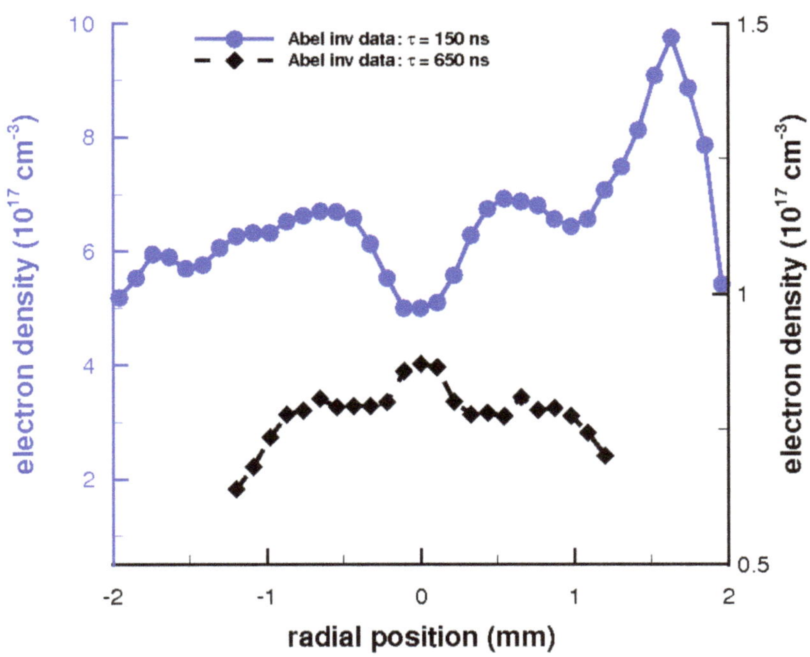

Figure 29.8. Electron densities versus radial position for 150 ns time delay (N_e-axis on the left-hand side) and for 650 ns time delay (N_e-axis on the right-hand side). Reproduced with permission from [21].

the electron density at line center reveals smaller electron density near $r = 0$ mm early in the plasma decay, with higher electron density away from center. While the determination of the decrease in electron density near center shows larger errors due to a value of the order of 5×10^{17} cm^{-3}, a diminished electron density at $r = 0$ mm position is inferred. In future experiments, the variation of the electron density for values larger than 3–5 $\times 10^{17}$ cm^{-3} should be investigated using Abel inverse transforms of the Balmer series hydrogen alpha line to further elucidate the electron densities in the center of the laser-induced plasma. In turn, the 650 ns data indicate a peak near the center position, yet the 650 ns Abel-inverted data lead to electron densities of the order of $\sim 1 \times 10^{17}$ cm^{-3} (N_e-axis on the right-hand side). For the 650 ns time delay data, only data recorded in the lateral range of $z = 1.1$ mm to $z = +3.3$ mm were considered for Abel inversion due to smaller line-profile FWHMs than for the 400 ns time delay. Furthermore, a decrease in the signal-to-noise ratio occurred for the Abel inverse transforms of the recorded data at the 650 ns time delay from optical breakdown.

29.4 Discussion

The determination of the electron density distribution across the plasma can be accomplished by use of an asymmetric Abel inversion. Filtering algorithms were not needed due to the choice of series solution that has been extensively tested to work without further preconditioning the wavelength and sensitivity corrected data. The results indicate radial variations that are of the order of a factor of ~ 2 early in the plasma decay, with a diminishing difference for later time delays from optical breakdown. Similar results are expected to be found following inversion of time-resolved hydrogen alpha spectra; however, this is of continued interest. The insights gained from the application of the Abel inverse transform include the extraction of spatially resolved spectra along the line of sight of a slice of the expanding plasma kernel. The spatial variation of the electron density can be associated with the plasma expansion following laser-induced optical breakdown.

References

[1] Miziolek A W, Palleschi V and Schechter I (ed) 2006 *Laser-Induced Breakdown Spectroscopy (LIBS)—Fundamentals and Applications* (New York: Cambridge University Press)

[2] Falcon R E 2012 *Cornell University Library, Astro-Ph* (Nash: Cornell University) http://arxiv.org/abs/1210.7197v1

[3] Falcon R E, Rochau G A, Bailey J E, Gomez T A, Montgomery M H, Winget D E and Nagayama T 2015 *Astrophys. J.* **806** 214

[4] Radon J 1917 *Ber. Sächs. Akad. Wissenschaft. Leipzig Math. Phys. Kl* **69** 262

[5] Radon J 1986 **5** 170

[6] Kunze H-J 2009 *Introduction to Plasma Spectroscopy* (Heidelberg: Springer)

[7] Cormack A M 1963 *J. Appl. Phys.* **34** 2722

[8] Cormack A M 1964 *J. Appl. Phys.* **35** 2908

[9] Deans S R 1983 *The radon transform and some of its applications*

[10] Merk S, Demidov A, Shelby D, Gornushkin I B, Panne U, Smith B W and Omenetto N 2013 *Appl. Spectrosc.* **67** 851
[11] Burger M, Skočić M and Bukvić S 2014 *Acta Part B.: At. Spectrosc.* **101** 51
[12] Pretzler G 1991 *Z. Naturforsch* **46a** 639
[13] Killer C 2014 Abel inversion algorithm http://www.mathworks.com/matlabcentral/fileexchange/43639-abel-inversion-algorithm
[14] Blades M W and Horlick G 1980 *Appl. Spectrosc.* **34** 696
[15] Blades M W 1983 *Appl. Spectrosc.* **37** 371
[16] Djurović S, Ćirišan M, Demura A V, Demchenko G V, Nikolić D, Gigosos M A and González M Á 2009 *Phys. Rev.* E **79** 046402
[17] Palomares J M, Torres J, Gigosos M A, van der Mullen J J A M, Gamero A and Sola A 2009 *Appl. Spectrosc.* **63** 1023
[18] Palomares J M, Torres J, Gigosos M A, van der Mullen J J A M, Gamero A and Sola A 2010 *J. Phys.: Conf. Ser.* **207** 012013
[19] Sobral H, Villagrán-Muniz M, Navarro-Gonzaález R and Raga A C 2000 *Appl. Phys. Lett.* **77** 3158
[20] Thiyagarajan M and Scharer J E 2008 *IEEE Trans. Plasma Sci.* **36** 2512
[21] Parigger C G, Gautam G and Surmick D M 2015 *Int. Rev. At. Mol. Phys.* **6** 43
[22] Arfken G B, Weber H J and Harris F E 2012 *Mathematical Methods for Physicists, A comprehensive Guide* seventh edition (New York: Academic)
[23] Oks E, Parigger C G and Plemmons D H 2003 *Appl. Opt.* **42** 5992
[24] Gautam G, Parigger C G, Surmick D M and Sherbini A M E L 2016 *J. Quant. Spect. Radiat. Transf.* **170** 189
[25] Parigger C G, Gautam G, Woods A C, Surmick D M and Hornkohl J O 2014 *Trends Appl. Spectrosc.* **11** 1
[26] Parigger C G, Woods A C, Surmick D M, Gautam G, Witte M J and Hornkohl J O 2015 *Spectrochim. Acta* B **107** 132
[27] Konjević N, Ivković M and Sakan N 2012 *Spectrochim. Acta* B **76** 16
[28] Ivković M, Konjević N and Pavlović Z 2015 *J. Quant. Spect. Radiat. Transf.* **154** 1
[29] Surmick D M and Parigger C G 2015 *Int. Rev. At. Mol. Phys.* **5** 71

Christian G Parigger and James O Hornkohl

Chapter 30

Hypersonic imaging and emission spectroscopy of hydrogen and cyanide following laser-induced optical breakdown

30.1 Introduction

This chapter communicates the connection of measured shadowgraphs from optically induced air breakdown with emission spectroscopy in selected gas mixtures. Laser-induced optical breakdown (LIBS) is generated using 850 mJ and 170 mJ, 6 ns pulses at a wavelength of 1064nm, the shadowgraphs are recorded using time-delayed 5 ns pulses at a wavelength of 532 nm and a digital camera, and emission spectra are recorded for typically a dozen of discrete time delays from optical breakdown by employing an intensified charge-coupled device. The symmetry of the breakdown event can be viewed as close-to spherical symmetry for time delays of several 100 ns. Spectroscopic analysis explores well-above hypersonic expansion dynamics using primarily the diatomic molecule cyanide (CN) and atomic hydrogen emission spectroscopy. Analysis of the air breakdown and selected gas breakdown events permits the use of Abel inversion for inference of the expanding species distribution. Typically, species are prevalent at higher density near the hypersonically expanding shock wave, measured by tracing CN and a specific carbon atomic line. Overall, recorded air breakdown shadowgraphs are indicative of laser-plasma expansion in selected gas mixtures, and optical spectroscopy delivers analytical insight into plasma expansion phenomena.

Laser-plasma research is experiencing remarkable interest in laser-induced optical breakdown (LIBS) [1], in part due to success in analytical chemistry, in a volley of engineering applications, or in dedicated diagnosis that may extend to technology-driven changes in the medical field. This work is concerned with experiments and analysis of phenomena associated with pulsed, nanosecond radiation. Optical breakdown is accomplished by focusing a laser beam to irradiances above threshold for local lightning or transient plasma generation in gaseous media. For plasma

doi:10.1088/978-0-7503-6204-7ch30

generation with focused nanosecond laser pulses, the initial portion of the laser pulse energy generates optical breakdown and the remaining portion interacts with the evolving plasma. Micro-plasma imaging is of general interest in the laser-induced breakdown experiments, this includes application of LIBS diagnostic in gases and near liquid or solid surfaces. The analysis and interpretation of observed expansion dynamics can be significantly alleviated when including symmetry considerations.

Optical radiation that is well-above optical breakdown threshold in air, hydrogen, and molar 1:1 CO_2:N_2 molar mixture at or near ambient laboratory conditions causes multiple breakdown spots in focus. One may associate these spots with peaks in computational maps of the focal area, especially when considering a single lens and spherical aberration [2]. As one reduces the irradiance to the threshold values for the gases of interest, the number of separate breakdown spots diminishes down to one. However, there is always a desire to obtain more diagnostic signal, and reasonably repeatable signals that favor use of radiation that is perhaps of the order of one magnitude, or more, above threshold. This chapter aims to investigate the occurrence of reasonable spherical symmetry in gas breakdown dynamics, and subsequently to apply Abel-inversion techniques for diagnostics of the expanding plasma. Of interest in this work is optical breakdown in air [3], early breakdown phenomena in hydrogen gas [4], and measurement of the diatomic molecule CN [5, 6], including spatial distribution for time delays of the order of one microsecond form initiation of optical breakdown. Experimental arrangements for spectroscopy and shadowgraph capture are communicated in chapter 16.

30.2 Shock waves

Laser-induced breakdown performed on solid, liquid, or gaseous materials produces a small explosion. This explosion, together with the excitation, plasma formation, and ablation of material, is accompanied by the surrounding material being fiercely displaced and the production of a shock wave. The geometry and the evolvement over time of a laser-induced shock wave are dependent on the energy of the laser and the shape of plasma produced. Blast waves due to nuclear explosions set the foundation for the study of shock wave production and propagation. The vast amounts of energy released in a fixed volume by nuclear explosions compared to normal explosions were examined by Bethe *et al* [7] at Los Alamos, NM, USA, in 1941 and Taylor [8] in the United Kingdom in 1950, yielding the theory of a point strong explosion. Studies by John von Neumann [7] and Sedov [9, 10], which assumed an adiabatic expansion and a sudden release of energy, E, in negligible volume and time, led to the development of the expansion law of the shock wave:

$$R(\tau) = \frac{1}{K}\left(\frac{E\tau^2}{\rho}\right)^{\frac{1}{5}}, \tag{30.1}$$

where $R(\tau)$ is the radius of the shock wave at time τ, K is a constant dependent on the adiabatic coefficient of the gas, and ρ is the density of the gas. For studies performed in standard ambient temperature and pressure (SATP) air, $K \approx 1$, which is consistent with shadowgraph studies performed by Gautam *et al* [3]. Comparisons of computed blast-wave radii, using equation (1) with $K = 1$, for SATP air and molar CN mixture are seen in table 30.1 and table 30.2. There is minimal variation in shock wave expansion in SATP air versus CN mixture, which would indicate that measured shadowgraphs in air provide an acceptable guide for the CN gaseous mixture.

Another important characteristic of the shock wave is the expansion velocity. The shock wave expansion velocity indicates whether the approximation used in the shock wave expansion law is accurate. Shock wave expansion velocity, $v(\tau)$, assuming $K \approx 1$, is determined by:

$$v(\tau) = \frac{2}{5}\tau^{-3/5}\left(\frac{E}{\rho}\right)^{\frac{1}{5}}. \tag{30.2}$$

Studies by Harith *et al* [11] discuss that the shock wave expansion law is a great approximation when the shock wave expansion velocity is around Mach numbers $\mathrm{Ma} \approx 2$. However, this work also discusses applicability of the shock wave radius equation, equation (30.1), for velocities with $\mathrm{Ma} \gg 2$, see section 30.6.3. Mach numbers, Ma, are calculated using:

Table 30.1. Computed shock wave radii for SATP air and for molar CN mixture, 160 mJ.

τ (ns)	R (mm) for Air [$\rho = 1.2$ kg m^{-3}]	R (mm) for CN [$\rho = 1.63$ kg m^{-3}]
200	1.40	1.31
450	1.93	1.82
700	2.31	2.17
950	2.61	2.45
1200	2.86	2.69
1450	3.09	2.90

Table 30.2. Computed shock wave radii for SATP air and for molar CN mixture, 200 mJ.

τ (ns)	R (mm) for Air [$\rho = 1.2$ kg m^{-3}]	R (mm) for CN [$\rho = 1.63$ kg m^{-3}]
200	1.46	1.37
450	2.02	1.90
700	2.41	2.27
950	2.73	2.56
1200	2.99	2.81
1450	3.23	3.04

$$\mathrm{Ma} = \frac{v(\tau)}{c}, \tag{30.3}$$

where c is the speed of sound in SATP, 343 m/s. Comparisons of computed shock wave expansion velocities and Mach numbers for energies 160 and 200 mJ, using equations (30.2) and (30.3), for SATP air and molar CN mixture are seen in table 30.3, table 30.4, table 30.5, and table 30.6. Shock wave expansion velocities in air for early time delays (450 ns or less) move at hypersonic speeds (Mach numbers 5 or greater), while at later time delays (greater than 450 ns), the shock wave expansion velocities move at supersonic speeds (Mach numbers 1.3–5, inclusive). For the CN mixture, the shock wave expansion velocities move at supersonic speeds except for early time delays of 200 ns or less where they move at hypersonic speeds. Therefore, as time elapses, the shock wave expansion law approximation improves [11] as Mach numbers approach Ma $\approx$ 2 and slower.

Table 30.3. Computed shock wave velocity for SATP air and for molar CN mixture, 160 mJ.

τ (ns)	v (km/s) for air [$\rho = 1.2$ kg m^{-3}]	v (km/s) for CN [$\rho = 1.63$ kg m^{-3}]
200	2.80	2.63
450	1.72	1.62
700	1.32	1.24
950	1.10	1.03
1200	0.95	0.90
1450	0.85	0.80

Table 30.4. Computed shock wave velocity for SATP air and for molar CN mixture, 200 mJ.

τ (ns)	v (km/s) for air [$\rho = 1.2$ kg m^{-3}]	v (km/s) for CN [$\rho = 1.63$ kg m^{-3}]
200	2.92	2.75
450	1.80	1.69
700	1.38	1.30
950	1.15	1.08
1200	1.00	0.94
1450	0.89	0.84

Table 30.5. Computed shock wave Mach numbers for SATP air and for molar CN mixture, 160 mJ.

τ (ns)	Ma for air [$\rho = 1.2$ kg m^{-3}]	Ma for CN [$\rho = 1.63$ kg m^{-3}]
200	8.15	7.67
450	5.01	4.71
700	3.84	3.61
950	3.20	3.01
1200	2.78	2.62
1450	2.48	2.34

Table 30.6. Computed shock wave Mach numbers for SATP air and for molar CN mixture, 200 mJ.

τ (ns)	Ma for air [$\rho = 1.2$ kg m^{-3}]	Ma for CN [$\rho = 1.63$ kg m^{-3}]
200	8.52	8.02
450	5.24	4.93
700	4.02	3.78
950	3.35	3.15
1200	2.91	2.74
1450	2.60	2.44

30.3 Electron density

30.3.1 Atomic carbon line interference

Laser-induced breakdown performed on the carbon dioxide and nitrogen gaseous mixture produces a variety of species, which includes C, C^+, C^-, CN^+, CN^-, CNN, CO, CO^+, CO_2, CO_2^+, C_2, C_2^+, C_2^-, CCN, CNC, C_2O, C_3, N, N^+, N^-, NCO, NO, NO^+, NO_2, N_2, N_2^+, N_2^-, NCN, N_2O, N_3, O, O^+, O^-, O_2, and O_2^+ [12]. This work focuses on the analysis of the CN violet band $\Delta v = 0$ system that has vibrational bands (0, 0), (1, 1), (2, 2), (3, 3), and (4, 4), which are 388.34, 387.14, 386.19, 385.47, and 385.09 nm, respectively. The Czerny–Turner spectrometer equipped with 1200 groove/mm grating is adjusted to the region of interest of 370 to 393.5 nm to observe the CN violet band $\Delta v = 0$ system via the Andor Solis software [13].

In LIBS, first order lines (m = 1) are of interest when performing analysis of LIBS produced spectra, but spectral lines have different orders or modes due to the use of diffraction gratings which follow the equation:

$$d\sin(\theta) = m\lambda, \tag{30.4}$$

where d is the distance between the center of two adjacent slits, θ is the angle at which maxima occur, m is the propagation mode of interest, and λ is wavelength of monochromatic light. For example, if a first order ($m = 1$) spectral line has a wavelength, λ, equivalent to 193 nm, then its wavelength, λ, measured in second order ($m = 2$) would be 386 nm when using equation (4). Even though the spectrometer is set for the desired region of interest and the CN band heads are well defined, there can be interference from the other species spectral lines in higher order that are produced by the laser-induced breakdown spectroscopy (LIBS). In the CN work, there is an overlap of the CI 193.09 nm atomic carbon line in second order and the vibrational band (2, 2) of 386.19 nm. Measurements included recording of data without and with the previously mentioned 309 nm cut-on filter (see section 2.2).

30.3.2 Line broadening and deconvolution

Spectral line broadening of observed plasma is caused by the effect of ions and electrons. Local conditions, such as local thermodynamic equilibrium, and extended conditions, such as the plasma radiations traversed path as viewed by an observer, cause the spectral lines to broaden. Doppler or thermal broadening, natural broadening, and pressure broadening are different types of local effect broadening. Doppler broadening is characterized as a Gaussian profile and is due to the position of the detector or observer relative to the velocity of the atoms or ions within a gas or plasma. Natural broadening is characterized as a Lorentzian profile and occurs due to the uncertainty associated with the lifetime of the excited states and its energy. Finally, pressure broadening is characterized as a Lorentzian profile and is caused by atoms or other ions neighboring the emitter atom or ion. One example of pressure broadening is Stark broadening, which is caused by the shifting and splitting of spectral lines due to the presence of an external electric field, known as the Stark effect. Electron number density, n_e, can be determined from the full width at half maximum (FWHM), $\Delta\lambda_{\text{Stark}}$, of the stark-broadened CI 193.09 nm atomic carbon line [14], measured in second order:

$$\Delta\lambda_{\text{Stark}}(\text{nm}) = 2w(\text{nm})n_e(10^{17}\,\text{cm}^{-3}), \tag{30.5}$$

where width parameter, w, was extrapolated [14, 15], to be $w \approx 0.0029$ nm or by the Stark shift of the CI 193.09 nm atomic carbon line [14]:

$$\delta\lambda_{\text{Stark}}(\text{nm}) = \text{d}(\text{nm})n_e(10^{17}\,\text{cm}^{-3}), \tag{30.6}$$

where the shift parameter, d, was extrapolated to be [14, 15] $\text{d} \approx 0.0029$ nm. In order to use equations (30.5) and (30.5) to determine n_e, deconvolution of measured Stark, FWHM and Stark shifts must be performed. This is due to the line broadening being largely influenced by Stark broadening, which is typically approximated using a Voigt profile [16]. The convolution of Gaussian and Lorentzian line profiles results in the Voigt profile. Therefore, a rough approximation between Gaussian, Lorentzian, and Voigt profile widths:

$$f_V = \frac{f_L}{2} + \sqrt{\left(\frac{f_L}{2}\right)^2 + f_G^2}, \tag{30.7}$$

where f_V is the FWHM of the Voigt profile, f_L is the FWHM of the Lorentzian profile, and f_G is the FWHM of the Gaussian profile, can be used to apply deconvolution to the measured line profile. In this paper, f_V represents the measured spectral line, f_L represents the Stark FWHM or Stark shift, and f_G represents the spectral resolution of the spectrometer. Therefore, rearranging equation (30.7) to determine f_L yields:

$$f_L = \frac{f_V}{2} - \frac{f_G^2}{f_V}, \tag{30.8}$$

which can be used in conjunction with equations (30.5) and (30.6) to determine n_e of the CI 193.09 nm atomic carbon line in second order.

30.3.3 Computation of electron density

Spectra that were unfiltered had an overlap of the (2, 2) CN band head of 386.19 nm and the second order CI 193.09 nm atomic carbon line, where spectra filtered with the cut-on filter only had the (2, 2) CN band head of 386.19 nm. A peak-fitting Matlab script [17] was applied to filtered and unfiltered spectra to evaluate CI 193.09 nm atomic carbon line in second-order Stark widths and shifts. A typical intermediate record of the five-peak fitting includes fitted background, a single profile for each of the five bands. Figure 30.1 illustrates recorded data and the overall 'peakfit.m version 9.0' result that is composed of background and sum of five Gaussians versus wavelength. Equation (30.8) was used on the extracted Stark FWHM and Stark shifts to apply

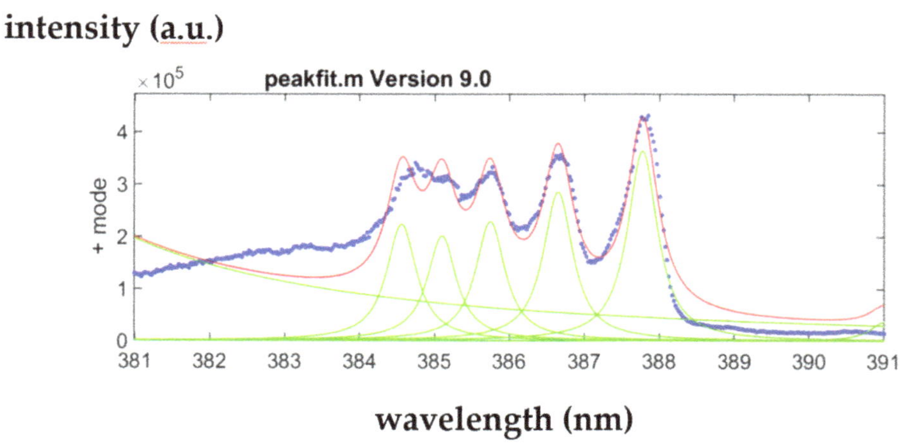

Figure 30.1. Typical result of the peakfit.m script applied to measured line-of-sight $\Delta v = 0$ CN spectra. Individual fitted peaks and the background variation (in green) are added up for the overall fit (in red) to the experimental data (dotted, in blue). Reproduced from [18]. CC BY 4.0.

deconvolution. The difference between deconvoluted Stark FWHM from unfiltered spectra and deconvoluted Stark FWHM from filtered spectra was performed to determine the FWHM contribution of the CI 193.09 nm atomic carbon line in second order only. The FWHM contribution of the CI 193.09 nm atomic carbon line in second order only was used in conjunction with equation (30.2) to determine n_e, where deconvoluted Stark shifts can also be used in conjunction with equation (30.6) to determine n_e. This communication reports results of n_e from only Stark widths.

30.4 Molecular spectra analysis

Plasma temperature from molecular spectra is dependent on vibrational and rotational elements of the molecular spectrum. Due to this dependence, temperature from molecular spectra can be used to evaluate the condition of the plasma. Temperature determination can be performed by fitting the measured spectra to the calculated theoretical diatomic spectrum. Diatomic line strength is used to calculate the needed theoretical diatomic spectrum for appropriate fitting. Diatomic line strength calculations are rather cumbersome to perform due to spectral line position requirements and the need for very accurate molecular rotational constants. Parigger and Hornkohl [19] describe the theoretical background and development of diatomic line strength tables necessary for temperature evaluation. Therefore, the CN line strength tables reported by Parigger *et al* [20] are used for the calculation of theoretical CN spectrum. The Nelder–Mead temperature (NMT) program [16] was used to fit measured CN spectral data. The NMT program utilizes the Nelder–Mead method [21], which is a numerical method used to find the minimum or maximum of an objective function in a multidimensional space. The NMT program requires initial fit parameter assumptions, which consist of the temperature of the molecular spectra, the line width of the molecular spectral line, and a linear baseline offset. Specifically, the Nelder–Mead algorithm creates a simplex established on the initial given fit parameters. Each vertex of the simplex represents a fit parameter and the size of the simplex is reduced by changing the vertex arrangement until the tolerance is achieved. During minimization, the first local minimum identified represents the minimum and the best fit parameters are established on final location of the simplex's vertices.

30.5 Abel inversion

The Abel transform, which analyzes spherically symmetric functions, can be applied to evaluate line-of-sight measurements from close to spherically symmetric plasmas. Figure 30.2 illustrates the line-of-sight experimental geometry. Specifically, the Abel-inversion [22–26] technique allows one to extract the radial distributions of electron densities of a close to spherically symmetric plasma directly from the recorded line-of-sight data. Letting $\varepsilon(r, \lambda)$ represent the radial emission coefficient, the Abel transform of $\varepsilon(r, \lambda)$ is shown to be [5, 12, 13, 27–30]:

$$I(z, \lambda) = 2\int_{z}^{P \gg R} \varepsilon(r, \lambda) \frac{r}{\sqrt{r^2 - z^2}} dr, \tag{30.9}$$

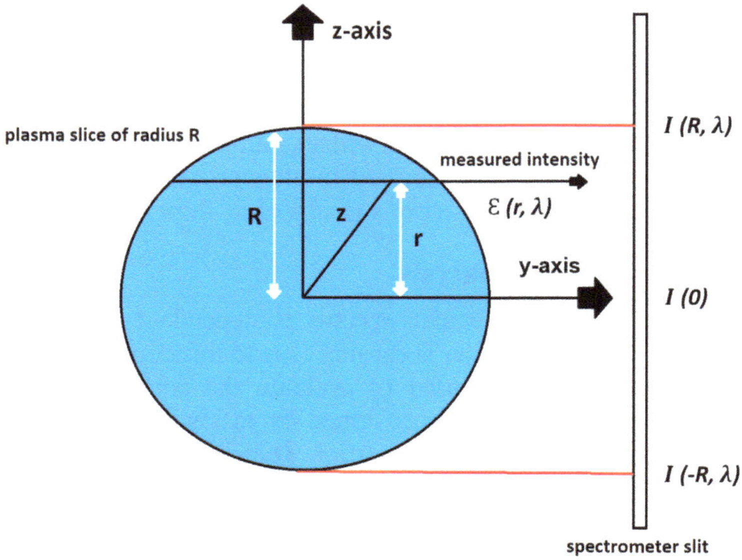

Figure 30.2. Line-of-sight geometry and Abel-inversion method. Reproduced from [18]. CC BY 4.0.

where z is the perpendicular distance from the origin of the line-of-sight, r is the radial distance from the center of the observed plasma at which electron density will be evaluated, R is the radius of the spherical object, and P is the upper integration limit which is established as being much greater than R, $P \gg R$. The recorded emission-intensity map contains spatial information along the slit dimension, and spectral information along the wavelength dimension. Abel inversion is performed on measured line-of-sight data of the emitted spectral intensity in order to determine the radial variation.

The emission intensity has subtle asymmetry along the direction perpendicular to the laser beam which may be a result of laser-plasma interaction that normally takes place in nanosecond laser-induced plasmas or may be a result of variations of the laser pulse profile. Figure 30.2 displays the geometry of the line-of-sight measurements utilized in this work, where line-of-sight measurements are along the y-axis, the direction of the laser beam is along the z-axis, and the x-axis is perpendicular to the y and z axes. Abel inversion is used to obtain radial dimension measurements of the plasma, which allows for the extraction of plasma radial information. Using the top half and bottom half of line-of-sight profiles, asymmetric data points are averaged utilizing the same symmetrization method for atomic hydrogen spectra [27–29] and application of Abel inversion is used.

Shadowgraph measurements of plasma kernels in hydrogen and hydrogen-nitrogen gaseous mixtures show a close to cylindrical symmetry or prolate spheroidal symmetry as compared to the close to spherically symmetric plasma kernels in standard ambient temperature and pressure in laboratory air for time delays of 100–10 000 ns [3, 28–30]. The deviation from close to spherically symmetric plasma kernels may be due to the laser energy employed for laser-induced breakdown;

therefore, as discussed in this work, shadowgraph measurements in standard ambient temperature and pressure are performed at laser energies observed in the carbon dioxide and nitrogen gaseous mixture, which are seen to be close to spherically symmetric [5]. Due to the spherically symmetric plasma requirement of the Abel-inversion technique, collected line-of-sight data are adjusted to allow subtle deviations from circular symmetry and modeled to be spherically symmetric [27–29].

Abel inversion of each wavelength determines the spatial distribution and subtle asymmetries present in the captured data are kept by applying an asymmetric factor, which then establish the asymmetric radial distributions [27–29]. Pretreatment of the captured data is not required when using the derivative free Abel-inverse transform method [22]. Inversion of equation (30.9) can be accomplished using an analytical method with derivatives, known as the Abel-inverse transform. Differentiation of spectra is rather challenging; therefore, coefficients are determined by using a complete set of orthogonal polynomials with a minimization method. The use of Chebyshev polynomials in conjunction with the available Matlab® script [22, 26] for Abel inversion of equation (30.9), allows the direct inversion of measured data. For this work, inversion was accomplished by choosing 15 polynomials [22, 23], which maintained fidelity of the spectra and was comparable to the use of a digital filter resulting in computed radial spectra. Smaller spectral resolution would occur with the selection of a smaller number of polynomials. Line-of-sight data along the spectrometer slit were captured and inverted for each wavelength to get the radial intensity distribution. Calibration and correction for system sensitivity using standard lamps is required for recorded data to undergo Abel inversion and curation of the spectra.

30.6 Results

30.6.1 Shadowgraphs

Shadowgraphs of plasma in SATP produced by infrared (IR) 1064nm radiation with excitation energy of 170 mJ and 6 ns pulse-width are shown below. A single 5 ns pulse-width 532 nm laser beam was used to capture the shadowgraphs. Shadowgraphs were taken in the range of 200 – 4200 ns time delays. Figure 30.3 displays typical results.

Further investigations of laser-induced laboratory air breakdown utilize pulse energies of 850 mJ per 6 ns, 1064nm pulses. Figures 30.4 and 30.5 illustrate recorded images in the range of 200–4000 ns.

Figures 30.4 and 30.5 display multiple breakdown events along the optical axis. Stagnation layers appear to be formed between individual breakdown spots, developing into vertical structures in the forward direction. Stagnation layers have been explored at the interface region of two colliding laser-induced plasmas [31]. The predicted initial plasma expansion speeds in laser-induced optical breakdown are of the order of 100 km/s [32], followed by a gas expansion that is analogous to that of a strong explosion [32]. Figures 30.4 and 30.5 also exhibit associated early expansion dynamics that occur at speeds well in excess of hypersonic speeds. The blue lines indicate the forward propagating shock wave boundaries that originate from

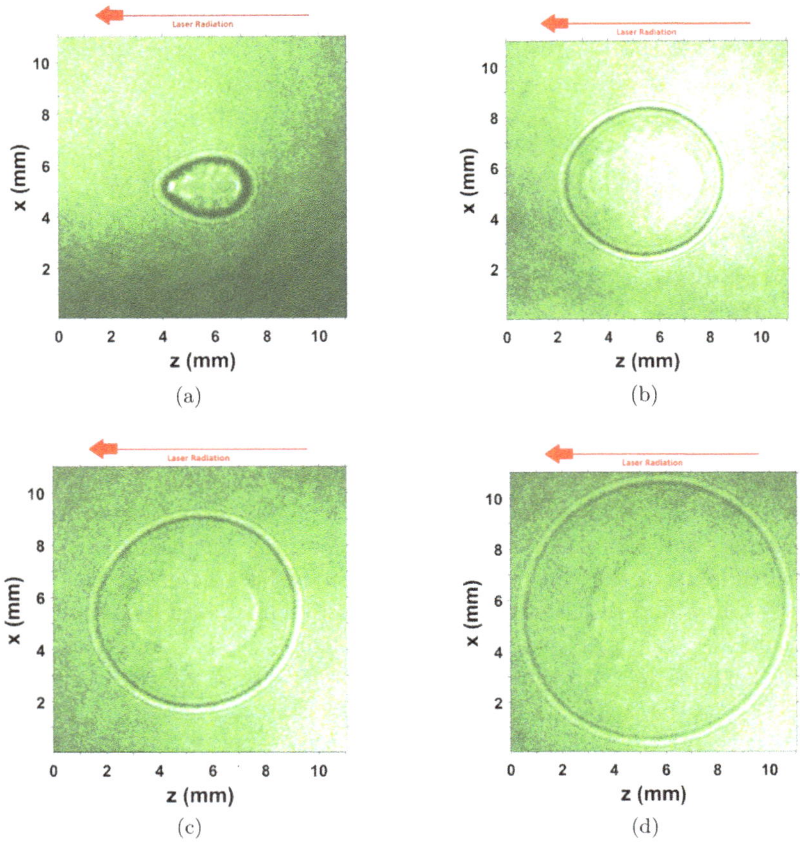

Figure 30.3. Single-shot shadowgraphs of expanding laser-induced plasma initiated with 170 mJ, 6 ns, 1064nm pulses, and imaged with 5 ns, 532 nm backlight, time delayed by (a) 200 ns, (b) 1200, (c) 2200 ns, and (d) 4200 ns. Reproduced from [18].

multiple breakdown spots appearing as beads. As indicated in Figure 30.3, the infrared 1064nm, 170 mJ, 6 ns laser beam is along the z-axis and moving from the right to left. The expanding shock wave and plasma kernel are clearly visible. At 0.2 s delay, the plasma kernel appears cylindrical and the expanding shock wave has a prolate spheroidal shape. As time delays approach 1 s, the plasma kernel and expanding shock wave become nearly spherical. As time elapses further, the plasma kernel and expanding shock wave continue to become close to spherical. The vertical extend is about a factor of 1.4 smaller for 170 mJ pulses than that for 850 mJ pulses, according to the Taylor–Sedov energy dependency, equation (30.1), for spherical expansion.

30.6.2 Emission spectra

The CN spectra captured by the spectrometer and two-dimensional intensified charge-coupled device for a fixed volume of the 1:1 molar CO_2:N_2 gas mixture held

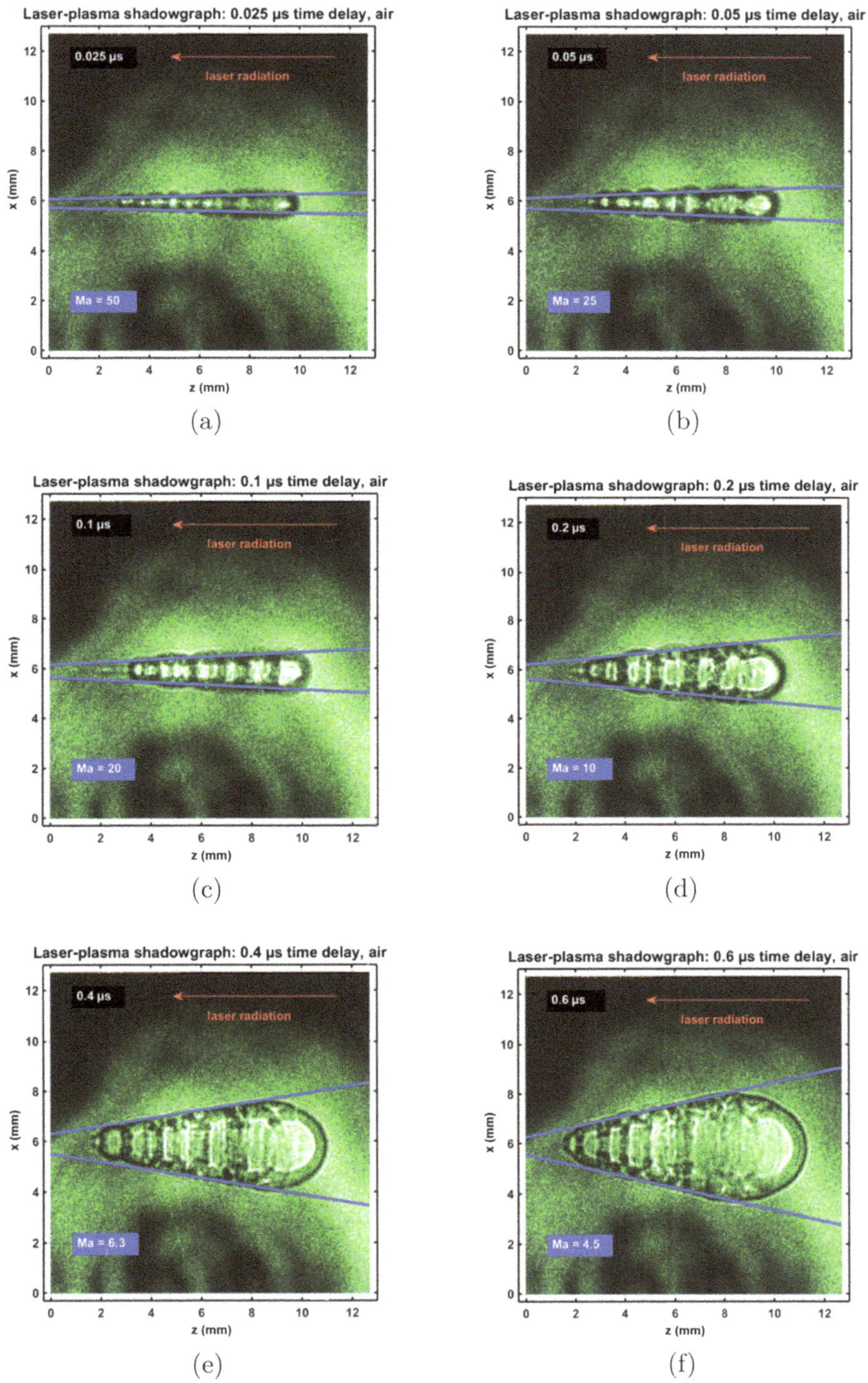

Figure 30.4. Shadowgraphs subsequent to laser-plasma generation with 850 mJ, 6 ns, 1064nm pulses. Time delays: (a) 25 ns, (b) 50 ns, (c) 100 ns, (d) 200 ns, (e) 400 ns, and (f) 600 ns. Reproduced from [18]. CC BY 4.0.

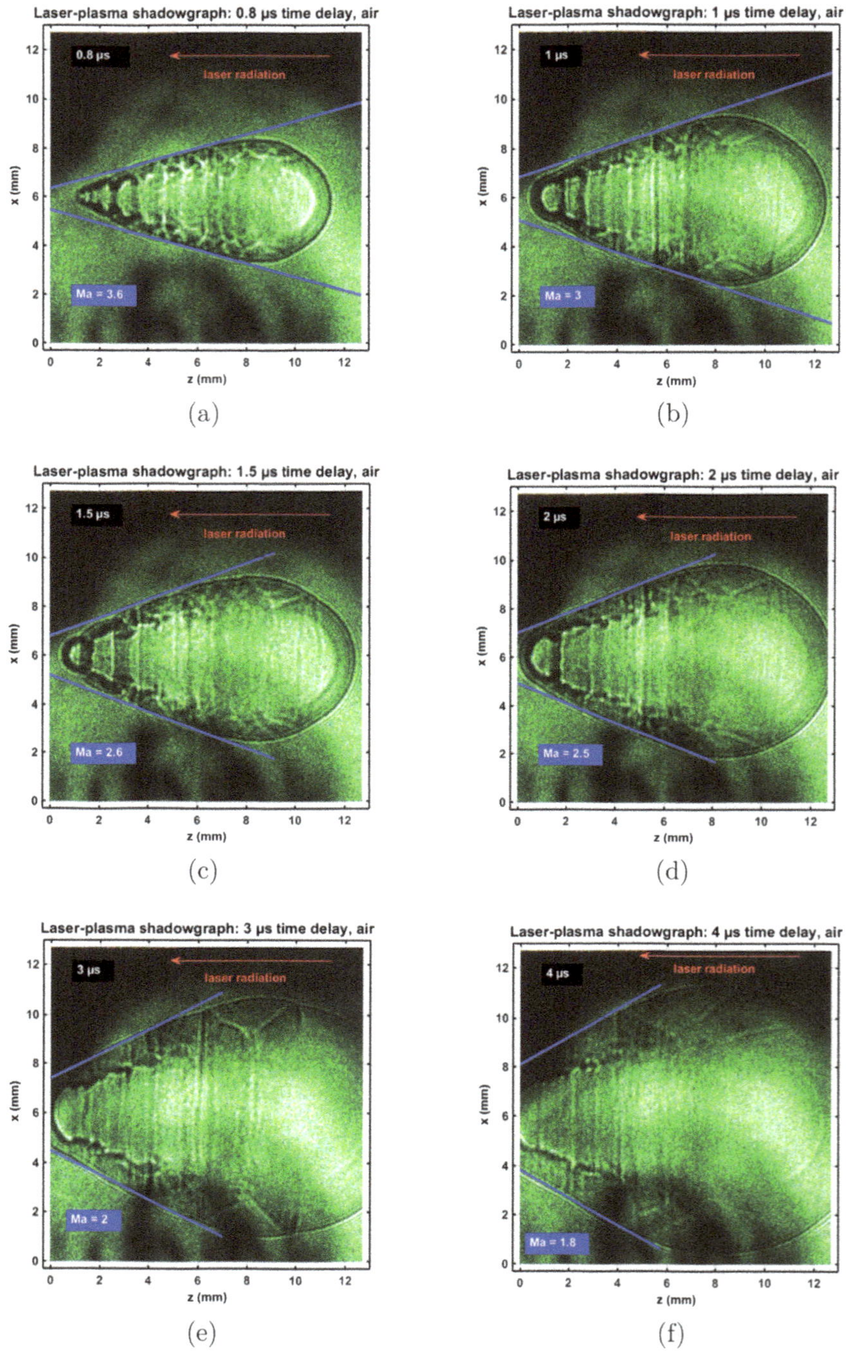

Figure 30.5. Shadowgraphs captured after laser-plasma generation with 850 mJ, 6 ns, 1064nm pulses. Time delays: (a) 800 ns, (b) 1000 ns, (c) 1500 ns, (d) 2000 ns, (e) 3000 ns, and (f) 4000 ns. Reproduced from [18]. CC BY 4.0.

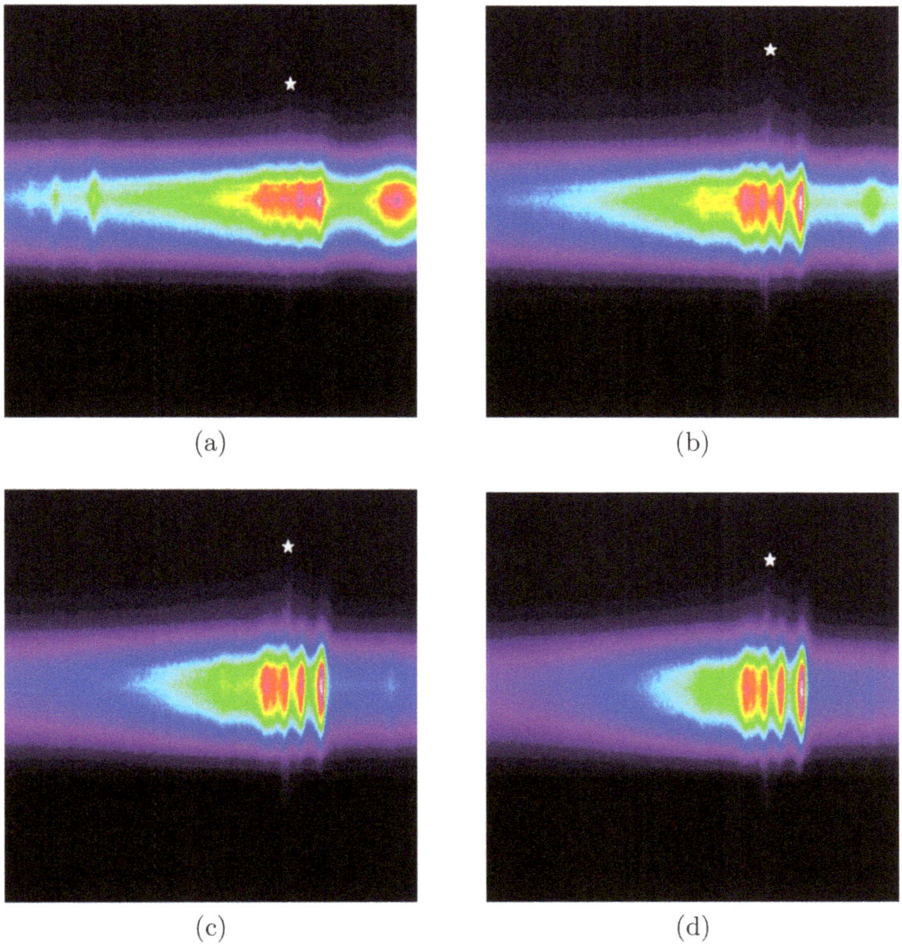

Figure 30.6. Optical breakdown CN spectra in a 1:1 molar CO_2:N_2 gas mixture held at atmospheric pressure for time delays of (a) 200 ns, (b) 450 ns, (c) 700 ns, and (d) 950 ns. Reproduced from [18]. CC BY 4.0.

at atmospheric pressure are shown in figure 30.6. As seen in figure 30.6, the CN violet system $B^2\Sigma^+ - X^2\Sigma^+$ vibrational bands of (0, 0), (1, 1), (2, 2), (3, 3), and (4, 4) are clearly visible and discernible. The overlap of the CI 193.09 nm atomic carbon line in second order and the (2, 2) CN band head is also seen in figure 30.6. At time delays greater than 2.5 s, the CI 193.09 nm atomic carbon line in second order appears to dissipate and does not overlap the (2, 2) CN band head.

Spectrometer-detector gate width: 125 ns. *, second-order atomic carbon line [5]. In separate experimental runs, CN spectra were captured for a fixed volume of the 1:1 molar CO_2:N_2 gas mixture held at atmospheric pressure with the use of a 309 nm cut-on wavelength filter. The 309 nm cut-on wavelength filter allows for the suppression of the CI 193.09 nm atomic carbon line in second order. Although it is advantageous to apply the 309 nm cut-on wavelength filter for the suppression of the CI 193.09 nm atomic carbon line in second order, the 309 nm cut-on wavelength filter causes a reduction in spectral intensity captured by the spectrometer and the intensified charge-coupled detector by $\approx$ 13%. Filtered and unfiltered spectra also show the CN plasma moving towards the laser as time elapses.

30.6.3 Shock wave and plasma expansion

The expanding shock wave radii results are shown in table 30.7, decreasing shock wave velocities and Mach number results are shown in table 30.8, and increasing plasma kernel radii are shown in table 30.9. The expanding shock wave radius for 0.2 s delay is not exactly consistent with the previously discussed shock wave

Table 30.7. Computed shock wave radii versus measured shock wave radii for SATP air, 170 mJ.

τ (ns)	Computed R (mm)	Measured R (mm)
200	1.41	1.00 ± 0.30
1000	2.69	2.67 ± 0.80
1200	2.90	2.83 ± 0.85
2200	3.69	3.57 ± 1.07
4200	4.78	4.95 ± 1.49

Table 30.8. Inferred shock wave velocities and Mach numbers for SATP air, 170 mJ.

τ (ns)	Velocity, v (km/s)	Mach number, Ma
200	4.03 ± 1.21	11.76 ± 1.30
1000	1.31 ± 0.39	3.82 ± 1.15
1200	1.08 ± 0.32	3.15 ± 0.95
2200	0.58 ± 0.17	1.67 ± 0.50
4200	0.30 ± 0.09	0.87 ± 0.26

Table 30.9. Measured plasma kernel radii for SATP air, 170 mJ.

τ (ns)	Measured r (mm)
200	0.45 ± 0.13
1000	2.25 ± 0.67
1200	2.40 ± 0.72
2200	3.00 ± 0.90
4200	4.04 ± 1.21

expansion law, equation (30.1), and this can be due to the velocity of the shock wave being greater than the Mach 2 maximum velocity requirement of the shock wave expansion law. For time delays approaching 1 μs and later, the expanding shock wave radii are consistent with the shock wave expansion law, with their shock wave expansion velocities, v(τ), being closer to Mach 2 and slower.

As previously discussed, shock wave expansions in SATP air appear similar in the CN mixtures used in this work. Therefore, the visualization of these shock wave expansions in SATP air provides a good model to the shock wave expansion behavior and plasma kernel geometry in the CN mixtures. The images recorded for 850 mJ laser-induced plasma generation are also analyzed using the Taylor–Sedov theory of blast-wave propagation from a point explosion yields the time dependent radius, R(t), of the shock front [27–29]:

$$R(\tau) = \xi(E/\rho)^{\frac{1}{n+2}}\tau^{\frac{2}{n+2}} \sim \tau^{\frac{2}{n+2}}. \tag{30.10}$$

Here, ξ ($\xi = 1/K$) is a constant in the range of 1.0 to 1.1 that depends on the specific heat capacity, E is the energy that is released during the explosion or the absorbed energy per laser pulse, ρ is the gas density, τ is the time delay, and n is the shape dependent parameter. The values of $n = 1$, 2, and 3 correspond to planar, cylindrical, and spherical shock waves, respectively. One can use equation (30.10) for computation of the blast-wave or shock front expansion generated from laser-induced optical breakdown. However, of primary interest is the dependence of the radius, R(τ), on time delay, τ. Figure 30.7 displays the maximum of the shock wave radius versus time delay measured perpendicular to the direction of the laser beam

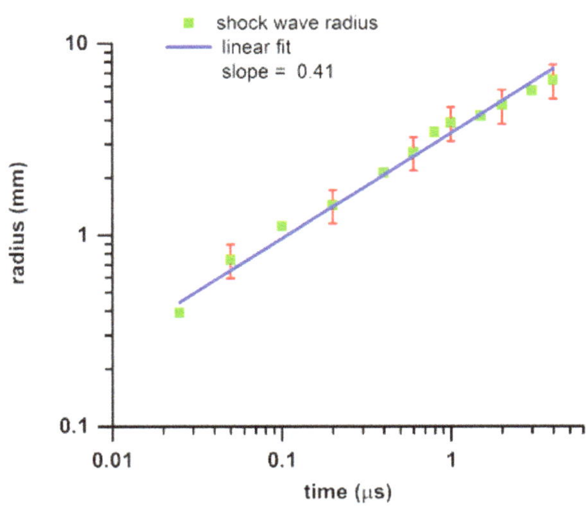

Figure 30.7. Log-log plot of shock wave expansion measured perpendicular to the laser-propagation direction when using 850 mJ, 6 ns, 1064nm pulses for optical breakdown in laboratory air. Reproduced from [18]. CC BY 4.0.

propagation. In view of figures 30.5 and 30.6, the maximum is determined from the images near z = 8 mm.

The linear fit (figure 30.7) reveals 0.41 for the slope, or $n = 2.9 \sim 3$. In other words, spherical expansion is inferred. This figure also shows 20% error bars. These error bars are estimated from the variations in the pulse energy for generation of optical breakdown, the trigger-jitter synchronization of the two laser beams (one for plasma generation, the other for backlight), and the shadowgraph analysis readout errors. One can also extract from the graph the approximate 1 mm per μs expansion velocity for time delays of $\sim$1 μ s, or Ma = 3. From equation (30.10), using $\xi = 1.0$ to 1.1, E = 800 mJ, $\rho = 1.225$ kg/m^3, and $n = 3$ yields for the radius R($\tau = 1$ μ s) = 3.7–4.1 mm, consistent with the measured value of 3.9 mm [4].

For ultra-high-pure hydrogen, and for time delays of the order of almost 0 to a few dozen ns, recorded spectral images are utilized for exploration of the well-above hypersonic expansion. Figure 30.8 illustrates two images captured from optical breakdown in near atmospheric hydrogen gas, i.e., at a cell pressure of (1.08 ± 0.033) $\times 10^5$ Pa (810 ± 25 Torr).

Figure 30.9 summarizes the expansion speed for early time delays. However, the speed of sound in hydrogen is approximately four times higher than in air, and the recorded air shadowgraphs can serve as a guide for shock wave appearances. For example, hydrogen expansion at a delay of 400 ns approximately corresponds to air shadowgraphs recorded at a delay of 1600 ns. Most importantly, if the irradiance is not significantly higher than that for optical breakdown thresholds, a spherically symmetric appearance of the shock wave for delays 10 × larger than indicated in figure 30.9 would be expected analogous to the 170 mJ shadowgraphs recorded in air. Indeed, captured shadowgraphs of optical breakdown in hydrogen [30] reveal a prolate spheroidal shock wave shape for a time delay of 400 ns (see figure 1 in [30]).

The determined expansion speeds for hydrogen shock wave expansion speeds are well-above hypersonic speed (hypersonic: Ma $\geqslant$ 5) or above reentry speeds (reentry: Ma $\leqslant$ 25) at time delays of 10 to 40 ns.

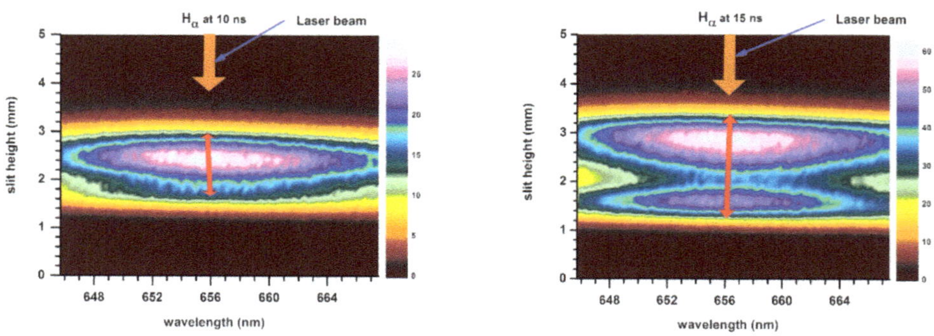

(a) (b)

Figure 30.8. (a) Hydrogen alpha plasma spectra images at 10 ns, and (b) 15 ns time delays. The red arrow indicates the measured plasma width. Reproduced from [18]. CC BY 4.0.

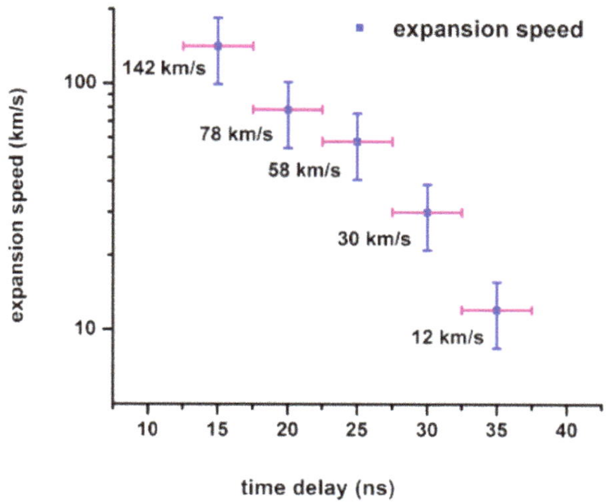

Figure 30.9. Plasma expansion speeds. The indicated time-delay error bars are due to the gate width of 5 ns. Reproduced from [18]. CC BY 4.0.

30.6.4 Electron density

The inferred Stark widths of the CI 193.09 nm carbon line in second order for the 1:1 molar CO_2:N_2 gaseous mixture held at atmospheric pressure were determined using the previously discussed peak-fitting Matlab® [17], deconvolution of the filtered and unfiltered measured peaks, and taking the difference between the filtered and unfiltered deconvoluted peaks.

The inferred Stark widths are plotted versus the slit height of the spectrometer, which can be seen in figure 30.10. Larger Stark widths are seen towards the edges of the plasma, while smaller Stark widths are seen in the center of the plasma.

For time delays of 450 and 950 ns, figure 30.10(a), (c), the Stark widths are between 0.4 and 0.5 nm and located towards the edges of the plasma. The Stark widths are used to calculate electron number density, n_e. The calculated n_e is plotted versus the slit height of the spectrometer, which can be seen in figure 30.10(b), (d). Peak electron densities are of the order of $n_e \approx 10^{19}\text{cm}^{-3}$, and values between the two peaks are of the order of $n_e \approx 10^{17}\text{cm}^{-3}$. Since n_e is directly proportional to the Stark width of the 193.09 nm carbon line in second order as previously shown in equation (5), n_e plots mimic the same behavior as the previously mentioned Stark width plots, where higher n_e is seen towards the edges of the plasma and lower ne is towards the center of the plasma. The higher electron densities toward the edges of the plasma appear to follow the previously discussed shock wave expansion law, equation (30.1), within the indicated error bars as seen in table 30.10.

The Stark shifts of the CI 193.09 nm carbon line in second order for the 1:1 molar CO_2:N_2 gaseous mixture held at atmospheric pressure were also determined using the peak-fitting Matlab® script [17]. Larger Stark shifts are seen towards the edges of the plasma, while smaller Stark widths are seen in the center of the plasma. However, similar results are found when using the Stark shifts, yet with the shock wave fronts more precisely demarcated when using the Stark widths.

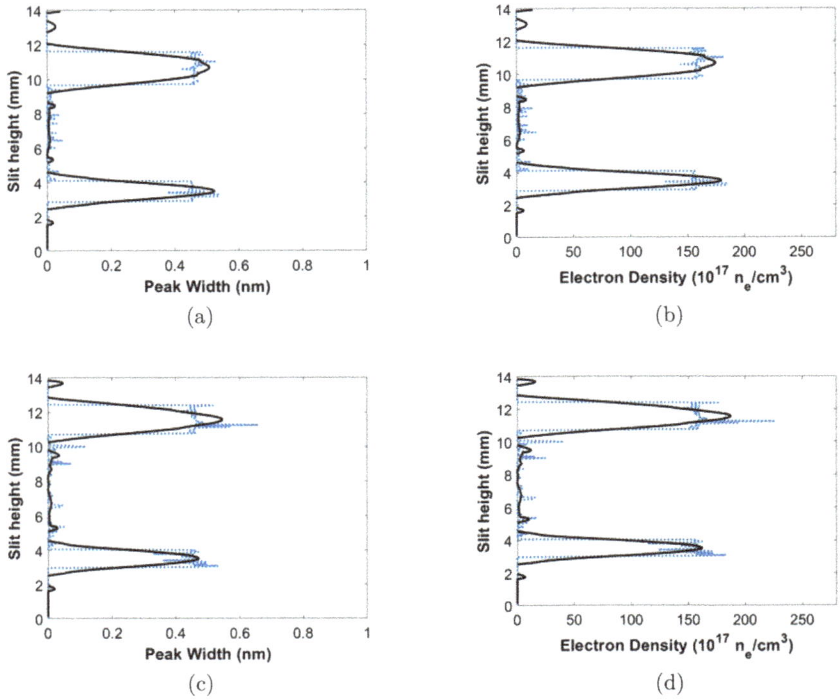

Figure 30.10. Inferred widths and calculated electron densities of C I 193.09 nm atomic carbon line in 2nd order versus slit height for 1:1 molar CO_2:N_2 gas mixture held at atmospheric pressure. Time delays: (a,b) 450 ns, and (c,d) 950 ns. Reproduced from [18]. CC BY 4.0.

Table 30.10. Computed shock wave radii versus plasma radius for 1:1 molar CO_2:N_2 gaseous mixture held at atmospheric pressure, 170 mJ.

τ (ns)	Computed R (mm)	Measured R_{Plasma}
450	1.84	2.90 ± 0.87
700	2.20	3.00 ± 0.90
950	2.48	3.30 ± 0.99
1200	2.72	3.35 ± 1.01
1450	2.94	4.00 ± 1.20
1700	3.13	4.30 ± 1.29
1950	3.31	4.40 ± 1.32
2200	3.47	4.30 ± 1.29

30.6.5 Cyanide temperature

Inferred temperatures of filtered line-of-sight CN spectra in the 1:1 molar CO_2:N_2 gaseous mixture held at atmospheric pressure are plotted versus slit height of the spectrometer, as shown in figure 30.11. The outgoing shock wave can be seen from

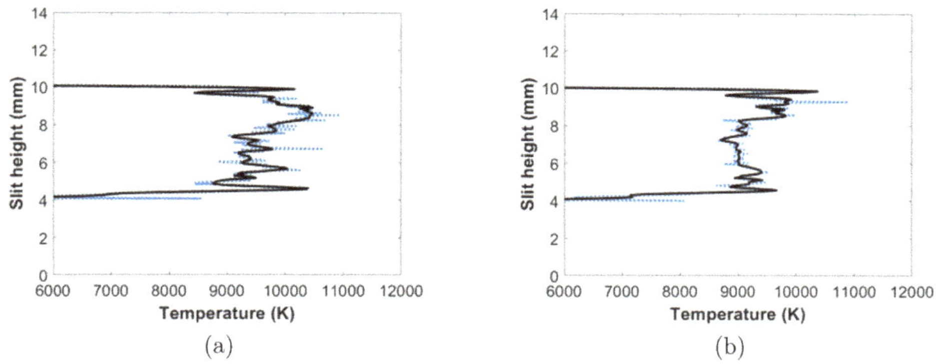

Figure 30.11. Temperature versus slit height for filtered line-of-sight CN spectra for fixed volume of 1:1 molar CO_2:N_2 gaseous mixture held at atmospheric pressure. Time delays: (a) 450 ns; (b) 950 ns. Reproduced from [18]. CC BY 4.0.

time delays of 450–950 ns. Figure 30.11 indicates that temperature variations occur in the central region, while increased temperatures are shown at the edges of the plasma. Higher temperatures are seen on the edge of the plasma towards the top of slit or towards the laser side. At a time delay of 450 ns, figure 30.11(a), the temperatures in the central region of the plasma are between 9500–10 000 K, while the temperatures at the edges of the plasma are more than 10 000 K. At time delays of 950 ns, figure 30.11b, the temperatures in the central region of the plasma cool to a range of 9000–9500 K, while temperatures at the edges of the plasma are between 9500–10 000 K. As time elapses further the plasma central region, temperatures cool even further to a range of 8500–9000 K for time delays of 1.2–1.7 s, while the edges of the plasma maintain a temperature range of 9500–10 000 K. From time delays of 1.95–2.2 μ s, the central region of the plasma sustains temperatures of 8500–9000 K and temperatures near the edge of plasma towards the bottom of the slit are around 9000 K, while temperatures near the edge of the plasma towards the top of the slit increase to greater than 11 000 K.

30.6.6 Abel inverted spectra

Abel inversion of the filtered 1:1 molar CO_2:N_2 gaseous mixture held at atmospheric pressure was performed by inverting measured line-of-sight data, $I(z, \lambda)$, for each wavelength, λ, to obtain the radial emissivity distribution, $\varepsilon(r, \lambda)$. Figures 30.12 and 30.13 illustrate the results. Previously measured shadowgraphs show the plasma generated in the 1:1 molar CO_2:N_2 gaseous mixture held at atmospheric pressure has a close to spherical shape, which would justify the use of Abel inversion.

The irradiance threshold for optical breakdown for the experimental arrangement (see section 2.1) is $\approx$ 20 mJ [5], or in terms of irradiance $\approx 3 \approx 10^{11}$ W cm $^{-}$2. When using pulse energies of up to about 170 mJ, or up to $\approx 10 \times$ breakdown threshold, close to symmetric shock wave expansion occurs. For pulse energies of 800 mJ, or

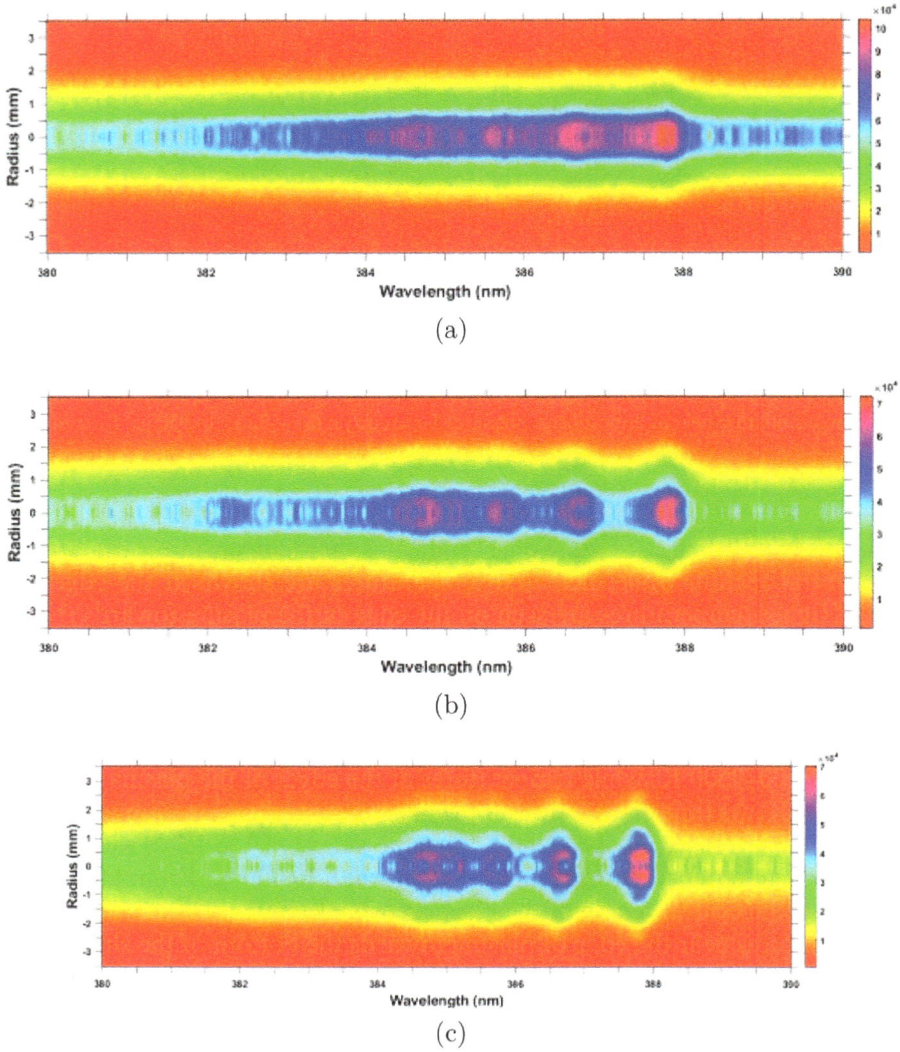

Figure 30.12. Abel-inverted CN spectra 1:1 molar CO_2:N_2 gaseous mixture held at atmospheric pressure. Time delay: (a) 200 ns, (b) 450 ns, and (c) 950 ns. Reproduced from [18]. CC BY 4.0.

40× breakdown threshold, the 'symmetric' expansion is seen in the region where most of the energy is absorbed (see figures 30.4 and 30.5).

At a time delay of 200 ns, figure 30.12(a), the CN distribution appears evenly distributed across the plasma. From time delays of 450–2200 ns, figures 30.12(b), (c) and 30.13(a)–(c), the CN signals begin to become stronger towards the edges of the plasma and weaker in the center of the plasma, which is consistent with the higher temperatures seen at the edges of the plasma as discussed previously. These results were projected to be similar to the shock wave results, but inside the plasma kernel and shock wave, the variations of the CN distribution were also expected.

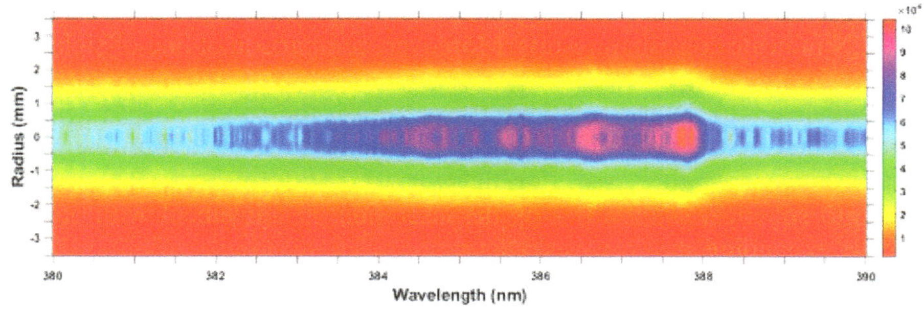

(a)

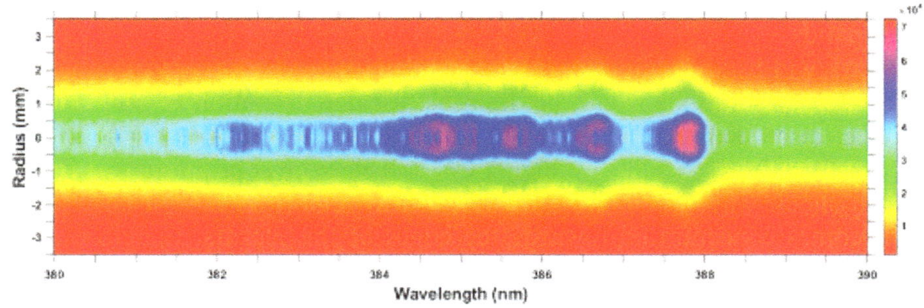

(b)

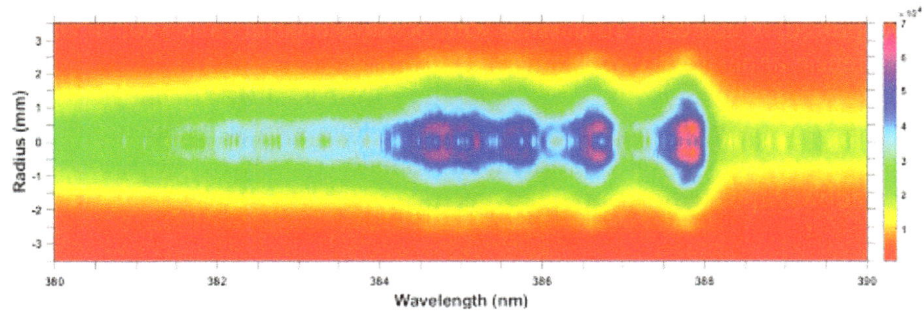

(c)

Figure 30.13. Abel-inverted CN spectra 1:1 molar CO_2:N_2 gaseous mixture held at atmospheric pressure. Time delay: (a) 1200 ns, (b) 1700 ns, and (c) 2200 ns. Reproduced from [18]. CC BY 4.0.

30.7 Discussion

The laser-induced optical breakdown studies of air and selected gases reveal that usually multiple breakdown spots occur along the optical axis when focusing radiation to above-threshold irradiance. Once optical breakdown is achieved, the absorbed radiation energy drives the shock wave towards the incoming radiation. The forward cone, or the initial asymmetry noticeable in the shadowgraphs, is a measure of how high above-threshold irradiance is employed. The species concentration near the shock wave is higher than in the center, especially well developed for time delays of the order of 1 s. Both atomic species and diatomic molecular species such as CN indicate consistent results. Moreover, comparison of line-of-sight and of Abel-inverted data show agreement, especially at the spatial location of a breakdown bead closest to the incoming beam one notices the development of the laser-induced plasma that is close to spherical for irradiance levels about one order of magnitude higher than threshold. For higher irradiances, the shock wave appears to expand into spherical shape towards the laser side, in agreement with computed spherically expanding shock waves.

The presented investigations of CN formation, especially near the expanding shock wave, are instrumental for potential medical and industrial CN diagnosis applications. But clearly, CN detection with LIBS shows the following conclusions:

- Shock wave expansion affects the formation of CN molecules as the plasma expands;
- Stark widths and shifts can be used to determine electron density, but higher spectral resolutions would be desirable for determination of accurate values of electron densities;
- For time delays around 1 s, higher CN and electron concentrations occur near the shock wave than those in the central region of the plasma. The CN becomes concentrated towards the edges of the plasma, therefore slit size, energy per pulse, and measurement acquisition time would need to be considered when capturing data especially for handheld design;
- The use of a 309 nm cut-on filter is an effective way to filter out unwanted atomic carbon line contributions but causes a 10 % reduction in the signal captured, which can cause issues with possible quantification for medical and forensic applications;
- As plasma expands and cools, radiation from excited CN molecules seems evenly distributed and indicates a close to homogenous temperature;
- Abel inversion is only justified for radially symmetric light sources, but shadowgraph studies support symmetrization to elucidate spatial dependence.

References

[1] Miziolek A W, Palleschi V and Schechter I (ed) 2006 *Laser-Induced Breakdown Spectroscopy (LIBS)—Fundamentals and Applications* (New York: Cambridge University Press)

[2] Singh J P and Thakur S N (ed) 2020 *Laser-Induced Breakdown Spectroscopy* (New York: Elsevier Science)

[3] Gautam G, Helstern C M, Drake K A and Parigger C G 2016 *Int. Rev. At. Mol. Phys.* **7** 45

[4] Gautam G and Parigger C G 2018 *Atoms* **6** 46

[5] Parigger C G, Helstern C M, Jordan B S, Surmick D M and Splinter R 2020 *Molecules* **25** 615

[6] Helstern C M 2020 Laser-induced breakdown spectroscopy and plasmas containing cyanide *PhD Dissertation* The University of Tennessee, Knoxville, Knoxville, TN

[7] Bethe H A, Fuchs K, Hirschfelder J O, Magee J L, Peierls R E and Neumann J 1947 *Blastwave*

[8] Taylor G I 1950 *Proc. Math. Phys. Eng. Sci.* **201** 159

[9] Sedov L I 1959 *Similarity and Dimensional Methods in Mechanics* (Cambridge, MA: Academic)
[10] Sedov L I 1946 *Prikl. Mat. Mech.* **10** 241
[11] Harith M A, Palleschi V, Salvetti A, Singh D P, Tropiano G and Vaselli M 1990 *Beams* **8** 247
[12] Parigger C G, Helstern C M and Gautam G 2019 *Atoms* **7** 74
[13] Parigger C G, Helstern C M and Gautam G 2017 *Int. Rev. At. Mol. Phys.* **8** 25
[14] Dackman M 2014 Laser-induced breakdown spectroscopy for analysis of high-density methane–oxygen mixtures *Master Thesis* The University of Tennessee, Knoxville, Knoxville, TN
[15] Griem H 1974 *Spectral Line Broadening by Plasma* (New York: Academic)
[16] Surmick D M 2016 Spectroscopic imaging of aluminum containing plasma *PhD Dissertation* The University of Tennessee, Knoxville, Knoxville, TN
[17] O'Haver T 2018 Peakfit algorithm https://www.mathworks.com/matlabcentral/fileexchange/23611-peakfit-m
[18] Parigger C G, Helstern C M and Gautam G 2020 *Symmetry* **12** 2116
[19] Parigger C G and Hornkohl J O 2020 *Quantum Mechanics of the Diatomic Molecule with Applications* (Bristol: IOP Publishing)
[20] Parigger C G, Woods A C, Surmick D M, Gautam G, Witte M J and Hornkohl J O 2015 *Spectrochim. Acta* B **107** 132
[21] Nelder J A and Mead R 1965 *Comput. J.* **7** 308
[22] Pretzler G and Naturforsch Z 1991 *Z. Naturforsch* **46a** 639
[23] Pretzler G, Jäger H, Neger T, Philipp H and Woisetschläger J 1992 *Z. Naturforsch.* A **47** 955
[24] Kandel Y P 2009 An experimental study of hydrogen balmer lines in pulsed laser plasma *PhD Dissertation* Wesleyan University, Middletown, CT
[25] Gornushkin I B, Shabanov S V and Panne U 2011 *J. Anal. At. Spectrom.* **26** 1457
[26] Killer C 2014 Abel inversion algorithm http://www.mathworks.com/matlabcentral/fileexchange/43639-abel-inversion-algorithm
[27] Parigger C G, Gautam G and Surmick D M 2015 *Int. Rev. At. Mol. Phys.* **6** 43
[28] Gautam G 2017 On laser-induced plasma containing hydrogen *PhD Dissertation* The University of Tennessee, Knoxville, Knoxville, TN
[29] Parigger C G, Surmick D M and Gautam G 2017 *J. Phys. Conf. Ser.* **810** 012012
[30] Gautam G, Parigger C G, Helstern C M and Drake K A 2017 *Appl. Opt.* **56** 9277
[31] Dardis J and Costello J T 2010 *Spectrochim. Acta* B **65** 535
[32] Zel'dovich Y B and Raizer Y P 1966 *Physics of Shock Waves and High-Temperature Hydrodynamic Phenomena* ed W D Hayes and R F Probstein Vol 1 (New York: Academic)

Part III

Appendices

IOP Publishing

Quantum Mechanics of the Diatomic Molecule (Second Edition)

Christian G Parigger and James O Hornkohl

Appendix A

Review of angular momentum commutators

The customary starting point for the quantum theory of angular momentum is the commutation formula for the Cartesian components of the angular momentum operator **J**,

$$\mathbf{J}_i\mathbf{J}_j - \mathbf{J}_j\mathbf{J}_i = i\varepsilon_{ijk}\mathbf{J}_k \tag{A.1}$$

where

$$\varepsilon_{ijk} = \begin{cases} +1 & i, j, k \text{ in cyclic order} \\ -1 & i, j, k \text{ not in cyclic order}. \\ 0 & \text{if any indices equal} \end{cases} \tag{A.2}$$

The above commutator property usually defines the angular momentum operator. Coordinate transformations leave the angular momentum operator definition invariant. The conservation law for angular momentum is fundamental. The definition of angular momentum, equation (A.1), is, of course, invariant under specifically spatial translations and rotations. Furthermore, equation (A.1) is invariant under coordinate inversion and time reversal.

Van Vleck's reversed angular momentum method starts with equation (A.1) but then utilizes change of sign of i for angular momentum when a transformation of coordinates to a system attached to a rotating molecule is made,

$$\mathbf{J}_{i'}\mathbf{J}_{j'} - \mathbf{J}_{j'}\mathbf{J}_{i'} = -i\varepsilon_{i'j'k'}\mathbf{J}_{k'}. \tag{A.3}$$

Here, the primed index denotes a rotated coordinate. This equation containing the reversed sign of i is known as Klein's [1] anomalous commutation formula.

Two approaches are debated in this book, namely an operator and an algebraic approach [2], without utilizing Klein's anomalous commutation formula. Each approach begins with the standard commutator formula (equation (A.1)). The operator approach is included in the text, the algebraic approach is discussed in detail in this appendix. In principle, equation (A.3) can be utilized in building angular momentum theory. However, in analogy to the distinction between

doi:10.1088/978-0-7503-6204-7ch31

right- and left-handed coordinate systems, different signs occur. We only use the *standard* sign as indicated in equation (A.1) in computation of a molecular diatomic spectrum [3], i.e., without resorting to use of Klein's anomaly and Van Vleck's reversed angular momentum method.

Here, the reversal of the sign in equation (A.1) is briefly investigated for a unitary and an anti-unitary transformation. The Euler rotation matrix is a real, unitary matrix (see equation (A.5)). The determinate of the Euler rotation matrix is +1 meaning that the sign of vectors is preserved under rotations. A spatial rotation of coordinates is a proper transformation. Conversely, the inversion or parity operator constitutes an improper rotation—this transformation cannot be described exclusively in terms of the Euler angles. However, angular momentum is a pseudo or axial vector, preserving the sign of **J** under improper rotations. The parity operator is also unitary and equation (A.1) is preserved by the parity operator. Time reversal (time inversion or reversal of motion) changes the sign of **J** and it complex-conjugates the imaginary unit due to time reversal being anti-unitary. Thus, equation (A.1) is invariant under time reversal. As shown in texts (e.g., Messiah [4]), the time reversal operator has been designed to be anti-unitary, consequently preserving the sign of i in commutation formulae.

An algebraic approach reveals that the commutator equation equation (A.1) remains invariant when proper rotation of coordinates is applied. Standard Cartesian coordinates are employed in the particular representation of the rotation matrix, the use of spherical polar coordinates would yield the same results. The laboratory referenced **J** is transformed to the rotated coordinate system by application of the rotation matrix $\mathbf{D}(\alpha\beta\gamma)$,

$$\mathbf{J}' = \mathbf{D}(\alpha\beta\gamma)\mathbf{J}, \tag{A.4}$$

where α, β, and γ are the Euler angles and $\mathbf{D}(\alpha\beta\gamma)$ is an orthogonal matrix whose determinant is +1, Goldstein [5],

$$\mathbf{D}(\alpha\beta\gamma) = \begin{pmatrix} \cos\alpha\cos\beta\cos\gamma - \sin\alpha\sin\gamma & \sin\alpha\cos\beta\cos\gamma + \cos\alpha\sin\gamma & -\sin\beta\cos\gamma \\ -\cos\alpha\cos\beta\sin\gamma - \sin\alpha\cos\gamma & -\sin\alpha\cos\beta\sin\gamma + \cos\alpha\cos\gamma & \sin\beta\sin\gamma \\ \cos\alpha\sin\beta & \sin\alpha\sin\beta & \cos\beta \end{pmatrix}. \tag{A.5}$$

The Euler angles and the matrix $\mathbf{D}(\alpha\beta\gamma)$ used here are those normally used in quantum mechanics, such as by Messiah [4], Davydov [6], Goldstein [5], Rose [7], Brink and Satchler [8], Tinkham [9], Gottfried [10], Baym [11], and Shore and Menzel [12]. This same set of Euler angles is also used by some authors of books on the theory of diatomic spectra, such as Judd [13] and Mizushima [14]. Evaluation of the angular momentum commutation formulae in the rotated system of coordinates gives

$$\mathbf{J}_{i'}\mathbf{J}_{j'} - \mathbf{J}_{j'}\mathbf{J}_{i'} = i\varepsilon_{i'j'k'}\mathbf{J}_{k'}. \tag{A.6}$$

This result is obtained from equations (A.1), (A.4), and (A.5). The calculation is simplified somewhat if one notes that for an orthogonal matrix the cofactors, i.e., signed minor determinants, are equal to the corresponding matrix elements of $\mathbf{D}(\alpha\beta\gamma)$ labeled m_{ij}, i.e., m_{ij} is its own cofactor. For example,

$$\begin{aligned} J_{x'}J_{y'} - J_{y'}J_{x'} = i\big[&(m_{12}m_{23} - m_{13}m_{22})(J_yJ_z - J_zJ_y) \\ &+ (m_{13}m_{21} - m_{11}m_{23})(J_xJ_z - J_zJ_x) \\ &+ (m_{11}m_{22} - m_{12}m_{21})(J_xJ_y - J_yJ_x)\big], \end{aligned} \tag{A.7}$$

and since

$$\begin{aligned} m_{12}m_{23} - m_{13}m_{22} &= m_{31}, \\ m_{13}m_{21} - m_{11}m_{23} &= - m_{32}, \\ m_{11}m_{22} - m_{12}m_{21} &= m_{33}, \end{aligned} \tag{A.8}$$

the right-hand side of equation (A.6) reduces to $iJ_{z'}$.

The review of Klein's [1] anomalous formula concludes that reversal of the sign of the sign is not required in diatomic molecular spectroscopy. The anomalous sign in the angular momentum commutators does not reveal a novel aspect of the nature of diatomic molecules. It is noteworthy that the anomalous commutation formula remains today a time-honored tradition in the theory of molecular spectra, as evidenced in several references [10–22]. Klein's anomalous commutators are means by which matrix elements of various operators in the molecular Hamiltonian are obtained, in particular those expressed in terms of angular momentum raising and lowering operators.

References

[1] Klein O 1929 *Zur Frage der Quantelung des asymmetrischen Kreisels* **58** 730
[2] Parigger C G and Hornkohl J O 2010 *Int. Rev. At. Mol. Phys.* **1** 25
[3] Hornkohl J O and Parigger C G 1996 *Am. J. Phys.* **64** 623
[4] Messiah A 1964 *Quantum Mechanics* (Amsterdam: North-Holland)
[5] Goldstein H, Poole C P and Safko J L 2001 *Classical Mechanics* 3rd edn (Reading, MA: Addison-Wesley)
[6] Davydov A S 1965 *Quantum Mechanics* (Oxford: Pergamon)
[7] Rose M E 1995 *Elementary Theory of Angular Momentum* (Mineola, NY: Dover)
[8] Brink D M and Satchler G R 1968 *Angular Momentum* (Oxford: Oxford University Press)
[9] Tinkham M 1964 *Group Theory and Quantum Mechanics* (New York: McGraw-Hill)
[10] Gottfried K 1989 *Quantum Mechanics* (Reading: Addison-Wesley)
[11] Baym G 1969 *Lectures on Quantum Mechanics* (London: Benjamin/Cummings)
[12] Shore B W and Menzel D H 1968 *Principles of Atomic Spectra* (Reading: Addison-Wesley)
[13] Judd B 1975 *Angular Momentum Theory for Diatomic Molecules* (New York: Academic)
[14] Mizushima M 1975 *The Theory of Rotating Diatomic Molecules* (New York: Wiley)
[15] Van Vleck J H 1951 The coupling of angular momentum vectors in molecules *Rev. Mod. Phys.* **23** 213
[16] Freed K F 1966 *J. Chem. Phys.* **45** 4214
[17] Kovács I 1969 *Rotational Structure in the Spectra of Diatomic Molecules* (New York: American Elsevier)
[18] Hougen J T 2001 *The Calculation of Rotational Energy Levels and Rotational Line Intensities in Diatomic Molecules* (Gaithersburg, MD: National Institute of Standards and Technology) http://physics.nist.gov/DiatomicCalculations. Originally published as *The Calculation of*

Rotational Energy Levels and Rotational Line Intensities in Diatomic Molecules, J. T. Hougen, NBS Monograph 115 (1970)

[19] Miller T A, Carrington A and Levy D H 1970 *Adv. Chem. Phys.* **18** 149

[20] Brown J M and Carrington A 2003 *Rotational Spectroscopy of Diatomic Molecules* (Cambridge: Cambridge University Press)

[21] Lefebvre-Brion H and Field R W 2004 *The Spectra and Dynamics of Diatomic Molecules* (New York: Elsevier/Academic)

[22] Bunker P R and Jensen P 1998 *Molecular Symmetry and Spectroscopy* 2nd edn (Ottawa: NRC))

Christian G Parigger and James O Hornkohl

Appendix B

Effects of raising and lowering operators

This appendix addresses details of the effects of raising and lowering operators on standard states $|JM\rangle$ and on elements of the rotation matrix $D^{J*}_{M\Omega}(\alpha\beta\gamma)$ [1]. The angular momentum raising and lowering operators have the following effects on the standard $|JM\rangle$ states,

$$J_{\pm}|JM\rangle = C_{\pm}(J,\, M)\ |J,\, M \pm 1\rangle, \tag{B.1}$$

where

$$\begin{aligned} C_{\pm}(J,\, M) &= \sqrt{J(J+1) - M(M \pm 1)} \\ &= \sqrt{(J \mp M)(J \pm M - 1)}\,. \end{aligned} \tag{B.2}$$

This general equation is, of course, applicable to the diatomic molecule. However, as a result of approximation, one deals with approximate diatomic eigenfunctions. Contained in Van Vleck's method is his discovery that the above standard results are not directly applicable to approximate diatomic eigenfunctions. Typically, two magnetic quantum numbers occur for approximate diatomic eigenfunctions, M and Ω in Hund's case (*a*) or M_N and Λ in case (*b*).

In modern notation, approximate diatomic angular momentum states are represented by elements of the rotation matrix, $D^{J*}_{M\Omega}(\alpha\beta\gamma)$, which carry two magnetic quantum numbers, one more than allowed by the nature of angular momentum. Only $\mathbf{J}^2$ and one of its components, by usual convention J_z, commute with the Hamiltonian. It will be important in our approach to find the effects of the raising and lowering operators on elements of the rotation matrix while applying standard theory.

The rotated raising operator, J'_{+},

$$J'_{+} = J_{x'} + iJ_{y'}, \tag{B.3}$$

lowers the Ω quantum number on Hund's case (*a*) states; see, for example, Van Vleck [2], Judd [3], Mizushima [4], Freed [5], Kovacs [6], Hougen [7], Carrington *et al* [8],

doi:10.1088/978-0-7503-6204-7ch32

Zare *et al* [9], Brown and Howard [10], and Lefebvre-Brion and Field [11]. Similarly, the rotated raising operator N'_+ lowers the Λ quantum number on case (*b*) kets. Agreement between eigenvalues of Hamiltonian matrices built using these results and experimentally measured term values has firmly established their correctness. Klein's anomalous commutators are often referenced in debating the reason why J'_+ lowers Ω and why N'_+ lowers Λ. However, of interest will be the following equation

$$J'_\pm \, D^{J*}_{M\Omega}(\alpha\beta\gamma) = -C_\mp(J, \Omega) \, D^{J*}_{M,\,\Omega\mp 1}(\alpha\beta\gamma), \tag{B.4}$$

which will be derived below. Note that J'_+ lowers Ω, that J'_- raises Ω, and that an unexpected minus sign occurs.

The nature of angular momentum does not allow M and Ω both to be rigorously *good* quantum numbers. This is equivalent to stating that J_z and $J_{z'}$ do not commute. According to definition of angular momentum, J_z does not commute with J_x or J_y, but it is, perhaps, not obvious that J_z and $J_{z'}$ fail to commute. One can easily show, Gottfried [12], that

$$[J_z, J_{z'}] = i\sin(\beta)J_\beta \tag{B.5}$$

$$= i\sin(\beta)[-\sin(\alpha)J_x + \cos(\alpha)J_y]. \tag{B.6}$$

where $J_\beta = \partial/\partial\beta$ is the angular momentum operator for rotation about the first intermediate y-axis. In general, J_z and $J_{z'}$ do not commute. Thus, $D^{J*}_{M,\,\Omega}(\alpha\beta\gamma)$ cannot represent a state of angular momentum of a molecule or any other system. The rotation matrix connects two different states of angular momentum.

B.1 Angular momentum operators

Angular momentum operator representations in terms of Euler angles are elaborated. A rotation provides a particularly simple way of expressing a component of angular momentum. The three Euler rotations give the following three components:

$$J_\alpha = -i\frac{\partial}{\partial\alpha} = J_z, \tag{B.7}$$

$$J_\beta = -i\frac{\partial}{\partial\beta}, \tag{B.8}$$

$$J_\gamma = -i\frac{\partial}{\partial\gamma} = J_{z'}. \tag{B.9}$$

Each of these operators is referenced to a different coordinate system, i.e., $J_\alpha = J_z$ in the laboratory system, $J_\beta = J_{y_1}$ in the first intermediate system, and $J_\gamma = J_{z'}$ in the fully rotated system. Applying the first Euler rotation to the vector operator $\mathbf{J}$ results in

$$\begin{pmatrix} J_{x_1} \\ J_{y_1} \\ J_{z_1} \end{pmatrix} = \begin{pmatrix} \cos\alpha & \sin\alpha & 0 \\ -\sin\alpha & \cos\alpha & 0 \\ 0 & 0 & 1 \end{pmatrix} \begin{pmatrix} J_x \\ J_y \\ J_z \end{pmatrix}. \tag{B.10}$$

From this one finds

$$J_{y_1} = J_\beta = -\sin\alpha\ J_x + \cos\alpha\ J_y, \tag{B.11}$$

giving J_β in terms of the laboratory coordinates of $\mathbf{J}$. Similarly, using the full rotation matrix, equation (A.4), one can express J_γ in laboratory coordinate system,

$$\mathbf{J}' = D(\alpha\beta\gamma)\mathbf{J}, \tag{B.12}$$

$$J_{z'} = J_\gamma = \cos\alpha\sin\beta\ J_x + \sin\alpha\sin\beta\ J_y + \cos\beta\ J_\alpha, \tag{B.13}$$

where the substitution $J_z = J_\alpha$ has been made. Equations (B.11) and (B.13) can be inverted for J_x and J_y:

$$J_x = -i\left(-\cos\alpha\cot\beta\frac{\partial}{\partial\alpha} - \sin\alpha\frac{\partial}{\partial\beta} + \frac{\cos\alpha}{\sin\beta}\frac{\partial}{\partial\gamma}\right), \tag{B.14}$$

$$J_y = -i\left(-\sin\alpha\cot\beta\frac{\partial}{\partial\alpha} + \cos\alpha\frac{\partial}{\partial\beta} + \frac{\sin\alpha}{\sin\beta}\frac{\partial}{\partial\gamma}\right), \tag{B.15}$$

$$J_z = -i\frac{\partial}{\partial\gamma}. \tag{B.16}$$

The method in obtaining these results included evaluation of J_β and J_γ in terms of the laboratory components of $\mathbf{J}$. Similarly, $J_{x'}$, $J_{y'}$, and $J_{z'}$ can be obtained by expressing J_α and J_β in terms of the components of $\mathbf{J}'$. The inverse of the full rotation matrix is applied to the rotated vector $\mathbf{J}'$,

$$\mathbf{J} = D^{-1}(\alpha\beta\gamma)\mathbf{J}' = D^{\dagger}(\alpha\beta\gamma)\mathbf{J}', \tag{B.17}$$

to find J_α in terms of the rotated coordinates of $\mathbf{J}$,

$$J_z = J_\alpha = -\sin\beta\cos\gamma\ J_{x'} + \sin\beta\sin\gamma\ J_{y'} + \cos\beta\ J_\gamma. \tag{B.18}$$

The Euler β-rotation is taken about the first intermediate y-axis meaning that the first intermediate and second intermediate y-axes coincide. Thus, J_β can be evaluated in fully rotated coordinates by applying the inverse of the γ rotation matrix to $\mathbf{J}'$,

$$\begin{pmatrix} J_{x_2} \\ J_{y_2} \\ J_{z_z} \end{pmatrix} = \begin{pmatrix} \cos\gamma & -\sin\gamma & 0 \\ \sin\gamma & \cos\gamma & 0 \\ 0 & 0 & 1 \end{pmatrix} \begin{pmatrix} J_{x'} \\ J_{y'} \\ J_{z'} \end{pmatrix}. \tag{B.19}$$

$J_{y_2} = J_{y_1} = J_\beta$; therefore, we find

$$J_\beta = \sin\gamma\ J_{x'} + \cos\gamma\ J_{y'}. \tag{B.20}$$

The two equations in two unknowns are inverted as before. We find for $J_{x'}$ and $J_{y'}$

$$J_{x'} = -i\left(\cos\gamma\cot\beta\frac{\partial}{\partial\gamma} + \sin\gamma\frac{\partial}{\partial\beta} - \frac{\cos\gamma}{\sin\beta}\frac{\partial}{\partial\alpha}\right), \tag{B.21}$$

$$J_{y'} = -i\left(-\sin\gamma\cot\beta\frac{\partial}{\partial\gamma} + \cos\gamma\frac{\partial}{\partial\beta} + \frac{\sin\gamma}{\sin\beta}\frac{\partial}{\partial\alpha}\right), \tag{B.22}$$

$$J_{z'} = -i\frac{\partial}{\partial\gamma}. \tag{B.23}$$

The raising and lowering operators are then constructed using the results above,

$$J_{+} = -ie^{i\alpha}\left(-\cot\beta\frac{\partial}{\partial\alpha} + i\frac{\partial}{\partial\beta} + \frac{1}{\sin\beta}\frac{\partial}{\partial\gamma}\right), \tag{B.24}$$

$$J_{-} = -ie^{-i\alpha}\left(-\cot\beta\frac{\partial}{\partial\alpha} - i\frac{\partial}{\partial\beta} + \frac{1}{\sin\beta}\frac{\partial}{\partial\gamma}\right), \tag{B.25}$$

$$J_{+'} = -ie^{-i\gamma}\left(\cot\beta\frac{\partial}{\partial\gamma} + i\frac{\partial}{\partial\beta} - \frac{1}{\sin\beta}\frac{\partial}{\partial\alpha}\right), \tag{B.26}$$

$$J_{-'} = -ie^{i\gamma}\left(\cot\beta\frac{\partial}{\partial\gamma} - i\frac{\partial}{\partial\beta} - \frac{1}{\sin\beta}\frac{\partial}{\partial\alpha}\right). \tag{B.27}$$

These general results also apply to systems composed of any number of particles. A modification or better *simplification* is required for a system consisting of a single particle (or two particles, since the two-body reduction can always be applied to a system of two particles). The third Euler angle, γ, is superfluous for a single particle, i.e., $\partial/\partial\gamma = 0$. Choosing the first Euler angle to be azimuthal angle ϕ and the second Euler angle to be the polar angle θ, the equations (B.14)-(B.16)) reduce to the familiar textbook equations for the angular momentum operators of a single particle.

Comparison of equations (B.14)–(B.16)) with equations (B.14–B.16) shows that the components of angular momentum are changed by a coordinate transformation. However, the defining commutators for angular momentum in terms of its Cartesian components remain invariant, although the individual components differ. We note that equations (B.7), (B.24), and (B.25) agree with Judd's [3] equation (1.22), but (B.9), (B.26), and (B.27) differ in sign from Judd's equation (1.23), presumably due to the use of the anomalous commutator formula. It appears that Judd obtained his rotated operators in a manner which guaranteed they would obey Klein's anomalous commutation formula.

B.2 Angular momentum commutators and rotation matrix elements

General relations for commutators are typically obtained by applying the commutator to an abstract ket describing a physical state. Alternatively, in the Schrödinger representation, the commutators are obtained by applying differential operators to physical eigenfunctions. This appendix demonstrates how two anomalous results may occur when applying $[J_{x'}, J_{y'}]$ and $[J_y, J_{y'}]$ to the rotation matrix elements $D^{J*}_{M\Omega}(\alpha\beta\gamma)$.

The commutator $[J_{x'}, J_{y'}]$ is evaluated using

$$[J_{x'}, J_{y'}] = \left[\frac{1}{2}(J'_+ + J'_-), \frac{1}{2i}(J'_+ - J'_-)\right] = -\frac{i}{2}\left[J'_-, J'_+\right], \tag{B.28}$$

and equation (B.4) which for convenience is repeated here

$$J'_\pm\ D^{J*}_{M\Omega}(\alpha\beta\gamma) = -C_\mp(J, \Omega)\ D^{J*}_{M,\ \Omega\mp 1}(\alpha\beta\gamma). \tag{B.29}$$

Successive application of the operators in the rotated frame of reference yields the intermediate result

$$-\frac{i}{2}\left[J'_-, J'_+\right]D^{J*}_{M\Omega}(\alpha\beta\gamma) = -\frac{i}{2}(C_+(J, \Omega - 1)C_-(J, \Omega) - C_-(J, \Omega + 1)C_+(J, \Omega))D^{J*}_{M\Omega}(\alpha\beta\gamma), \tag{B.30}$$

which after inserting (compare equation (B.2))

$$C_\pm(J, \Omega) = \sqrt{(J \mp \Omega)(J \pm \Omega + 1)}, \tag{B.31}$$

leads to

$$[J_{x'}, J_{y'}]D^{J*}_{M\Omega}(\alpha\beta\gamma) = -i\Omega D^{J*}_{M\Omega}(\alpha\beta\gamma). \tag{B.32}$$

It might be tempting to conclude anomalous commutator relations in the rotated molecular frame from this identity. However, the rotation matrix elements contain two quantum numbers M and Ω, one too many to represent a physical eigenfunction.

Similarly, $[J_y, J_{y'}]$ is evaluated using

$$[J_y, J_{y'}] = \left(-\frac{1}{4}\left[J_+, J'_+\right] - \frac{1}{4}\left[J_-, J'_-\right] + \frac{1}{4}\left[J_+, J'_-\right] + \frac{1}{4}\left[J_-, J'_+\right]\right), \tag{B.33}$$

and applying it to the rotation matrix elements $D^{J*}_{M\Omega}(\alpha\beta\gamma)$. Note $J_{y'}$ acts on Ω (see equations (B.29)) while J_y acts on M,

$$J_\pm\ D^{J*}_{M\Omega}(\alpha\beta\gamma) = C_\pm(J, M)\ D^{J*}_{M\pm 1,\ \Omega}(\alpha\beta\gamma). \tag{B.34}$$

The intermediate step is given here

$$
\begin{aligned}
[J_y, J_{y'}]D^{J*}_{M\Omega}(\alpha\beta\gamma) = \\
&+\frac{1}{4}C_-(J,\,\Omega)J_+D^{J*}_{M\ \Omega-1}(\alpha\beta\gamma) + \frac{1}{4}C_+(J,\,M)J'_+D^{J*}_{M+1\ \Omega}(\alpha\beta\gamma) \\
&+\frac{1}{4}C_+(J,\,\Omega)J_-D^{J*}_{M\ \Omega+1}(\alpha\beta\gamma) + \frac{1}{4}C_-(J,\,M)J'_-D^{J*}_{M-1\ \Omega}(\alpha\beta\gamma) \\
&-\frac{1}{4}C_+(J,\,\Omega)J_+D^{J*}_{M\ \Omega+1}(\alpha\beta\gamma) - \frac{1}{4}C_+(J,\,M)J'_-D^{J*}_{M+1\ \Omega}(\alpha\beta\gamma) \\
&-\frac{1}{4}C_-(J,\,\Omega)J_-D^{J*}_{M\ \Omega-1}(\alpha\beta\gamma) - \frac{1}{4}C_-(J,\,M)J'_+D^{J*}_{M-1\ \Omega}(\alpha\beta\gamma),
\end{aligned}
\tag{B.35}
$$

which reduces to

$$
[J_y, J_{y'}]D^{J*}_{M\Omega}(\alpha\beta\gamma) = 0. \tag{B.36}
$$

Again, one might be tempted to infer from equation (B.36) a general commutator relation that also applies to angular momentum states.

References

[1] Parigger C G and Hornkohl J O 2010 *Int. Rev. At. Mol. Phys.* **1** 25

[2] Van Vleck J H 1951 The coupling of angular momentum vectors in molecules *Rev. Mod. Phys.* **23** 213

[3] Judd B 1975 *Angular Momentum Theory for Diatomic Molecules* (New York: Academic)

[4] Mizushima M 1975 *The Theory of Rotating Diatomic Molecules* (New York: Wiley)

[5] Freed K F 1966 *J. Chem. Phys.* **45** 4214

[6] Kovács I 1969 *Rotational Structure in the Spectra of Diatomic Molecules* (New York: American Elsevier)

[7] Hougen J T 2001 *The Calculation of Rotational Energy Levels and Rotational Line Intensities in Diatomic Molecules* (Gaithersburg, MD: National Institute of Standards and Technology) http://physics.nist.gov/DiatomicCalculations. Originally published as *The Calculation of Rotational Energy Levels and Rotational Line Intensities in Diatomic Molecules*, J. T. Hougen, NBS Monograph 115 (1970)

[8] Miller T A, Carrington A and Levy D H 1970 *Adv. Chem. Phys.* **18** 149

[9] Zare R N, Schmeltekopf A L, Harrop W J and Albritton D L 1973 *J. Mol. Spectrosc.* **46** 37

[10] Brown J M and Howard B J 1976 *Mol. Phys.* **31** 1517

[11] Lefebvre-Brion H and Field R W 2004 *The Spectra and Dynamics of Diatomic Molecules* (New York: Elsevier/Academic)

[12] Gottfried K 1989 *Quantum Mechanics* (Reading: Addison-Wesley)

IOP Publishing

Quantum Mechanics of the Diatomic Molecule (Second Edition)

Christian G Parigger and James O Hornkohl

Appendix C

Modified Boltzmann plots

This appendix elaborates on the use of Boltzmann plots [1] for diatomic molecular spectroscopy inferences of temperature from a measured emission spectra. The Boltzmann plot method evaluates temperature by finding the slope from a semilog graph of recorded spectral lines, e.g., the integrated line shape in determination of hydrogen excitation temperature in atomic spectroscopy. In molecular spectroscopy, usually there are many lines within a spectral resolution of the order of 0.1nm, which are typical for laser-plasma emission spectroscopy that uses resolving powers of the order of 5000. Consequently, a modified Boltzmann plot approach [2] is developed for analysis of measured molecular spectra.

C.1 Boltzmann plots

The formal approach starts with an equation containing an exponential,

$$f(x) = A\, e^{-\alpha\, x}, \tag{C.1}$$

one evaluates the logarithm of both sides, and puts the result into point-slope form for a straight line,

$$\begin{aligned} \ln[f(x)] &= -\,\alpha\, x + \ln A \\ y(x) &= \ \ m\, x + b. \end{aligned} \tag{C.2}$$

In spectroscopy, spontaneous emission from a gas adheres to the spectral radiance, or frequently labeled 'intensity', originating from an upper level, u, to a lower level, l,

$$I_{ul} = \frac{64\,\pi^4\,(a_0 e)^2\, c\, C_{\mathrm{abs}}\, N_0}{3\,Q}\, C_{\tilde{\nu}}\, \tilde{\nu}_{ul}^{\,4}\, S_{ul}\, e^{-h\,c\,F_u/(k_B\,T)}. \tag{C.3}$$

doi:10.1088/978-0-7503-6204-7ch33

The individual terms are listed below.

- $a_0\ e$: Bohr radius ×electronic charge.
- c, h, k_B: speed of light, Planck constant, Boltzmann constant.
- C_{abs}: absolute intensity calibration factor and $C_{\tilde{\nu}}$—relative spectral sensitivity calibration factor.
- N_0: number density, molecules / cm^3.
- Q: partition function of gas.
- $\tilde{\nu}$: wave number of spectral line produced by $u \to l$ transition.
- S_{ul}: electric dipole line strength.
- F_u: upper term value (i.e., energy eigenvalue divided by $h\ c$).
- T: absolute temperature.

The linearized spontaneous emission equation,

$$\ln\left(\frac{I_{ul}}{C_{\tilde{\nu}}\ \tilde{\nu}^4\ S_{ul}}\right) = -\frac{h\ c}{k_B\ T}\ F_u - \ln\left(\frac{64\ \pi^4\ (a_0 e)^2\ c\ C_{\mathrm{abs}}\ N_0}{3\ Q}\right), \tag{C.4}$$

$$= y\ m \quad x + \qquad\qquad b,$$

is solved to find the slope m. A Boltzmann plot is a graph of $\ln\left(\frac{I_{ul}}{C_{\tilde{\nu}}\ \tilde{\nu}^4\ S_{ul}}\right)$ versus the upper term value F_u. The slope is proportional to $-1/T$, of course provided that thermodynamic equilibrium is reached,

$$T = -\frac{h\ c}{k_B\ m} \tag{C.5}$$

Since the points would deviate from straight line, one labels the graph as a nonlinear Boltzmann plot, which would imply departure from thermal equilibrium. Figure C.1 illustrates an example of a Boltzmann plot.

Figure C.2 displays the spectrum from which the Boltzmann plot is constructed.

A Boltzmann plot can be made using the peak intensity or (better) the intensity integrated under each spectral line, but in either case the spectral lines must be fully resolved. The following shows how to make a Boltzmann plot when the spectral lines are not fully resolved. Table C.1 shows an example of a line list.

An emission spectral line is broadened by physical processes in the radiating gas and the properties of the spectrometer used to record the spectral line. A widely-used measure of spectral line width is the full width at half maximum (FWHM.) A spectrum is effectively broken into a finite number of pixels between the minimum λ_{min} and the maximum λ_{max} wavelengths in the spectrum (or the minimum or maximum frequencies ν or wave numbers $\tilde{\nu}$), and the FWHM can be used as the pixel width.

An emission spectrum is computed by evaluating expressions indicated in equation (C.3). One selects a line shape function (i.e., a Gaussian) and an FWHM, and computes the contributions of each line. A snippet of a program code utilizes a Gaussian line shape.

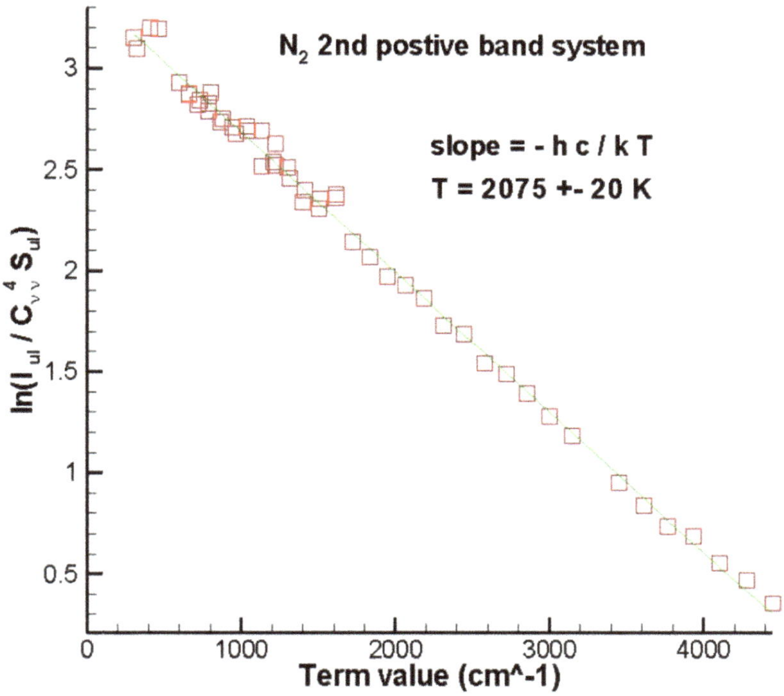

Figure C.1. A Boltzmann plot for a spectrum from the N_2 second positive system.

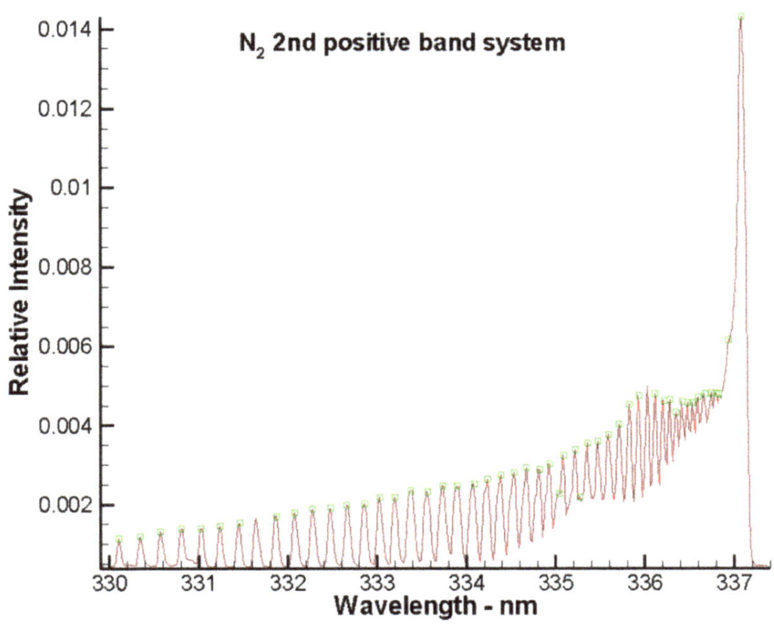

Figure C.2. A Boltzmann plot constructed a N_2 second positive system.

Table C.1. Spectral line lists: an example downloaded from NIST.

Wavelength air (Å)	A_{ul} (s^{-1})	E_u (cm −1)	E_l (cm −1)	g_u
2118.312	1.03×10^7	0	47 192.38	2
2123.362	1.22×10^7	112.061	47 192.38	4
2129.663	1.52×10^7	0	46 940.97	2
2134.733	1.81×10^7	112.061	46 941.55	4
2145.555	2.33×10^7	0	46 593.32	2
2150.699	2.79×10^7	112.061	46 593.95	4
2168.805	3.06×10^7	0	46 093.424	2
2174.028	3.65×10^7	112.061	46 094.312	4
2199.150	1.75×10^6	0	45 457.244	2
2204.590	3.49×10^6	112.061	45 457.244	4
2204.660	4.53×10^7	0	45 344.165	2
2210.046	5.40×10^7	112.061	45 345.594	4
2257.999	3.77×10^6	0	44 273.133	2

```
!     Guassian line shape program segment
      delwn = (wn_lo-wn_hi) / (npts-1)
      con = 2.0 * sqrt(alog(2.0))
      do 70 k=1, kmax
         ndel = nint(2.5*FWHM_wn/delwn)
         if (wn(k).lt.wn_lo.or.wn(k).gt.wn_hi) go to 70
         n0 = nint((wn(k)-wnmin)/delwn) + 1
         nmin = n0 - ndel
         if (nmin.lt.1)  nmin = 1
         nmax = n0 + ndel
         if (nmax.gt.npts)  nmax = npts
         do n=nmin, nmax
            u = con * (wn(k)-x(n)) / FWHM_wn
            if (u.lt.9.21)  then
               y(n) = y(n) + peak(k) * exp(-u*u)
               if (y(n).gt.ymax)  ymax = y(n)
            end if
         end do
   70 end do
```

C.2 Modified Boltzmann plot

The intensity I_p falling on a pixel (*i.e*, a small wave number range $\Delta\tilde{\nu}$) is the sum of all of the intensities of the spectral lines that contribute to the intensity range $\Delta\tilde{\nu}$,

$$I_p = \sum_{i=1}^{\text{all lines}} \text{const.}\ f(\tilde{\nu}_i,\, \Delta\tilde{\nu})\ C_i\ \tilde{\nu}_i^4\ S_i\ e^{-h\,c\,F_i/(k\,T)} \tag{C.6}$$

where $f(\tilde{\nu}_i,\, \Delta\tilde{\nu})$ is the line shape function. Because the term value F_i varies greatly from line to line, the exponential cannot be taken outside the summation. Therefore, an arbitrary exponential $e^{h\,c\,F_j/(k\,T)}$ is inserted,

$$I_p = e^{-h\,c\,F_j/(k\,T)} \sum_{i=1}^{\text{all lines}} \text{const.}\ f(\tilde{\nu}_i,\, \Delta\tilde{\nu})\ C_i\ \tilde{\nu}_i^4\ S_i\ e^{-h\,c\,(F_j\,-\,F_i)/(k\,T)}. \tag{C.7}$$

The exponential $e^{-h\,c\,F_j/(k\,T)}$ can be isolated, meaning that for assumed values of the term value F_j and temperature T, one can create a Boltzmann plot. The value of F_j is chosen to be F_i for the spectral line making the biggest contribution to I_p. The value of

$$ln\left(\frac{I_p}{C_i\ \tilde{\nu}^4\ S_i}\right) / \sum_{i=1}^{\text{all lines}} \text{const.}\ f(\tilde{\nu}_i,\, \Delta\tilde{\nu}_i)\ e^{-h\,c\,(F_j\,-\,F_i)/(k\,T)} \tag{C.8}$$

is plotted versus F_j for a trial T, but there is no reason why the temperature inferred from this plot will equal the trial temperature. However, the new temperature can replace the trial temperature, and the process is repeated, i.e., the procedure must be iterated until the trial temperature and the temperature from the Boltzmann plot agree.

An example output from modified Boltzmann plot routine displays the effectiveness of the iteration to find the excitation temperature.

Trial temperature = 6000.
1 2078.
2 2073.
3 2073.
4 2073.
5 2073.
Std.Dev. Boltzmann plot = 5.1378×10^{-2}
$a0 = 3.364 \quad \pm 1.2980 \times 10^{-2}$
Second radiation constant=1.438 79
$a1 = -6.9404 \times 10^{-4} \quad \pm 6.314 \times^1 0 - 6$
Boltzmann plot temperature = 2073. Std.dev. = $\pm 18.9K$

References

[1] Wiese W L 1991 *Spectrochim. Acta* B **46** 831

[2] Hornkohl J O, Parigger C G and Lewis J W L 1991 *J. Quant. Spectrosc. Radiat. Transfer* **46** 405

IOP Publishing

Quantum Mechanics of the Diatomic Molecule (Second Edition)

Christian G Parigger and James O Hornkohl

Appendix D

Aspects of nitric oxide computations

This appendix communicates a collection of notes on the molecular parameters of nitric oxide (NO). These notes contain considerable discussion of the quantum mechanics of the diatomic molecule. The interesting topic of hyperfine structure is ignored.

In the present state of the applied science, the molecular parameters are found by fitting term value differences computed from an approximate Hamiltonian to the best experimental values for the line positions. The computation begins with trial values of the parameters, which are then iteratively refined until the differences between the computed and experimental line positions are minimized in the least squares sense.

This appendix discusses results found from fitting some of the best available NO data. The literature on NO is too vast for a comprehensive review. It can only be hoped that the following offers a representative sampling.

D.1 Matrix elements of the Hamiltonian

This section gives a review of the methods of applied quantum mechanics in diatomic spectroscopy. As an example of high quality experimental spectra, reference will be made to the NO rotation-vibration data reported by Amiot *et al* [1]. Their measured line positions are not reported in the journal article, but their data are still available from the Canadian Depository of Unpublished Data. The Macki and Wells [2] data for the (1,0) band are more accurate than the data of Amiot *et al*, but Macki and Wells report only smoothed line positions instead of measured values.

For their analysis of their NO spectra, Amiot *et al* reference the important paper by Zare *et al* [3], which describes the computation of term values by diagonalization of the Hamiltonian. The Zare *et al* 's method consists of four steps, which will be further discussed.

In the first step, trial values of the molecular parameters and analytical methods are used to evaluate the matrix elements of the Hamiltonian in the Hund's case (a) basis.

doi:10.1088/978-0-7503-6204-7ch34

The second step is an analytical approximation, called the Van Vleck transformation, which reduces the dimension of the square Hamiltonian matrix to $2(2S+1) \times 2(2S+1)$, where S is the total electronic spin quantum number for the dominant state in the basis—if the latter is a Σ state, then the result of the Van Vleck transformation has the dimensions $(2S+1) \times (2S+1)$. In addition to reducing the dimensions of the Hamiltonian, the Van Vleck transformation also splits the Hamiltonian into two independent submatrices of opposite parity.

The ground state of NO is usually modeled as a dominant $^2\Pi$ state, which contributes a 4×4 submatrix to the Hamiltonian, weakly mixed with a $^2\Sigma^+$ state, which contributes a 2×2 submatrix. Thus, in this model the Hund's case (a) representation of the Hamiltonian is a 6×6 matrix. Van Vleck's approximate transformation reduces the Hamiltonian to a 4×4 matrix composed of two independent 2×2 matrices of opposite parity. The third step in Zare *et al*'s method is numerical diagonalization of the two submatrices. The fourth step is a matrix computation which gives corrections to the trial values of the molecular parameters. If the computed corrections are not negligibly small, steps 1–4 are iteratively repeated until the corrections become negligibly small.

The following determinations of NO parameters differ from Zare *et al*'s method in one way. The Van Vleck transformation is skipped. Thus, for the ground state of NO the present computations involve numerical diagonalization of a single 6×6 matrix instead of numerical diagonalization of two 2×2 submatrices.

Various different Hamiltonian models for NO ($X\,^2\Pi$) will be investigated below but the conclusion is that the accepted model of a dominant $^2\Pi$ state with a small component of a $^2\Sigma^+$ state adequately describes the available experimental observations. Table D.1 shows the fit of this model to the NO (1,0) band data of Amiot *et al* [1].

Most of the symbols in the following tables have their standard meanings. $T_v = T_e + G_v$ where T_e is the electronic term value and G_v is the vibrational term value. $\Delta\tilde{\nu}$ is the search tolerance used by the fitting program when it searches for an experimental spectral line to match a predicted line.

An experimental line for which there is no predicted line having the same J' and J'' whose vacuum wave number is within the search tolerance $\Delta\tilde{\nu}$ is eliminated from the fit. The table caption entry '405 lines of 419' indicates that 14 experimental lines were rejected from the fit, and σ is the standard deviation of the predicted line positions (vacuum wave number) with respect to the 405 accepted experimental line positions. Values in parenthesis represent 1 standard deviation expressed in the last digits. Values in square brackets were held fixed during the fitting process. Other symbols in the table will be defined below. The ratio of the standard deviation of the fit to the band origin, $\sigma/\tilde{\nu}_{10} \approx 2 \times 10^{-7}$.

Most of the richness of the typical diatomic spectrum is produced by the rotational Hamiltonian, the term in the diatomic Hamiltonian representing the kinetic energy of rotation of the nuclei, given by

$$H_{\text{rot}} = B(r)\,\mathbf{R}^2 \tag{D.1}$$

Table D.1. The molecular parameters for the $v = 0$ and $v = 1$ states of the ground electronic state of NO obtained by fitting the model of a dominant $^2\Pi$ state weakly mixed with a $^2\Sigma^+$ state to the (1,0) band data of Amiot *et al.* [1]

$X\ ^2\Pi(v = 1)$	$T_v = 2824.6331(5)$
$B_v = 1.678\ 5715(27)$	$D_v = 5.4893(26) \times 10^{-6}$
$A_v = 122.7026(21)$	$A_{vJ} = 3.475(38) \times 10^{-4}$
$<AL> = -177.8(1.0)$	$<AL>_J = 4.2(1.8) \times 10^{-3}$
$<BL> = -1.3922(76)$	$<BL>_J = -4.2(1.3) \times 10^{-5}$
$X\ ^2\Pi(v = 0)$	$[T_v = 948.66]$
$B_v = 1.696\ 1445(30)$	$D_v = 5.4735(29) \times 10^{-6}$
$A_v = 122.9495(23)$	$A_{vJ} = 3.565(42) \times 10^{-4}$
$<AL> = -180.4(1.1)$	$<AL>_J = 3.4(2.0) \times 10^{-3}$
$<BL> = -1.433(84)$	$<BL>_J = -3.1(1.5) \times 10^{-5}$
$A\ ^2\Sigma^+(v = 0)$	$[T_v = 45087.65]$
$[B_v = 1.986\ 312]$	$[D_v = 5.575 \times 10^{-6}]$
$[\gamma_v = -1.34 \times 10^{-4}]$	

The parameters for the $^2\Sigma^+$ state were held fixed to those values found from a fit to the $A\ ^2\Sigma^+(v = 0) \leftrightarrow X\ ^2\Pi(v = 0)$ γ-system data of Engleman *et al* [4]. The upper state extends to $0.5 \leqslant J' \leqslant 40.5$, includes 405 of 429 lines, with an accuracy of $\Delta\tilde{\nu} = 0.001$ cm^{-1} and $\sigma = 0.000\ 38$ cm^{-1}. The term value difference equals $\tilde{\nu}_{10} = 1875.9731$.

in which **R** is the orbital angular momentum operator for the nuclei whose motion has been reduced to that of a single, fictitious particle of reduced mass μ,

$$\mu = \frac{m_a m_b}{m_a + m_b} \tag{D.2}$$

of the nuclei whose masses are m_a and m_b, and

$$B(r) = \frac{\hbar}{4\pi c \mu r^2} \tag{D.3}$$

where r is the internuclear distance. The rotational parameter $B(r)$ has the units of energy, but the right-hand side of the above equation has been divided by hc (Planck's constant times the speed of light) to give $B(r)$ the spectroscopist's unit of energy which has the units of reciprocal length.

When, as has been assumed here, the normally very small influence of nuclear spin on the term values can safely be ignored, the total angular momentum,

$$\mathbf{J} = \mathbf{L} + \mathbf{R} + \mathbf{S} \tag{D.4}$$

is the sum of the total electronic orbital angular momentum **L**, the total nuclear orbital angular momentum **R**, and the total electronic spin **S**. The rotational Hamiltonian becomes [5]

$$\begin{aligned} H_{\text{rot}} &= B(r)(\mathbf{J} - \mathbf{L} - \mathbf{S})^2 \\ &= B(r)\Big[\mathbf{J}^2 + \frac{L_+L_- + L_-L_+}{2} + L_z^2 + \mathbf{S}^2 - (J_+L_- + J_-L_+ + 2J_zL_z) \\ &\quad - (J_+S_- + J_-S_+ + 2J_zS_z) + (L_+S_- + L_-S_+ + 2L_zS_z)\Big] \end{aligned} \tag{D.5}$$

$$\begin{aligned} &= B(r)\Big[\mathbf{J}'^2 + \frac{L'_+L'_- + L'_-L'_+}{2} + L_{z'}^2 + \mathbf{S}'^2 - (J'_+L'_- + J'_-L'_+ + 2J_{z'}L_{z'}) \\ &\quad - (J'_+S'_- + J'_-S'_+ + 2J_{z'}S_{z'}) + (L'_+S'_- + L'_-S'_+ + 2L_{z'}S_{z'})\Big]. \end{aligned} \tag{D.6}$$

where unprimed components are in the laboratory coordinate system, primed components are in a coordinate system attached to the molecule, and the raising and lowering operators are defined by, for example,

$$J_\pm = J_x \pm J_y, \tag{D.7}$$

with similar equations for $S_\pm$ and $L_\pm$. The analytical form of each term in the Hamiltonian is preserved under a coordinate transformation, compare equations (D.5) and (D.6). Matrix elements of Hamiltonian terms are much easier to analytically evaluate in molecular coordinates than in the laboratory coordinates.

As explained in many quantum mechanics texts, specification of the states of a quantum system is normally done with the standard $|nJM\rangle$ states of the system, where J is the quantum number for the total angular (the quantity which obeys the conservation law for angular momentum), M is the quantum number for the laboratory z component of $\mathbf{J}$, and n denotes all other required quantum numbers. The exact equation for the standard $|nJM\rangle$ states of the diatomic molecule is [5]

$$\langle \mathbf{r}_1, \mathbf{r}_2, \ldots, \mathbf{r}_\mathcal{N}, \mathbf{r}|nJM\rangle = \sum_{\Omega=-J}^{J} \langle \mathbf{R}'_e r|n\rangle \, D^{J*}_{M\Omega}(\alpha\beta\gamma) \tag{D.8}$$

where $\mathbf{r}_1, \mathbf{r}_2, \ldots, \mathbf{r}_\mathcal{N}$ are the laboratory coordinates of the $\mathcal{N}$ electrons; $\mathbf{r}$ is the internuclear vector; α, β, and γ are the Euler angles; R'_e represents $3\mathcal{N} - 1$ rotated electronic coordinates (γ is the missing electronic coordinate); and $D^J_{M\Omega}(\alpha\beta\gamma)$ is the matrix element of the rotation operator (note that its complex conjugate appears in the above equation). The Hund's case (a) basis is defined by [5]

$$\langle \mathbf{r}_1, \mathbf{r}_2, \ldots, \mathbf{r}_\mathcal{N}, \mathbf{r}|nJM\Omega\Lambda S\Sigma\rangle = \sqrt{\frac{2J+1}{8\pi^2}} \langle \mathbf{R}'_e r|n\rangle \, |S\Sigma\rangle \, D^{J*}_{M\Omega}(\alpha\beta\gamma). \tag{D.9}$$

in which Ω is the quantum number for the z' component of J and Σ is the quantum number for the z' component of S. The relationship

$$J_{z'} = L_{z'} + S_{z'} \tag{D.10}$$

holds because the z'-axis lies on the internuclear axis and the nuclei have no internuclear component of orbital angular momentum (i.e., $R_{z'} = 0$). Thus,

$$\Omega = \Lambda + \Sigma \tag{D.11}$$

and Λ, the quantum number for $L_{z'}$ is also a case (a) quantum number, although it does not explicitly appear on the right-hand side of equation (D.9).

The Born–Oppenheimer approximation has not been applied in either the exact equation for the $|nJM\rangle$ diatomic states, equation (D.8), or the case (a) basis, equation (D.9). The smallness of the electronic mass in comparison to the nuclear mass produces an approximate separation of the electronic and nuclear motions. In the usual exposition of the Born–Oppenheimer approximation, the approximate separation of the electronic and nuclear motions involves all of the electronic and nuclear internal coordinates. However, in the present formulation the Euler angles play no part in the Born–Oppenheimer separation, which is concerned exclusively with the eigenfunction $\langle \mathbf{R}'_e r|n\rangle$. The quantum number v, the quantum number associated with the internuclear distance r, is extracted from the collection of quantum numbers represented by the symbol n, and the Born–Oppenheimer approximate separation of $\langle \mathbf{R}'_e r|nv\rangle$ is then made,

$$\begin{aligned} \langle \mathbf{R}'_e r|nv\rangle &\approx \langle \mathbf{R}'_e;\, r|n\rangle\, \langle r|v\rangle \\ &= \psi_n(\mathbf{R}'_e;\, r)\, \psi_v(r) \end{aligned} \tag{D.12}$$

in which the semicolon is a notational device indicating that the electronic eigenfunction $\psi_e(\mathbf{R}'_e;\, r)$ is a parametric function of the internuclear distance r. This means that the electronic Schrödinger equation has a different solution for each value of the internuclear distance. Because the electronic eigenfunction is a parametric function of r, all electronic matrix elements are functions of r.

Application of the Born–Oppenheimer approximation to the case (a) basis gives

$$\langle \mathbf{r}_1, \mathbf{r}_2, \ldots, \mathbf{r}_\mathcal{N}, \mathbf{r}|nvJM\Omega\Lambda S\Sigma\rangle = \sqrt{\frac{2J+1}{8\pi^2}}\langle \mathbf{R}'_e;\, r|n\rangle\, \langle r|v\rangle |S\Sigma\rangle\, D^{J*}_{M\Omega}(\alpha\beta\gamma). \tag{D.13}$$

Matrix elements of the rotational Hamiltonian can be calculated using equations (D.6) and (D.13). Zare *et al* [3] give tables of case (a) matrix elements. They are also given by Lefebvre-Brion and Field [6]. Sign ambiguities between Zare *et al* and Lefebvre-Brion and Field can be resolved through application of equations (D.6) and (D.9) [5]. Table D.5 gives example matrix elements of $H_{\rm rot}$ in a state consisting of a mixture of $^2\Pi$ and $^2\Sigma$ case (a) basis states. The diagonal matrix elements of the rotational Hamiltonian, $\langle nJM\Omega\Lambda S\Sigma|H_{\rm rot}|nJM\Omega\Lambda S\Sigma\rangle$, are given by

$$\begin{aligned} &B_v\, \langle nJM\Omega\Lambda S\Sigma|\mathbf{J}^2 + L_{z'}^2 - 2J_{z'}L_{z'} - 2J_{z'}S_{z'} + 2L_{z'}S_{z'}|JM\Omega\Lambda S\Sigma\rangle \\ &= B_v\, [J(J+1) - \Omega^2 + S(S+1) - \Sigma^2], \end{aligned} \tag{D.14}$$

where the rotational constant B_v is defined by

$$B_v = \frac{\hbar}{4\pi c\mu}\langle nv|r^{-2}|nv\rangle = \frac{\hbar}{4\pi c\mu}\langle v|r^{-2}|v\rangle. \tag{D.15}$$

Because the internuclear distance is held constant for evaluation of an electronic matrix element, the rotational constant is diagonal with respect to the electronic quantum numbers n.

The operations involving **J** and **S** are straightforward. Operation of $\mathbf{J}^2$, J_z, and $J_{z'}$ on the standard $|nJM\rangle$ and $|nJ\Omega\rangle$ states obey,

$$\mathbf{J}^2\,|nJM\rangle = J(J+1)\,|nJM\rangle \tag{D.16}$$

$$J_z\,|nJM\rangle = M\,|nJM\rangle \tag{D.17}$$

$$J_{z'}\,|nJ\Omega\rangle = \Omega\,|nJ\Omega\rangle \tag{D.18}$$

The only possible point of confusion here is the standard state $|nJ\Omega\rangle$, which is referenced to the rotated, molecule fixed, z-axis (i.e., the z'-axis) instead of the laboratory z-axis. For any system, the standard $|nJM\rangle$ and $|nJ\Omega\rangle$ states are related by

$$\langle \mathbf{r}_1, \mathbf{r}_2, \ldots, \mathbf{r}_{\mathcal{N}}|nJM\rangle = \sum_{\Omega=-J}^{J} \langle \mathbf{r}'_1, \mathbf{r}'_2, \ldots, \mathbf{r}'_{\mathcal{N}}|nJ\Omega\rangle\; D^{J*}_{M\Omega}(\alpha\beta\gamma) \tag{D.19}$$

where, for the moment, $\mathcal{N}$ is the number of particles having spatial coordinates in the system. This is a general result holding for any quantum system, see, e.g., Thompson's [7] equation (6.19) or equation (xx) of [5].

The operations involving **L** are more complicated. The operators $\mathbf{J}^2$ and $J_\pm$ act on the rotation matrix element but not $\langle \mathbf{R}'_e r|nv\rangle$ or the electronic spin states $|S\Sigma\rangle$. The operator $\mathbf{S}^2$ affects nothing but the spin ket $|S\Sigma\rangle$. The total electronic orbital angular momentum,

$$\mathbf{L} = \sum_{i=1}^{\mathcal{N}} \mathbf{l}_i \tag{D.20}$$

in which l_i is the orbital angular momentum of the ith electron is a function of $3\mathcal{N}$ coordinates where $\mathcal{N}$ is the number of electrons. The symbol $\mathbf{R}'_e$ represents only $3\mathcal{N}-1$ electronic coordinates, and the third Euler angle γ is the remaining electronic coordinate. Thus, $3\mathcal{N}-1$ components of L operate on $\langle \mathbf{R}'_e|nv\rangle$ and the remaining component l_γ acts on $D^{J*}_{M,\,\Omega}(\alpha\beta\gamma)$. A Clebsch–Gordon expansion of the rotation matrix element gives

$$\begin{aligned}\langle \mathbf{r}_1, \mathbf{r}_2, \ldots, \mathbf{r}_{\mathcal{N}}, \mathbf{r}|nJM\Omega\Lambda S\Sigma\rangle &= \sqrt{\frac{2J+1}{8\pi^2}}\,\langle \mathbf{R}'_e r|nv\rangle\;|S\Sigma\rangle \\ &\times \sum_{M_N=-N}^{N}\sum_{M_S=-S}^{S} \langle NM_N SM_S|JM\rangle\;\langle N\Lambda S\Sigma|J\Omega\rangle\; D^{N*}_{M_N\Lambda}(\alpha\beta\gamma)\; D^{S*}_{M_S\Sigma}(\alpha\beta\gamma).\end{aligned} \tag{D.21}$$

in which N, the laboratory referenced M_N, and the molecule referenced Λ are the quantum numbers associated with the total orbital angular momentum **N**,

$$\mathbf{N} = \mathbf{L} + \mathbf{R} \tag{D.22}$$

For the one electron for which γ is the angular coordinate of rotation about the internuclear axis,

$$\begin{aligned} l_{z'}\, D^{N*}_{M_N\Lambda}(\alpha\beta\gamma) &= -\, i\, \frac{\partial}{\partial\gamma} D^{N*}_{M_N\Lambda}(\alpha\beta\gamma) \\ &= -\, i\, \frac{\partial}{\partial\gamma} e^{iM_N\alpha}\, d^{N}_{M_N\Lambda}(\beta)\, e^{i\Lambda\gamma} \\ &= \Lambda\, D^{N*}_{M_N\Lambda}(\alpha\beta\gamma). \end{aligned} \tag{D.23}$$

Thus, operation of l_γ in the case (a) basis is given by

$$l_\gamma\, |nJM\Omega\Lambda S\Sigma\rangle = \Lambda\, |nJM\Omega\Lambda S\Sigma\rangle. \tag{D.24}$$

Calculation of off-diagonal matrix elements of the rotational Hamiltonian is complicated by the nonstandard behavior of the case (a) states under operation of the raising and lowering operators. The rotation matrix element carries two magnetic quantum numbers, one more than mathematically admitted by the definition of angular momentum. The familiar behavior of standard $|nJM\rangle$ under raising and lowering operators,

$$\begin{aligned} J_\pm\, |nJM\rangle &= \sqrt{J(J+1) - M(M \pm 1)}\, |nJ,\, M \pm 1\rangle \\ &= C_\pm(JM)\, |nJ,\, M \pm 1\rangle \end{aligned} \tag{D.25}$$

does not hold for the rotation matrix element, even though it looks like an angular momentum eigenfunction because it is specified in terms of angular coordinates and quantum numbers. The rotation matrix elements obey [5]

$$J_\pm D^{J}_{M\Omega}(\alpha\beta\gamma) = -\, C_\mp(JM)\, D^{J}_{M\mp1,\,\Omega}(\alpha\beta\gamma), \tag{D.26}$$

$$J'_\pm D^{J}_{M\Omega}(\alpha\beta\gamma) = C_\pm(J\Omega)\, D^{J}_{M,\,\Omega\pm1}(\alpha\beta\gamma), \tag{D.27}$$

$$J_\pm D^{J*}_{M\Omega}(\alpha\beta\gamma) = C_\pm(JM)\, D^{J*}_{M\pm1,\,\Omega}(\alpha\beta\gamma), \tag{D.28}$$

$$J'_\pm D^{J*}_{M\Omega}(\alpha\beta\gamma) = -\, C_\pm(J\Omega)\, D^{J*}_{M,\,\Omega\mp1}(\alpha\beta\gamma). \tag{D.29}$$

These equations show that the rotation matrix element does not have the mathematical properties of an angular momentum eigenfunction. One can conclude, with the support of mathematical certainty, that $D^{J}_{M\Omega}(\alpha\beta\gamma)$ is not an angular momentum eigenfunction. Armed with the ability to count to two, one can draw the same conclusion from the symbol $D^{J}_{M\Omega}(\alpha\beta\gamma)$. Nevertheless, essentially without exception in diatomic literature, the rotation matrix element is treated as if it were an angular momentum eigenfunction, and this is the source of considerable confusion.

Using equation (D.29) for the nonstandard case (a) basis states but using the standard result

$$S_\pm\, |S\Sigma\rangle = C_\pm(S\Sigma)\, |S,\, \Sigma \pm 1\rangle \tag{D.30}$$

for the electronic spin states because $|S\Sigma\rangle$ is standard angular momentum state of the type given in equation (D.25), one finds (table D.2)

Table D.2. Matrix elements of the rotational Hamiltonian (equation (D.6)).

$\langle nvJM\Omega\Lambda S\Sigma \| H_{\mathrm{rot}} \| nvJM\Omega\Lambda S\Sigma\rangle$	$=\langle nvJM\Omega\Lambda S\Sigma \| B(r)(\mathbf{J}^2 + L_{z'}^2 + \mathbf{S}^2 - 2J_{z'}L_{z'} - 2J_{z'}S_{z'} + 2L_{z'}S_{z'}) \| nvJM\Omega\Lambda S\Sigma\rangle$ $=\langle nv \| B(r) \| nv\rangle[J(J+1) - \Omega^2 + S(S+1) - \Sigma^2]$ $=B_v[J(J+1) - \Omega^2 + S(S+1) - \Sigma^2]$
$\langle nvJM\Omega\Lambda S\Sigma \| H_{\mathrm{rot}} \| nvJM, \Omega \pm 1, \Lambda S, \Sigma \pm 1\rangle$	$=\langle nvJM\Omega\Lambda S\Sigma \| -B(r)(J_+'S_-' + J_-'S_+') \| nvJM, \Omega \pm 1, \Lambda S, \Sigma \pm 1\rangle$ $=\langle nv \| B(r) \| nv\rangle\ C_\pm(J\Omega)\ C_\pm(S\Sigma)$ $=B_v\sqrt{J(J+1) - \Omega(\Omega \pm 1)}\sqrt{S(S+1) - \Sigma(\Sigma \pm 1)}$
$\langle nvJM\Omega\Lambda S\Sigma \| H_{\mathrm{rot}} \| n'v'JM, \Omega \pm 1, \Lambda \pm 1, S\Sigma\rangle$	$=\langle nvJM\Omega\Lambda S\Sigma \| -B(r)(J_+'L_-' + J_-'L_+') \| n'v'JM, \Omega \pm 1, \Lambda \pm 1, S\Sigma\rangle$ $=\langle nv \| B(r)(L_+' + L_-') \| n'v'\rangle\ C_\pm(J\Omega)$ $=< BL > \sqrt{J(J+1) - \Omega(\Omega \pm 1)}$
$\langle nvJM\Omega\Lambda S\Sigma \| H_{\mathrm{rot}} \| n'v'JM\Omega, \Lambda \pm 1, S, \Sigma \mp 1\rangle$	$=\langle nvJM\Omega\Lambda S\Sigma \| B(r)(L_+'S_-' + L_-'S_+') \| n'v'JM\Omega, \Lambda \pm 1, S, \Sigma \mp 1\rangle$ $=\langle nv \| B(r)(L_+' + L_-') \| n'v'\rangle\ C_\mp(S\Sigma)$ $=< BL > \sqrt{S(S+1) - \Sigma(\Sigma \mp 1)}$

$$\begin{aligned}
&\langle nvJM\Omega\Lambda S\Sigma|H_{\mathrm{rot}}|nvJM,\,\Omega\mp 1,\,\Lambda S,\,\Sigma\pm 1\rangle\\
&= -\ \langle nJM\Omega\Lambda S\Sigma|J'_{+}S'_{-} + J'_{-}S'_{+}|nJM,\,\Omega\mp 1,\,\Lambda S,\,\Sigma\pm 1\rangle\\
&=B_{v}\sqrt{J(J+1)-\Omega(\Omega\pm 1)}\sqrt{S(S+1)-\Sigma(\Sigma\pm 1)}\,.
\end{aligned} \tag{D.31}$$

$$\begin{aligned}
&\langle nvJM\Omega\Lambda S\Sigma|H_{\mathrm{rot}}|n'v'JM\Omega,\,\Lambda\mp +1,\,S,\,\Sigma\pm 1\rangle\\
&= -\ \langle nvJM\Omega\Lambda S\Sigma|L'_{+}S'_{-} + L'_{-}S'_{+}|nJM,\,\Omega\mp 1,\,\Lambda S,\,\Sigma\pm 1\rangle\\
&=\langle nv|B(r)(L'_{+} + L'_{-})|n'v'\rangle\sqrt{S(S+1)-\Sigma(\Sigma-1)}\\
&\qquad+\langle nv|B(r)(L'_{+} + L'_{-})|n'v'\rangle\sqrt{S(S+1)-\Sigma(\Sigma+1)}
\end{aligned} \tag{D.32}$$

$$=\langle BL\rangle[\sqrt{S(S+1)-\Sigma(\Sigma-1)}+\sqrt{S(S+1)-\Sigma(\Sigma+1)}] \tag{D.33}$$

where

$$\langle BL\rangle = \langle nv|B(r)(L'_{+} + L'_{-})|n'v'\rangle \tag{D.34}$$

Equation (D.31) and the example given in table D.3 show that the Hund's case (a) matrix representation of $\mathbf{R}^2$ is nondiagonal. Therefore, the matrix of the rotational Hamiltonian, equation (D.1), is also nondiagonal. Strictly speaking, the Hund's case (a) basis is not a valid eigenfunction. At best, it is a poor physical approximation. This is no way detracts from its use as a basis. Before the widespread availability of digital hardware and numerical algorithms, diagonalization of the 6 × 6 matrix of table D.3 presented an essentially impossible task, but today numerical diagonalization of this matrix is trivial with even a very modest computer. In current practice, in table D.4 the fact that the Hund's case (a) basis does not yield a diagonal Hamiltonian matrix adds only the single line of code, `CALL JACOBI`, to a computer program.

Table D.5 gives the results of fitting the same data of Amiot *et al* used in table D.1 but with a different $^2\Sigma$ state. A comparison between tables D.1 and D.5 shows that

Table D.3. Matrix elements of $\mathbf{R}^2$ in a state which is a mixture of $^2\Pi$ and $^2\Sigma$ case (a) basis states.

Λ'			−1	−1	1	1	0	0
	Σ'		−1/2	1/2	−1/2	1/2	−1/2	1/2
		Ω'	−3/2	−1/2	1/2	3/2	−1/2	1/2
Λ	Σ	Ω						
−1	−1/2	−3/2	119.000	10.954	0	0	0	0
−1	1/2	−1/2	10.954	121.000	0	0	0	0
1	−1/2	1/2	0	0	121.000	10.954	0	0
1	1/2	3/2	0	0	10.954	121.000	0	0
0	−1/2	−1/2	0	0	0	0	121.000	11.000
0	1/2	1/2	0	0	0	0	11.000	121.000

Table D.4. Matrix elements of $\mathbf{R}^2$ in a state which is a mixture of $^3\Pi$ and $^3\Sigma$ case (a) basis states.

Λ'			−1	−1	−1	1	1	1	0	0	0
	Σ'		−1	0	1	−1	0	1	−1	0	1
		Ω'	−2	−1	0	0	1	2	−1	0	1
Λ	Σ	Ω									
−1	−1	−2	107.000	14.697	0	0	0	0	0	0	0
−1	0	−1	14.697	111.000	14.832	0	0	0	0	0	0
−1	1	0	0	14.832	111.000	0	0	0	0	0	0
1	−1	0	0	0	0	111.000	14.832	0	0	0	0
1	0	1	0	0	0	14.832	111.000	14.697	0	0	0
1	1	2	0	0	0	0	14.697	107.000	0	0	0
0	−1	−1	0	0	0	0	0	0	110.000	14.832	0
0	0	0	0	0	0	0	0	0	14.832	112.000	14.832
0	1	1	0	0	0	0	0	0	0	14.832	110.000

Table D.5. Another fit to the data of Amiot, Bacis, and Guelachvili [1] with a different model Hamiltonian.

$X\,^2\Pi(v=1)$	$T_v = 2824.6331(5)$
$B_v = 1.678\,5715(27)$	$D_v = 5.4893(26) \times 10^{-6}$
$A_v = 122.7029(21)$	$A_{vJ} = 3.471(38) \times 10^{-4}$
$AL_+ = -223.8(1.3)$	$AL_{+J} = 6.9(2.3) \times 10^{-3}$
$BL_+ = -1.7549(95)$	$BL_{+J} = -4.0(1.6) \times 10^{-5}$
$X\,^2\Pi(v=0)$	$[T_v = 948.66]$
$B_v = 1.696\,1444(29)$	$D_v = 5.4735(29) \times 10^{-6}$
$A_v = 122.9498(23)$	$A_{vJ} = 3.561(42) \times 10^{-4}$
$AL_+ = -225.4(1.4)$	$AL_{+J} = 5.8(2.6) \times 10^{-3}$
$BL_+ = -1.79(10)$	$BL_{+J} = -3.9(1.8) \times 10^{-5}$
$A^2\Sigma^+(v=0)$	$[T_v = 70000.0]$
$[B_v = 1.2]$	$[D_v = 5.0 \times 10^{-6}]$

Notes. In this example the weak $^2\Sigma^+$ contribution comes from a fictitious $^2\Sigma^+$ state. Angular momentum range $0.5 \leqslant J' \leqslant 40.5$, includes 405 of 419 lines, accuracy $\Delta\tilde{\nu} = 0.001$ cm^{-1} and $\sigma = 0.000\,38$ cm^{-1}.

only the mixing parameters AL_+ and BL_+ and the associated corrections for centrifugal stretching AL_{+J} and BL_{+J} are significantly influenced by the change in the parameters of the $^2\Sigma$ state.

Zare *et al* give the equations

$$o_v^\Sigma = \sum_{n'v'} \frac{\langle n'\ ^{2S+1}\Pi\ v'\ J'|\frac{1}{2}AL_+|n\ ^{2S+1}\Sigma\ v\ J\rangle^2}{E_{nvJ} - E_{n'v'J'}} \tag{D.35}$$

$$p_v^\Sigma = 4\sum n'v' \frac{\langle n'\ ^{2S+1}\Pi\ v'\ J'|\frac{1}{2}AL_+|n^{2S+1}\Sigma vJ\rangle\langle n'\ ^{2S+1}\Pi\ v'\ J'|BL_+|n\ ^{2S+1}\Sigma\ v\ J\rangle}{E_{nvJ} - E_{n'v'J'}} \tag{D.36}$$

$$q_v^\Sigma = 2\sum_{n'v'} \frac{\langle n'\ ^{2S+1}\Pi\ v'\ J'|AL_+|n\ ^{2S+1}\Sigma\ v\ J\rangle^2}{E_{nvJ} - E_{n'v'J'}} \tag{D.37}$$

for the Van Vleck transformation.

References

[1] Guelachvili G, Amiot C and Bacis R 1978 *Can. J. Phys.* **56** 251
[2] Macki A G and Wells J S 1991 *Wavenumber Calibration tables from Heterodyne Frequency Measurements* NIST Special Publication 821 (Washington DC: NIST)
[3] Zare R N, Schmeltekopf A L, Harrop W J and Albritton D L 1973 *J. Mol. Spectrosc.* **46** 37

[4] Engleman R Jr+, Rouse P E, Peek H M and Baiamonte V D 1970 *Eta and Gamma Band Systems of Nitric Oxide* Los Alamos Sci. Lab. Report LA-4364, Los Alamos, NM
[5] Hornkohl J O and Parigger C G 1996 *Am. J. Phys.* **64** 623
[6] Lefebvre-Brion H and Field R W 2004 *The Spectra and Dynamics of Diatomic Molecules* (New York: Elsevier/Academic)
[7] Thompson W J 1994 *Angular Momentum* (New York: Wiley)

IOP Publishing

Quantum Mechanics of the Diatomic Molecule (Second Edition)

Christian G Parigger and James O Hornkohl

Appendix E

Parity in diatomic molecules

This appendix communicates the application of the parity operator to the general diatomic eigenfunction [1]. The parity eigenvalue is a product of two factors, one that depends on the total angular momentum quantum number and a second constant factor that can be interpreted as the intrinsic parity of the molecule. These new results allow one to rigorously design an algorithm for the computation of diatomic spectra by utilizing that allowed transitions have nonvanishing rotational line strengths.

E.1 Introduction

The diatomic Hamiltonian matrix is historically parity-partitioned, thereby giving parity a more important role in diatomic spectroscopy than in atomic spectroscopy, e.g., see Zare *et al* [2]. Typically, a diatomic line list will include rotational parity designations for the lower levels [3]. Several authors, e.g., Hougen [4], Røeggen [5], Judd [6], and Larsson [7], have presented treatments of diatomic parity using the approximate Born–Oppenheimer separation of the diatomic eigenfunction into rotational, vibrational, and electronic factors.

In this work, operation of the parity operator on the general Wigner–Witmer [8] diatomic eigenfunction is used to yield the parity eigenvalues that are composed of a constant and an angular momentum dependent part. The computation of diatomic molecular spectra is accomplished without the need to explicitly include parity selection rules. The fundamental Wigner–Witmer diatomic eigenfunction simplifies the determination of rotational line strengths, i.e., Hönl–London factors. Allowed transitions are governed by nonzero rotational line strengths.

E.2 Parity operator

The discrete parity operation can be accomplished with a rotation and a reflection. The parity operator, $\mathcal{P}$, can be written as a product,

$$\mathcal{P} = C_2\, \sigma_v. \tag{E.1}$$

doi:10.1088/978-0-7503-6204-7ch35

The determinant of the matrix representations $\sigma_v(y, z)$ and $C_2(x)$ in laboratory xyz-coordinates are -1 and $+1$, respectively. The C_2 operator is a proper rotation that can be expressed as a discrete transformation of Euler angles. The Euler angles are the arguments of the Wigner D-function used to formulate rotational symmetry; consequently, the eigenvalues of C_2 are controlled by the angular momentum, J. The $\sigma_v(y, z)$ operation results in a constant factor, and the C_2 operation yields the angular momentum dependent part of the parity eigenvalue.

E.3 Rotation operator and Wigner *D*-function

Molecular eigenfunctions are normally expressed in rotated coordinates. The representations of the eigenfunction in original and rotated coordinate systems are connected with the rotation operator, $\mathcal{R}(\alpha, \beta, \gamma)$; Euler angles α, β and γ; and the Wigner D-functions,

$$\begin{aligned}\langle \mathbf{r}_1, \mathbf{r}_2, \dots, \mathbf{r}_N \,|JM\rangle &= \sum_{\Omega=-J}^{J} \langle \mathbf{r}_1, \mathbf{r}_2, \dots, \mathbf{r}_N \,|\mathcal{R}(\alpha, \beta, \gamma)|J\Omega\rangle\langle J\Omega \,|\mathcal{R}^{\dagger}(\alpha, \beta, \gamma)|JM\rangle \\ &= \sum_{\Omega=-J}^{J} \langle \mathbf{r}'_1, \mathbf{r}'_2, \dots, \mathbf{r}'_N \,|J\Omega\rangle\, D^{J*}_{M\Omega}(\alpha, \beta, \gamma).\end{aligned} \tag{E.2}$$

Angular momentum is the generator of rotations; therefore, one can expect that application of the discrete operator C_2 to the arguments of the D-function would yield a relationship between angular momentum and parity.

In terms of spatial and angular coordinates appropriate to the diatomic molecule, equation (E.2) can be written as

$$\begin{aligned}&\langle \rho, \zeta, \chi, \mathbf{r}_2, \dots, \mathbf{r}_n, \mathrm{r}, \theta, \phi \,|nvJM\rangle \\ &= \sum_{\Omega=-J}^{J} \langle \rho, \zeta, \chi', \mathbf{r}'_2, \dots, \mathbf{r}'_N, \mathrm{r}, \theta', \phi' \,|nvJ\Omega\rangle\, D^{J*}_{M\Omega}(\alpha, \beta, \gamma).\end{aligned} \tag{E.3}$$

Here, ρ is the distance of one electron (the electron arbitrarily labeled 1 but it could be any one of the electrons), ζ is the distance of that electron above or below the plane that passes through the center of mass of the two nuclei (the coordinate origin), and χ is the angle of rotation of that electron about the internuclear vector $\mathbf{r}(\mathrm{r}, \theta, \phi)$. The vibrational quantum number, v, has been extracted from the quantum numbers collection, n, which represents all required quantum numbers except J, M, Ω, and v.

The variables ρ, ζ, and r are scalars, which are unaffected by rotations. The physical rotation ϕ and the angle of coordinate rotation α are about the same axis, namely the z-axis. The physical rotation θ and the angle of coordinate rotation β are also about the same axis, namely the first intermediate y-axis of the full coordinate rotation. The angles χ and γ are both rotations about the z'-axis. Thus,

$$\phi' = \phi - \alpha, \quad \theta' = \theta - \beta, \quad \chi' = \chi - \gamma. \tag{E.4}$$

In coordinate rotations, one is at liberty to choose α, β, and γ. If one chooses for the angles $\alpha = \phi$, $\beta = \theta$, and $\gamma = \chi$, then all angular dependence of $\langle \rho, \zeta, \chi', \mathbf{r}'_2, \ldots, \mathbf{r}'_N, \mathrm{r}, \theta', \phi' \,|nvJ\Omega\rangle$ is removed. This yields the Wigner and Witmer [8] diatomic eigenfunction,

$$\begin{aligned} &\langle \rho, \zeta, \chi, \mathbf{r}_2, \ldots, \mathbf{r}_n, \mathrm{r}, \theta, \phi \,|nJM\rangle \\ &= \sum_{\Omega=-J}^{J} \langle \rho, \zeta, \mathbf{r}'_2, \ldots, \mathbf{r}'_N, \mathrm{r} \,|nv\rangle \, D^{J*}_{M\Omega}(\phi, \theta, \chi). \end{aligned} \tag{E.5}$$

The values of the quantum numbers, J and Ω, influence the electronic–vibrational eigenfunction $\langle \rho, \zeta, \mathbf{r}'_2, \ldots, \mathbf{r}'_N, r \,|nv\rangle$, but the electronic–vibrational eigenfunction is not an angular momentum state vector.

E.4 Parity of diatomic states

Parity is rotationally invariant. Inversion of the signs of all rotated coordinates inverts the signs of all unrotated coordinates, and vice versa. Therefore, the parity operator can be represented by $C_2(x')\, \sigma_v(y', z')$. The application to the right-hand side of the Wigner–Witmer diatomic eigenfunction (E.5) yields the parity eigenvalues,

$$p = -p_{\Sigma}\,(-1)^J \qquad J \text{ half-integer}, \tag{E.6}$$

$$p = p_{\Sigma}\,(-1)^J \qquad J \text{ integer}. \tag{E.7}$$

The constant part of the parity eigenvalue, p_Σ, is labeled in accord with standard spectroscopic notation. The imaginary values of $(-1)^J$ occurring when J is half-integer can be avoided if one adopts the convention [3] to always subtract 1/2 from J when J is half-integer. With this convention, equation (E.6) is replaced by

$$p = -p_{\Sigma}\,(-1)^{J-1/2} \qquad J \text{ half-integer}. \tag{E.8}$$

The value of p_Σ does not depend upon quantum numbers. It is a global value applying to all states of a given molecule. If the diatomic molecule can be said to have an intrinsic parity, then it is clearly p_Σ. One would expect the product of the intrinsic parities of the fundamental particles composing the molecule to equal p_Σ.

E.5 Parity in an algorithm for computing diatomic spectra

The following describes an algorithm in which equations (E.7) and (E.8) become practical equations for computing diatomic parity.

Consider the algorithm for computation of the wavelengths and intensities in the spectrum of a molecule from the first principles of quantum mechanics. The upper H' and lower H Hamiltonian matrices are computed and numerically diagonalized by unitary matrices, U' and U. The upper $F'_{n'v'J'}$ and lower F_{nvJ} terms are the eigenvalues of the Hamiltonians,

$$F'_{n'v'J'} = U'^{\dagger} H'\, U' \tag{E.9a}$$

$$F_{nvJ} = U^{\dagger}H\ U, \tag{E.9b}$$

and the vacuum wave numbers, $\tilde{\nu}$, of the predicted spectral lines,

$$\tilde{\nu} = F'_{n'v'J'} - F_{nvJ}, \tag{E.10}$$

are term differences. Of the very large number of computed term differences, only those for which the Condon and Shortley [9] line strength does not vanish are spectral lines. The line strength, $S(nvJ,\ n'v'J')$, is the sum over all M and M' of the irreducible tensor $T_k^{(q)}$ expectation values, $\langle nvJM\ |T_k^{(q)}|n'v'J'M'\rangle$. The exact separation of the total angular momentum in the Wigner–Witmer diatomic eigenfunction results in a diatomic line strength composed of two parts,

$$S(nvJ,\ n'v'J') = S(nv,\ n'v')\ S(J,\ J'), \tag{E.11}$$

the electronic–vibrational strength, $S(nv,\ n'v')$, and the unitless rotational line strength or Hönl–London factor, $S(J,\ J')$. The Born–Oppenheimer approximation separates the electronic–vibrational strength into electronic and vibrational parts. In the Hund's case (a) basis built from the Wigner–Witmer eigenfunction, the third Euler angle, $\chi = \gamma$, appears in the Wigner D-function,

$$\begin{aligned}|a\rangle &= |nvJM\Omega S\Sigma\rangle \\ &= \sqrt{\frac{2J+1}{8\pi^2}}\langle \rho,\ \zeta,\ \mathbf{r}'_2,\ \dots,\ \mathbf{r}'_N,\ \mathrm{r}\ |nv\rangle\ |S\Sigma\rangle\ D^{J*}_{M\Omega}(\phi,\ \theta,\ \chi).\end{aligned} \tag{E.12}$$

The algorithm for computation of the vacuum wave numbers, $\tilde{\nu}$, and diatomic spectral line strengths, $S(nvJ,\ n'v'J')$, is a straightforward application of quantum mechanics, but except for the very simplest molecules is also very far removed from the realm of the possible. However, with two very stringent caveats, the algorithm can be implemented for the diatomic molecule. The first caveat is that the vacuum wave numbers, $\tilde{\nu}$, for many spectral lines in many bands of a band system must have been experimentally measured with high accuracy, such as that provided by Fourier transform spectroscopy. The second caveat is that using semiempirical molecular constants one must be able to build upper and lower Hamiltonian matrices whose eigenvalue differences accurately predict the measured vacuum wave numbers. A fitting process is required [2]. One assumes trial values for the molecular constants, computes the spectral lines positions, $\tilde{\nu}$, and from the differences between $\tilde{\nu}-\tilde{\nu}_{\mathrm{exp}}$ finds the corrections to the molecular parameters. The difference between computed and measured line positions will typically equal the measurement error margins.

As a specific example, the line position data of Faris and Cosby [10] are used for the NO beta (3,0) band for the purpose of then creating a complete line list for the band with line strengths. Figure E.1 illustrates a spectrum generated from the NO line list.

A multiphoton 1+1 excitation was used to observe 10 of the 12 possible branches [10], with particular attention given to the parity designations of the numerous Λ doublets. These data are particularly suited for testing applications of the algorithm for the calculation of diatomic spectra. A total of 428 lines were fitted with a

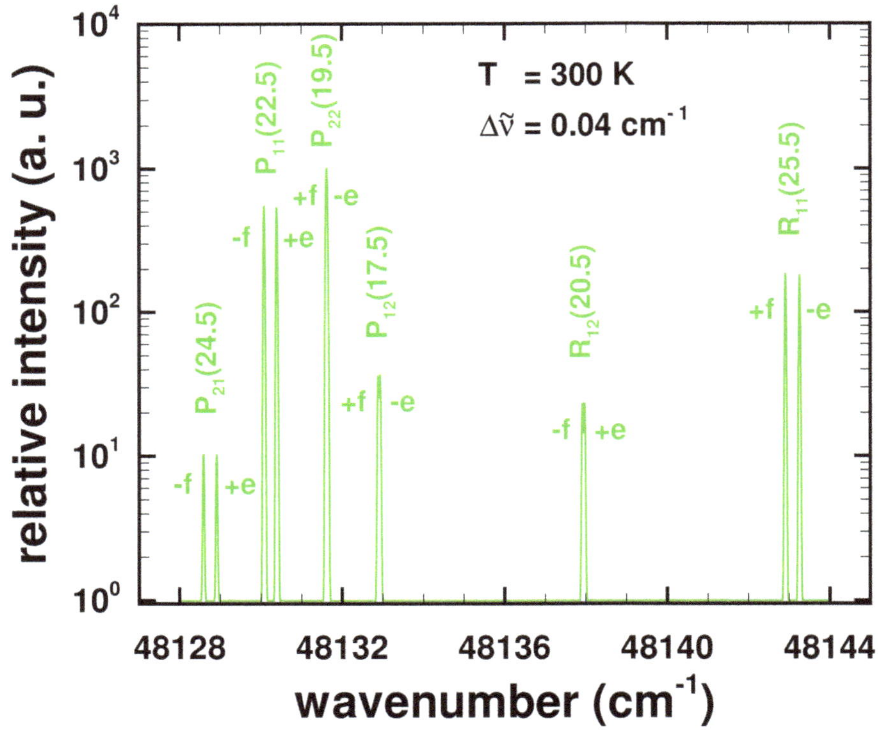

Figure E.1. Section of computed spectrum of the NO $B\,^2\Pi - X\,^2\Pi$ (3,0) band. The lines are further described in table E.1 [1].

standard deviation of 0.030 cm^{-1}, and a line list having no missing lines for the range of upper and lower J values was computed. Table E.1 provides details of the lines displayed in figure E.1. The computed parity eigenvalues agree with those assigned by Faris and Cosby [10].

The Hund's case (a) basis is mathematically complete. A sum of basis functions, $|a\rangle$, can be quantitatively very accurate. The parity operator, $\mathcal{P}$, commutes with the Hamiltonian. Thus, the orthogonal matrix that diagonalizes the case (a) representation of the Hamiltonian will also diagonalize the case (a) representation of $\mathcal{P}$.

The exact separation of the coordinates of the total angular momentum in the Wigner D-function greatly simplifies implementation of the algorithm, which uses nonvanishing line strengths to determine if a computed term difference represents an allowed spectral line. The Hönl–London factors are computed from the Hund's case (a) transition moment and the matrices, U and U', which diagonalize the upper and lower Hamiltonians.

A single selection rule handles all types of diatomic spectra. If the Hönl–London factor, $S(J,\, J')$, is nonvanishing, then the transition is allowed. Parity plays no part in the fitting process that determines the molecular parameters, but the parity eigenvalues are computed from the finalized values of the molecular parameters. The presented algorithm can be used to predict molecular spectra for the purpose of fitting measured data [11].

Table E.1. Lines of the NO $B\ ^2\Pi - X\ ^2\Pi$ (3,0) band (see figure E.1).

J		p	$F_{J'}$ (cm^{-1})	F_J (cm^{-1})	$\tilde{\nu}$ (cm^{-1})	$S(J, J')$	$\tilde{\nu}$-$\tilde{\nu}_{\rm exp}$ (cm^{-1})
24.5	P_{21}	$-f$	49 111.456	982.866	48 128.590	0.813	0.021
24.5	P_{21}	$+e$	49 111.501	982.588	48 128.914	0.812	0.078
22.5	P_{11}	$-f$	48 952.618	822.538	48 130.080	21.749	0.014
22.5	P_{11}	$+e$	48 952.673	822.282	48 130.390	21.805	−0.020
19.5	P_{22}	$+f$	48 876.118	744.504	48 131.615	18.832	−0.007
19.5	P_{22}	$-e$	48 876.146	744.505	48 131.641	18.786	
17.5	P_{12}	$+f$	48 747.194	614.306	48 132.888	0.515	−0.013
17.5	P_{12}	$-e$	48 747.244	614.307	48 132.936	0.527	−0.012
20.5	R_{12}	$-f$	48 952.618	814.699	48 137.919	0.881	0.034
20.5	R_{12}	$+e$	48 952.673	814.701	48 137.971	0.880	−0.077
25.5	R_{11}	$+f$	49 210.930	1068.021	48 142.908	25.424	0.034
25.5	R_{11}	$-e$	49 210.985	1067.731	48 143.254	25.317	−0.018

Notes. The Hönl–London factors, $S(J, J')$, and parity eigenvalues, p, are derived from numerical diagonalization of Hamiltonians in Hund's case (a) basis. The $P_{22}(19.5)$ Λ doublet is not resolved in the experiments [10].

References

[1] Hornkohl J O and Parigger C G 2017 *Int. J. Mol. Theor. Phys.* **1** 00103

[2] Zare R N, Schmeltekopf A L, Harrop W J and Albritton D L 1973 *J. Mol. Spectrosc.* **46** 37

[3] Brown J M, Hougen J T, Huber K P, Johns J W C, Kopp I, Lefebvre-Brion H, Merer A J, Ramsay D A, Rostas J and Zare R N 1975 *J. Mol. Spectrosc.* **55** 500

[4] Hougen J T 2021 *The Calculation of Rotational Energy Levels and Rotational Line Intensities in Diatomic Molecules* (Gaithersburg, MD: National Institute of Standards and Technology) http://physics.nist.gov/DiatomicCalculations, 2001. Originally published as *The Calculation of Rotational Energy Levels and Rotational Line Intensities in Diatomic Molecules*, J. T. Hougen, NBS Monograph 115 (1970)

[5] Røeggen I 1971 *Theor. Chim. Acta* **21** 398

[6] Judd B 1975 *Angular Momentum Theory for Diatomic Molecules* (New York: Academic)

[7] Larsson M 1981 *Phys. Scr.* **23** 835

[8] Wigner E and Witmer E E 1928 *Z. Phys.* **51** 859 H. Hettema, *Quantum Chemistry: Classic Scientific Papers*, p. 287, (Singapore: World Scientific) 2000

[9] Condon E U and Shortley G 1953 *The Theory of Atomic Spectra* (Cambridge: Cambridge University Press)

[10] Faris G W and Cosby P C 1992 *J. Chem. Phys.* **97** 7073

[11] Parigger C G, Woods A C, Surmick D M, Gautam G, Witte M J and Hornkohl J O 2015 *Spectrochim. Acta* B **107** 132

IOP Publishing

Quantum Mechanics of the Diatomic Molecule (Second Edition)

Christian G Parigger and James O Hornkohl

Appendix F

Rotational line strengths for the CN BX (5,4) band

This appendix communicates rotational line strengths for the cyanide (CN) $B\,^2\Sigma^+ - X\,^2\Sigma^+$ (5,4) band [1]. Rotational line strengths, computed from eigenvectors of Hund's case (a) matrix representations of the upper and lower Hamiltonians using Wigner–Witmer basis functions, show a larger than expected influence from the well-known perturbation in the (5,4) band. Comparisons with National Solar Observatory (NSO) experimental Fourier transform spectroscopy data reveal nice agreement of measured and predicted spectra.

F.1 Introduction

The CN violet $B\,^2\Sigma^+ - X\,^2\Sigma^+$ band system is one of the most studied band systems. Ram *et al* [2] and Brooke *et al* [3] have summarized the available experimental and theoretical information. Of the many known bands in the violet system, only the (5,4) band is considered here. This band exhibits a weak, quantitatively understood perturbation [4] caused by mixing of the $v = 17$ level of $A^2\Pi$ with the $v = 5$ level of $B\,^2\Sigma^+$. The particular perturbation of the CN (5,4) band is not considered in the study by Brooke *et al* [3] but is evaluated in this work by isolating the spectral features of this band that is part of the CN violet system. Numerical diagonalizations of upper and lower Hamiltonians with and without the perturbation are investigated and compared with available experimental spectra. The simulations rely on determining rotational strengths without parity-partitioned Hamiltonians. It is anticipated that the investigated (5,4) band modifications can be possibly confirmed with the new PGOPHER program that was recently released by Western [5].

F.2 CN (5,4) band spectra

For the computation of rotational spectra, the square of transition moments are numerically computed using the eigenvectors of upper and lower Hamiltonians. This approach can also be selected in the new PGOPHER program [5]. For the diatomic molecule, the results effectively yield the Hönl–London factors, yet we do not utilize

doi:10.1088/978-0-7503-6204-7ch36

tabulated Hönl–London factors that are available in standard textbooks. Table F.1, and figures F.1 and F.2 compare results obtained with and without taking into account the mixing. Results of modeling the angular momentum states of the upper $v = 5$ vibrational level as a mixture of $^2\Sigma$ and $^2\Pi$ Hund's case (a) basis functions, a so-called 'de-perturbation" or perturbation analysis, agree well that of Ito *et al* [4] whose used the line position measurements of Engleman [6]. The 100 lines of the more recent data of Ram *et al* [2] were fitted with a standard deviation of 0.025 cm^{-1}. Failure to include spin-orbit mixing of the $B^2\Sigma^+$ and $A^2\Pi$ basis states increased the standard deviation to 0.25 cm^{-1}.

The table and synthetic spectra reveal that the changes caused by spin-orbit mixing are relatively very much larger for the rotational line strengths, $S(J', J)$, than for the line positions, $\tilde{\nu}$. The simulation results compare nicely with measured spectra [2] available from the NSO at Kitt Peak [7]. Figure F.3 displays the recorded and simulated spectra for a resolution of 0.03 cm^{-1}.

The influence of $^2\Sigma^+ + {}^2\Pi$ mixing on the rotational line strengths, $S(J', J)$, was recognized because computation of $S(J', J)$ is an integral part of the unique line position fitting algorithm. Upper and lower Hamiltonian matrices in the Hund's case (a) basis are numerically diagonalized, and the spectral line vacuum wave number $\tilde{\nu}$ is the difference between upper and lower Hamiltonian eigenvalues.

Table F.1. Lines in the CN $B^2\Sigma^+X^2\Sigma^+$ (5,4) band near the perturbation.

J'	J		p'	$\tilde{\nu}$	$S_{J'J}$	$\Delta\tilde{\nu}$	$S^{(0)}_{J'J}$	$\Delta\tilde{\nu}^{(0)}$
9.5	8.5	R_{11}	$-e$	28 013.117	9.474	−0.010	9.474	0.337
9.5	8.5	R_{22}	$+f$	28 017.421	9.474	0.001	9.474	−0.059
10.5	9.5	R_{11}	$+e$	28 016.992	9.1988	−0.004	10.476	0.600
10.5	9.5	R_{22}	$-f$	28 021.651	11.171	−0.000	10.476	−0.067
11.5	10.5	R_{11}	$-e$	28 020.540	7.868	−0.041	11.478	1.193
11.5	10.5	R_{22}	$+f$	28 025.866	12.240	0.006	11.478	−0.067
12.5	11.5	R_{22}	$-f$	28 030.125	13.288	0.007	12.480	−0.072
12.5	11.5	R_{11}	$+e$	28 030.431	13.812		12.480	
13.5	12.5	R_{11}	$-e$	28 032.081	17.455	−0.053	13.481	−1.870
13.5	12.5	R_{22}	$+f$	28 034.428	14.325	0.011	13.481	−0.073
14.5	13.5	R_{11}	$+e$	28 035.672	17.919	−0.005	14.483	−1.102
14.5	13.5	R_{22}	$-f$	28 038.773	15.356	0.013	14.483	−0.076
15.5	14.5	R_{11}	$-e$	28 039.742	18.442	0.007	15.484	−0.807
15.5	14.5	R_{22}	$+f$	28 043.161	16.383	0.009	15.484	−0.084
16.5	15.5	R_{11}	$+e$	28 043.989	19.132	0.011	16.485	−0.655
16.5	15.5	R_{22}	$-f$	28 047.590	17.405	0.006	16.485	−0.091

Notes. $\tilde{\nu}$ are the fitted line positions, $S(J', J)$ are the rotational line strengths computed in the fitting algorithm. $S^{(0)}(J', J)$ and $\Delta\tilde{\nu}^{(0)}$ are the line strengths and errors in the fitted line positions, respectively, when the off-diagonal spin-orbit coupling constants $\langle AL+\rangle$ and $\langle BL+\rangle$ are set equal to 0. Spin-orbit mixing of $B^2\Sigma^+$ and $A^2\Pi$ shifts the upper e parity levels. An error in the $\tilde{\nu}(J', J)$ associated with these upper e parity levels is produced if the mixing is ignored. A relatively large fractional error, e.g., -3.974/17.455 versus -1.870/28032 for $R_{11}(12.5)$, can occur in the rotational line strengths, $S(J', J)$.

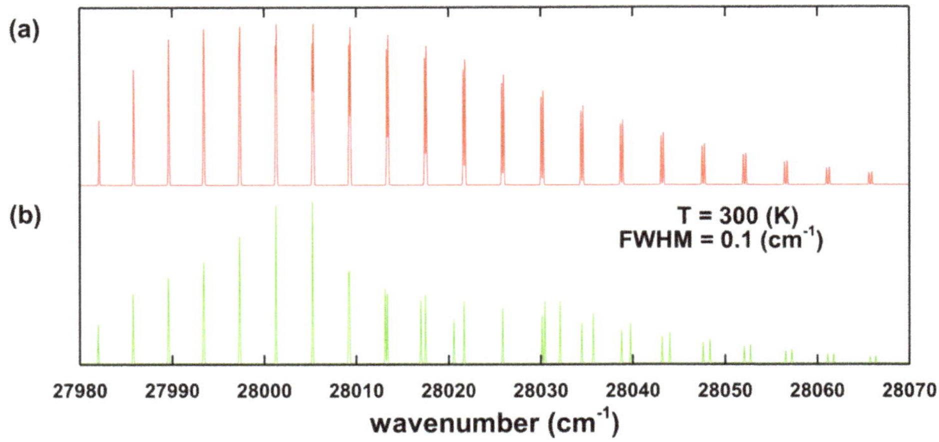

Figure F.1. Synthetic emission spectra showing the influence of inclusion of the $v = 17$, $A\,^2\Pi$ basis in the upper $v = 5$ state of the CN violet (5,4) band. In the upper spectrum, (a), the upper states are pure $^2\Sigma^+$. The $v = 17$, $A\,^2\Pi$ energy eigenvalues lie very near the $v = 5$, $B\,^2\Sigma^+$ eigenvalues, and this explains the large influence of the $A\,^2\Pi$ basis. In the lower spectrum, (b), the upper states are treated as the sum $c_{\Sigma}{}^2\Sigma^+ + c_{\Pi}{}^2\Pi$ with $c_{\Sigma} \gg c_{\Pi}$. Only R-branch lines are shown here, including those given in table F.1 [1].

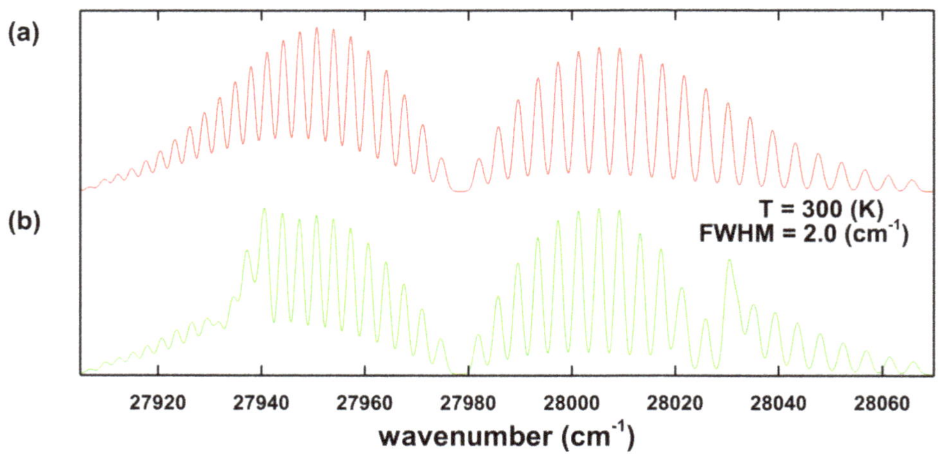

Figure F.2. The lower resolution spectra include both the P and R branches. (a) pure, (b) addition of a small amount of $^2\Pi$ to the upper basis affects the lower spectrum of the violet (5,4) band, even at low resolution [1].

To determine which of the many eigenvalue differences represent allowed spectral lines, the factor $S(J', J)$ is computed from the upper and lower eigenvectors for each eigenvalue difference. A nonvanishing $S(J', J)$ denotes an allowed diatomic spectral line. Parity-partitioned effective Hamiltonians are not used. Parity and branch designation are not required in the fitting algorithm. Input data to the fitting program is a table of vacuum wave number $\tilde{\nu}$ versus J' and J. The nonvanishing of the rotational strength is the only selection rule used. Applications of this rule lead to the establishment of spectral databases for diatomic molecular spectroscopy of

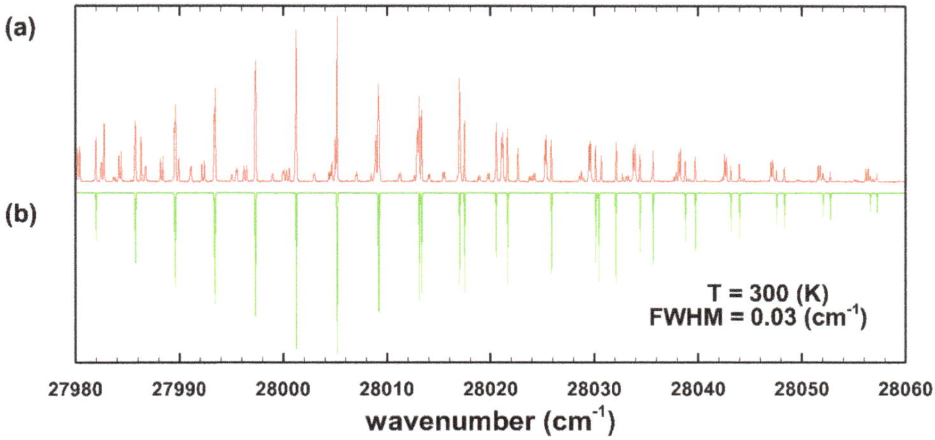

Figure F.3. Comparison of measured and simulated spectra. (a) Segment of the recorded [2] Fourier transform spectrum 920 212R0.005 [7], (b) computed spectrum for a temperature of 300 K and a spectral resolution of 0.03 cm^{-1}. The computed (5.4) band is flipped vertically to show how the predicted line positions of the R-branch match the vacuum wave numbers of the experimental spectrum [1].

selected transitions [8]. Over and above the PGOPHER program [5], there are other extensive efforts in predicting diatomic molecular spectra including, such as the so-called DUO program [9] for diatomic spectroscopy.

F.3 Wigner–Witmer diatomic eigenfunction

The Hund's case (a) basis functions were derived from the Wigner and Witmer [10] diatomic eigenfunction,

$$\langle \rho, \zeta, \chi, \mathbf{r}_2, \dots, \mathbf{r}_N, r, \theta, \phi \,|nvJM\rangle = \sum_{\Omega=-J}^{J} \langle \rho, \zeta, \mathbf{r}'_2, \dots, \mathbf{r}'_N, r \,|nv\rangle\, D^{J*}_{M\Omega}(\phi, \theta, \chi). \quad \text{(F.1)}$$

The coordinates are ρ the distance of one electron (the electron arbitrarily labeled 1 but it could be any one of the electrons) from the internuclear vector $\mathbf{r}(r, \theta, \phi)$, the distance ζ of that electron above or below the plane perpendicular to $\mathbf{r}$ and passing through the center of mass of the two nuclei (the coordinate origin), the angle χ for rotation of that electron about the internuclear vector $\mathbf{r}$, and the remaining electronic coordinates $\mathbf{r}_2, \dots, \mathbf{r}_N$ in the fixed and $\mathbf{r}'_2, \dots, \mathbf{r}'_N$ in the rotating coordinate system. The vibrational quantum number v has been extracted from the quantum numbers collection n which represents all required quantum numbers except J, M, Ω, and v. The Wigner–Witmer diatomic eigenfunction has no application in polyatomic theory, but for the diatomic molecule the exact separation of the Euler angles is a clear advantage over the Born–Oppenheimer approximation for the diatomic molecule in which the angle of electronic rotation, χ, is unnecessarily separated from the angles describing nuclear rotation, θ and ϕ. Equation (F.1) can be derived by writing the general equation for coordinate (passive) rotations α, β, and γ of the eigenfunction, replacing two generic coordinate vectors with the diatomic vectors $\mathbf{r}(r, \theta, \phi)$ and $\mathbf{r}'(\rho, \zeta, \chi)$, and equating the angles of coordinate rotation to the angles of physical rotation ϕ, θ, and ϕ. The general equation for coordinate rotation holds in isotropic space, and therefore the quantum numbers J, M, and Ω in the Wigner–Witmer eigenfunction include all electronic and nuclear spins. If nuclear spin were to be included, J, M, and Ω would be replaced by F, M_F, and Ω_F, but hyperfine structure is not resolved in the (5, 4) band data reported by [2], and equation (F.1) is written with the appropriate spectroscopic quantum numbers.

It is worth noting that the rotation matrix element $D^J_{M\Omega}(\phi, \theta, \chi)$ and its complex conjugate $D^{J*}_{M\Omega}(\phi, \theta, \chi)$ do not fully possess the mathematical properties of quantum mechanical angular momentum. It is well known that a sum of Wigner D-functions is required to build an angular momentum state. The equation

$$J'_{\pm}\, D^{J*}_{M\Omega}(\phi, \theta, \chi) = \sqrt{J(J+1) - \Omega(\Omega \mp 1)}\; D^{J*}_{M,\,\Omega\mp 1}(\phi, \theta, \chi) \tag{F.2}$$

is not a phase convention [11–13] but a mathematical result readily obtained from equation (F.1) and

$$J'_{\pm}\, |J\Omega\rangle = \sqrt{J(J+1) - \Omega(\Omega \pm 1)}\; |J, \Omega \pm 1\rangle, \tag{F.3}$$

in which the prime on the operator $J'_{\pm}$ indicates that it is written in the rotated coordinate system where the appropriate magnetic quantum number Ω.

F.4 Hund's basis functions

The Hund's case (a) basis function based upon the Wigner–Witmer diatomic eigenfunction is

$$\begin{aligned} |a\rangle &= \langle \rho, \zeta, \chi, \mathbf{r}'_2, \ldots, \mathbf{r}'_N, r, \theta, \phi\, |nvJMS\Lambda\Sigma\Omega\rangle \\ &= \sqrt{\frac{2J+1}{8\pi^2}}\, \langle \rho, \zeta, \mathbf{r}'_2, \ldots, \mathbf{r}'_N, r\, |nv\rangle\, |S\Sigma\rangle\, D^{J*}_{M\Omega}(\phi, \theta, \chi). \end{aligned} \tag{F.4}$$

As noted above, a sum of $|a\rangle$ basis functions is required to build an eigenstate of angular momentum. The basis function would also not be an eigenstate of the parity operator. The case (a) matrix elements, $p_{ij}^{(a)}$, of the parity operator $\mathcal{P}$,

$$p_{ij}^{(a)} = p_{\Sigma}(-)^J \delta(J_i J_j)\, \delta(\Omega_i, -\Omega_j)\, \delta(\Lambda_i, -\Lambda_i)\, \delta(n_i n_j), \tag{F.5}$$

show that a single $|a\rangle$ basis function is not an eigenstate of parity. The procedure called parity symmetrization adds $|JM\Omega\rangle$ and $|JM, -\Omega\rangle$ basis functions, thereby destroying the second magnetic quantum number Ω and yielding a function which at least possesses the minimal mathematical properties of an eigenstate of angular momentum, parity, and the other members of the complete set of commuting operators. The general procedure would be to continue adding basis functions to the upper and lower bases until eigenvalue differences between the upper and lower Hamiltonians accurately predict measured line positions.

F.5 The upper Hamiltonian matrix for the (5,4) band

Electronic spin **S** interactions with electronic orbital momentum **L** and nuclear orbital momentum **R** produce both diagonal and off-diagonal matrix elements in the Hund's case (a) representation of the Hamiltonian. The off-diagonal elements connect different basis states. For example, both of the mentioned spin-orbit interactions connect $^2\Sigma^+$ and $^2\Pi$. Because Van Vleck transformed Hamiltonians are not used, the appropriate parameters for the strength of these interactions are $\langle AL+\rangle$ and $\langle BL+\rangle$. Table F.2 lists the molecular parameters used in the Hamiltonian matrices. Tables F.3 and F.4 show the Hamiltonian matrices without and with spin-orbit interactions, respectively.

F.6 A diatomic line position fitting algorithm

A basic tool for the diatomic spectroscopist is a computer program that accepts a table of experimentally measured vacuum wave numbers $\tilde{\nu}_{\text{exp}}$ versus J' and J, and outputs a set of molecular parameters with which one can reproduce the $\tilde{\nu}_{\text{exp}}$ with a standard deviation comparable to the estimated experimental error. In practice, an experimental line list frequently shows gaps, vis. spectral lines are missing. Following a successful fitting process, one can use the molecular parameters to predict all lines. A computed line list is especially useful when it includes the Condon and Shortley [15] line strength from which the Einstein coefficients and oscillator strength [16, 17] and the HITRAN line strength [18] can be calculated. A feature of the line fitting program described below is its use of nonzero rotational strengths (see equation (F.8)) to mark which of the many computed differences between upper and lower term values represents the vacuum wave number of an allowed spectral line.

Table F.2. Molecular parameters used in this work, which relies on Hamiltonians that are not parity-partitioned.

	$X^2\Sigma^+$	$B^2\Sigma^+$	$A^2\Pi$
	$v=4$	$v=5$	$v=17$
B_v	1.820 866(13)	1.845 727(13)	1.404 833
D_v	6.172(36) $\times 10^{-6}$	8.003(38) $\times 10^{-6}$	5.66 $\times 10^{-6}$
A_v			−50.5253
γ_v	−1.98(43) $\times 10^{-4}$	−1.921(44) $\times 10^{-2}$	
γ_{D_v}	−1.98(43) $\times 10^{-4}$		
T_v	8011.7871	35 990.1780(25)	36 010.5732
$\langle AL+\rangle$		4.25(0.03)	
$\langle BL+\rangle$		0.0205(0.001)	

Notes. Values not followed by a number in parenthesis were held fixed or an error estimate was not computed. A value in parenthesis is the standard deviation in the fitted value. Parameters for the $A\,^2\Pi$ state were fitted by the Nelder–Mead minimization algorithm using values given by Brooke *et al* [3] as trial values. Error estimates were not computed, and the values of Brooke *et al* [3] were only very slightly changed.

Table F.3. Hamiltonian matrix for states modeled as the mixture of $^2\Sigma^+$ and $^2\Pi$ basis states.

			v	5	5	17	17	17	17
			Λ	0	0	−1	−1	1	1
			Σ	−0.5	0.5	−0.5	0.5	−0.5	0.5
v	Λ	Σ	Ω	−0.5	0.5	−1.5	−0.5	0.5	1.5
5	0	−0.5	−0.5	36 351.6409	−25.6707	0	0	0	0
5	0	0.5	0.5	−25.6707	36 351.6409	0	0	0	0
17	−1	−0.5	−1.5	0	0	36 257.6340	−19.5866	0	0
17	−1	0.5	−0.5	0	0	−19.5866	36 310.9646	0	0
17	1	−0.5	0.5	0	0	0	0	36 310.9646	−19.5866
17	1	0.5	1.5	0	0	0	0	−19.5866	36 257.6340
		E_{nvJ}		36 377.3116	36 325.9702	36 251.2135	36 317.3851	36 317.3851	36 251.2135

Off-diagonal spin-orbit coupling has been removed. Consequently, the 2×2 matrices along the main diagonal are independent, and could be individually diagonalized. The bottom row contains the energy eigenvalues. Using matrices like these to model upper states of the CN violet (5,4) band, the 100 experimental spectral lines reported by Ram *et al* [2] were fitted with a standard deviation of 0.25 cm^{-1}. This Hamiltonian was computed for $\langle AL+\rangle = \langle BL+\rangle = 0$ but is otherwise identical to the Hamiltonian in table F.4. Standard Hund's case (a) matrix elements [11, 13] were used.

Table F.4. Off-diagonal spin-orbit coupling 6×6 matrix.

			v	5	5	17	17	17	17
			Λ	0	0	−1	−1	1	1
			Σ	−0.5	0.5	−0.5	0.5	−0.5	0.5
v	Λ	Σ	Ω	−0.5	0.5	−1.5	−0.5	0.5	1.5
5	0	−0.5	−0.5	36 351.6409	−25.6707	2.8566	2.3274	2.8639	0
5	0	0.5	0.5	−25.6707	36 351.6409	0	2.8639	2.3274	2.8566
17	−1	−0.5	−1.5	2.8566	0	36 257.6340	−19.5866	0	0
17	−1	0.5	−0.5	2.3274	2.8639	−19.5866	36 310.9646	0	0
17	1	−0.5	0.5	2.8639	2.3274	0	0	36 310.9646	−19.5866
17	1	0.5	1.5	0	2.8566	0	0	−19.5866	36 257.6340
		E_{nvJ}		36 377.3957	36 327.7869	36 250.9625	36 317.3525	36 315.8194	36 251.1620

Consequently, the fitting process creates a complete line list including rotational factors. Parity plays no part in the fitting process, but the same orthogonal matrix that diagonalizes the case (a) from the three independent 2×2 matrices of table F.3, the off-diagonal matrix elements mix the Hund's case (a) basis states, and the standard deviation of the spectral line fit mentioned in table F.3 is reduced by a factor of 10 to 0.025 cm^{-1}. The spin-orbit coupling constants $\langle AL+\rangle = 4.25(0.03)$ and $\langle BL+\rangle = 0.205(0.001)$ were used in computation of this Hamiltonian. This single 6×6 matrix describing ${}^2\Pi -{}^2\Sigma^+$ mixing can be compared with the two 3×3 parity-partitioned matrices of Brown and Carrington [14].

The Hamiltonian matrix will also diagonalize the case (a) parity matrix whose elements are given in equation (F.5). The $p = \pm 1$ parity eigenvalue becomes a computed quantity, and the *e*/*f* parity designation is established from the parity eigenvalue using the accepted convention Brown *et al* [19].

Trial values of upper and lower state molecular parameters, typically taken from previous works by other for the band system in question, are used to compute upper H' and lower H Hamiltonian matrices in the case (a) basis given by equation (F.4) for specific values of J' and J. The upper and lower Hamiltonians are numerically diagonalized,

$$T' = \tilde{U}'\, H'\, U' \tag{F.6a}$$

$$T = \tilde{U}\, H\, U \tag{F.6b}$$

giving the upper T' and lower T term values. The vacuum wave number $\tilde{\nu}$ is determined,

$$\tilde{\nu}_{ij} = T_i' - T_j, \tag{F.7}$$

and the rotational strength is evaluated,

$$S_{ij}(J', J) = (2J+1)\left| \sum_n \sum_m \tilde{U}'_{in} \langle J\Omega;\, q,\, \Omega' - \Omega\, | J'\Omega' \rangle\, U_{mj}\, \delta(\Sigma_n' \Sigma_m) \right|^2. \tag{F.8}$$

The degree of the tensor operator, q, responsible for the transitions amounts to $q = 1$ for electric dipole transitions. For a nonzero rotational factors, $S(J', J)$, the vacuum wave number $\tilde{\nu}_{ij}$ is added to a table of computed line positions to be compared with the experimental list $\tilde{\nu}_{\mathrm{exp}}$ versus J' and J. The Clebsch–Gordan coefficient, $\langle J\Omega;\, q,\, \Omega' - \Omega\, | J'\Omega' \rangle$, is the same one appearing in the pure case (a)—case (a) formulae for $S(J', J)$. For a specific values of J' and J, one constructs tables for $\tilde{\nu}_{\mathrm{exp}}$ and computed $\tilde{\nu}_{ij}$. The errors $\Delta\tilde{\nu}_{ij}$,

$$\Delta\tilde{\nu}_{ij} = \tilde{\nu}_{ij} - \tilde{\nu}_{\mathrm{exp}}, \tag{F.9}$$

are computed where each $\tilde{\nu}_{ij}$ is the one that most closely equals one of the $\tilde{\nu}_{\mathrm{exp}}$. Once values of $\tilde{\nu}_{ij}$ and $\tilde{\nu}_{\mathrm{exp}}$ are matched, each is marked unavailable until a new list of $\tilde{\nu}_{ij}$ is computed. The indicated computations are performed for all values of J' and J in the experimental line list, and corrections to the trial values of the molecular parameters are subsequently determined from the resulting $\Delta\tilde{\nu}_{ij}$. The entire process is iterated until the parameter corrections become negligibly small. As this fitting process successfully concludes, one obtains a set of molecular parameters that predict the measured line positions $\tilde{\nu}_{\mathrm{exp}}$ with a standard deviation that essentially equals the experimental estimates for the accuracy of the $\tilde{\nu}_{\mathrm{exp}}$.

F.7 Discussion

The influence on intensities in the (5,4) band of the CN violet system caused by the weak spin-orbit mixing, figures F.1 and F.2, is significantly larger than initially anticipated. This was noticed because computation of the rotational strengths is an integral part of our line position fitting program. The eigenvectors that diagonalize the Hamiltonian to yield fitted line position $\tilde{\nu}$ also yield $S(J', J)$. The eigenvectors of Van Vleck transformed effective Hamiltonians are rarely discussed in the literature; equally, the calculations of Hönl–London factors appear problematic. In established diatomic molecular practice, Hönl–London factors are determined independently of line positions. Analytical approximations utilize the parameter $Y = A/B$ to account for the influence of spin-orbit interaction on $S(J', J)$. Kovács [20] gives many examples, Li *et al* [21] give a more recent application. These analytical approximations can accurately account for intermediate spin-orbit coupling which smooth transitions between case (a) and case (b) with increasing J' and J, but show limited sensitivity to abrupt changes in $S(J', J)$ near perturbations such as those seen the (5, 4) band in the CN violet system.

It is noted in passing that the $S^{(0)}(J', J)$ in table F.1 that describe rotational strengths without off-diagonal spin-orbit constants, are exactly equal to pure case (b) Hönl–London factors for $^2\Sigma^+ - {}^2\Sigma^+$ transitions, even though all computations were carried out in the Hund's case (a) basis. This observation merely makes the point that the Hund's case (b) is defined due to the physical absence of spin-orbit coupling.

F.8 Conclusion

The Wigner–Witmer diatomic eigenfunction makes it possible to form an exact, mathematical connection between computation of $\tilde{\nu}$ and $S(J', J)$ in a single algorithm. The concept of the nonvanishing rotational strengths as the omnipotent selection rule initially conceived as a simplifying convenience in a computer algorithm is now seen to be more valuable, as evidenced in this work's analysis of the CN (5,4) band perturbations by isolating a specific branch. Future work is planned for comparisons of the CN (10,10) band spectra that include perturbation and that show promising agreements with experiments and PGOPHER predictions.

References

[1] Hornkohl J O and Parigger C G 2017 *Int. J. Mol. Theor. Phys.* **1** 00102

[2] Ram R S, Davis S P, Wallace L, Englman R, Appadoo D R T and Bernath P F 2006 *J. Mol. Spectrosc.* **237** 225

[3] Brooke J S A, Ram R S, Western C M, Li G, Schwenke D W and Bernath P F 2014 *J. Quant. Spectrosc. Radiat. Transfer* **210** Astrophys. J. Supp. Series

[4] Ito H, Fukuda Y, Ozaki Y, Kondow T and Kuchitsu K 1987 *J. Mol. Spectrosc.* **121** 84
[5] Western C M 2017 *J. Quant. Spectrosc. Radiat. Transfer* **186** 221
[6] Engleman R 1974 *J. Mol. Spectrosc.* **49** 106
[7] Kroto H W 2016 National Solar Observatory (NSO) at Kitt Peak, McMath-Pierce Fourier Transform Spectrometer (FTS) data, (accessed December 20, 2016). ftp://vso.nso.edu/FTS_cdrom/FTS30/920212R0.005.
[8] Parigger C G, Woods A C, Surmick D M, Gautam G, Witte M J and Hornkohl J O 2015 *Spectrochim. Acta* B **107** 132
[9] Yurchenko S N, Lodi L, Tennyson J and Stolyarov A V 2016 *Comput. Phys. Commun.* **202** 262
[10] Wigner E and Witmer E E 1928 *Z. Phys.* **51** 859
Hettema H 2000 *Quantum Chemistry: Classic Scientific Papers* p 287 (Singapore: World Scientific)
[11] Zare R N, Schmeltekopf A L, Harrop W J and Albritton D L 1973 *J. Mol. Spectrosc.* **46** 37
[12] Howard B J and Brown J M 1976 *J. Mol. Phys.* **31** 1517
[13] Lefebvre-Brion H and Field R W 2004 *The Spectra and Dynamics of Diatomic Molecules* (New York: Elsevier/Academic)
[14] Brown J M and Carrington A 2003 *Rotational Spectroscopy of Diatomic Molecules* (Cambridge: Cambridge University Press)
[15] Condon E U and Shortley G 1953 *The Theory of Atomic Spectra* (Cambridge: Cambridge University Press)
[16] Hilborn R C 1982 *Am. J. Phys.* **50** 982 http://arxiv.org/ftp/physics/papers/0202/0202029.pdf
[17] Thorne A P 1988 *Spectrophysics* 2nd edn (London: Chapman and Hall)
[18] Rothman L S *et al* 1998 *J. Quant. Spectrosc. Radiat. Transfer* **60** 665
[19] Brown J M, Hougen J T, Huber K P, Johns J W C, Kopp I, Lefebvre-Brion H, Merer A J, Ramsay D A, Rostas J and Zare R N 1975 *J. Mol. Spectrosc.* **55** 500
[20] Kovács I 1969 Rotational Structure in the Spectra of Diatomic Molecules *Rotational Structure in the Spectra of Diatomic Molecules* (American Elsevier: New York)
[21] Li G, Harrison J J, Ram R S, Western C M and Bernath P F 2012 *J. Quant. Spectrosc. Radiat. Transfer* **113** 67

IOP Publishing

Quantum Mechanics of the Diatomic Molecule (Second Edition)

Christian G Parigger and James O Hornkohl

Appendix G

Intrinsic parity of the diatomic molecule

This appendix discusses appearances of selected homonuclear spectra. In quantum field theory, the parity of a state is broken into two parts: one depending on the angular momentum and the second part, called the *intrinsic parity*, that is independent of angular momentum. Particles currently viewed as fundamental (e.g., the photon and electron) are assigned an intrinsic parity. Since parity is multiplicative, the intrinsic parity of a collection of fundamental particles is the product of the individual intrinsic parities. Thus, nucleons have an intrinsic parity of $+1$, and the intrinsic parity of an atomic or molecular system having N electrons is simply $(-)^N$. Does intrinsic parity have any relevance in atomic and molecular physics? By comparing experimental spectra recorded by others with synthetic spectra computed by us using intrinsic parity $(-)^N$ in the computations, we conclude that intrinsic parity is relevant to the spectra of homonuclear diatomic molecules.

The unresolved hyperfine structure (unresolved nuclear spin states) of a diatomic molecule state effectively give the state a degeneracy factor (statistical weight) $g = (2I_\mathrm{a} + 1)(2I_\mathrm{b} + 1)$, where I_a is the nuclear spin of one nucleus and I_b the nuclear spin of the second nucleus. When the two nuclei are identical, the total nuclear spin statistical weight remains $(2I + 1)^2$ where $I = I_\mathrm{a} = I_\mathrm{b}$, but the number of nuclear spin states g_+ for which the parity is $+1$ and the number of states g_- for which the parity is -1 are unequal. This leads to the alternation of intensity in the spectrum of a homonuclear diatomic molecule. The experimental spectra given in the bottom trace in figures (G.1)–(G.3) are examples. In each figure above the experimental spectrum is a computed diatomic spectrum, which agrees agrees reasonably well with the experimental spectrum, and the top trace in each figure is another computed spectrum which is in qualitative disagreement with the experimental spectrum. The only difference between the two computed spectra are the parity values used in their computation.

doi:10.1088/978-0-7503-6204-7ch37

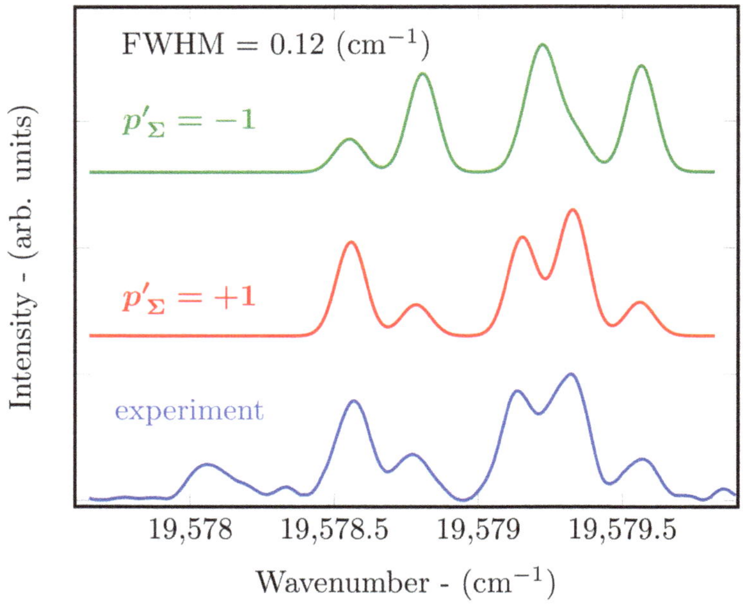

Figure G.1. Two synthetic $^{13}C_2$ $d^3\Pi_g - a^3\Pi_u$ spectra compared with an experimental spectrum [1]. The Λ-doublets are partially resolved. The intrinsic parity of an neutral diatomic molecule is +1 (i.e., the number of electrons is even). For the upper $d^3\Pi_g$ state, $p'_{\rm gu} = +1$, while for the lower $a^3\Pi_u$ state, $p_{\rm gu} = -1$. Thus, $p'_\Sigma = p_0\, p'_{\rm gu} = +1$ and $p_\Sigma = p_0\, p_{\rm gu} = -1$. The molecular parameters were found by fitting the line position data of [1]

The synthetic spectra in Figs (G.1)–(G.3) were computed using the following algorithm.

- For given upper total angular momentum quantum J' and lower quantum number J, the upper and lower Hamiltonian matrices H' and H are computed in the Hund's case (a) basis and are numerically diagonalized to obtain the eigenvalues F' and F and eigenvectors U' and U,

$$F' = U'^{\dagger}\, H'\, U' \tag{G.1}$$

$$F = U^{\dagger}\, H\, U \tag{G.2}$$

- The Hönl–London factors $S_{J'J}$ are computed,

$$S_{J'J} = (2J+1)\left| \sum_{a'}\sum_{a} \langle J\, |a\rangle\, \langle a|C^{J'\Omega'}_{J\Omega q\kappa}|a'\rangle\, \langle a'\, |J'\rangle \right|^2 \tag{G.3}$$

  and the term difference (i.e., line position) is computed for each nonvanishing Hönl–London factor,

$$\tilde{\nu} = F' - F, \qquad \text{if } S_{J'J} \neq 0 \tag{G.4}$$

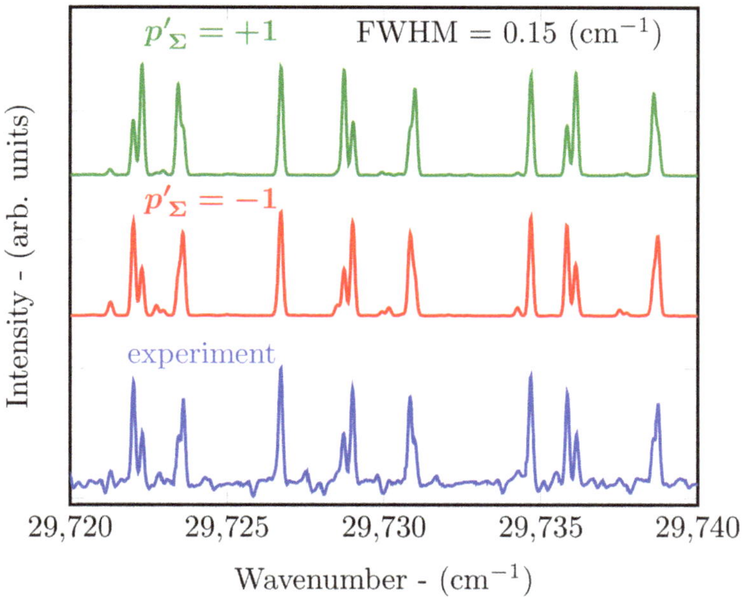

Figure G.2. Two synthetic $C^3\Pi_u - B^3\Pi_g$ N_2 second positive system spectra compared with an experimental spectrum [2]. The Λ-doublets are partially resolved. This spectrum reverses the situation of figure G.1. The intrinsic parity is again +1, but here $p'_{gu} = -1$ and $p_{gu} = +1$, and one finds $p'_\Sigma = p_0\, p'_{gu} = -1$ and $p_\Sigma = p_0\, p_{gu} = +1$. The molecular parameters were found by fitting the line position data [2].

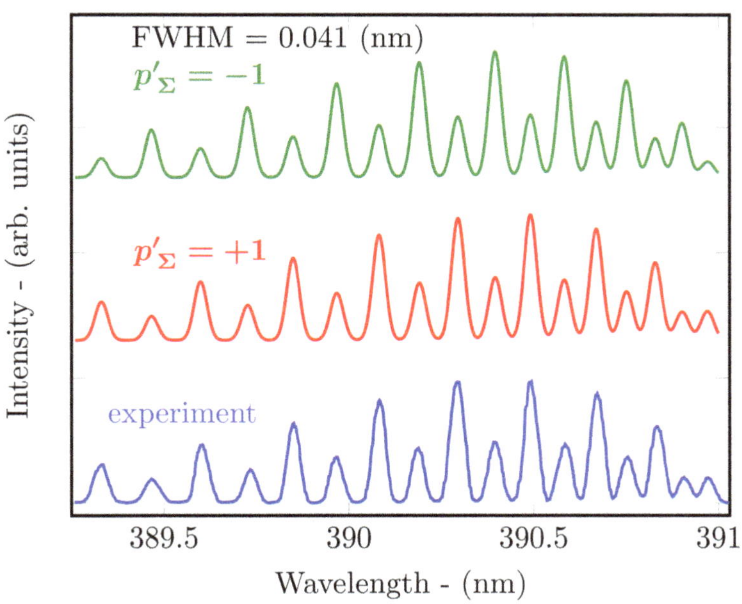

Figure G.3. There are no Λ-doublets in a ${}^2\Sigma_u^+ - {}^2\Sigma_g^+$ spectrum, but the parity alternates from line to line causing an alternation of intensity. The two synthetic spectra are compared with an experimental spectrum [3]. The intrinsic parity of the singly ionized N_2^+ molecule is −1. Thus, $p'_\Sigma = p_0\, p'_{gu} = 1$ and $p_\Sigma = p_0\, p_{gu} = +1$. The molecular parameters were found by fitting the line position data of [4].

- A line list is created by repeating the above for all required J' and J.
- A thermal distribution of excited states and a Gaussian spectrometer slit function having a specified full width at half maximum are assumed.
- The spectrum was broken into a number of pixels, and the the contribution of each line in the computed line list to each pixel is computed.

The matrix elements of parity in the Hund's case (a) basis are the product of a constant p_Σ and $(-)^J$,

$$p_{ij} = p_\Sigma\,(-)^J\,\delta_{J_iJ_j}\,\delta_{\Omega_i,-\Omega_j}\,\delta_{\Lambda_i,-\Lambda_j}\,\delta_{n_in_j}. \tag{G.5}$$

The quantum number n denotes the electronic basis, and other symbols have the standard meanings given them in diatomic spectroscopy. Parity is a member of the complete set of commuting observables. Therefore, the orthogonal matrices U' and U that diagonalize the upper and lower Hamiltonians computed in the case (a) basis will also diagonalize the upper and lower parity matrices computed in the case (a) basis.

The only difference between the two computed spectra in each of figures (G.1)–(G.3) are the values of the upper p'_Σ and lower p_Σ. When computing the synthetic spectra given in figures (G.1)–(G.3), we imposed the selection rule $p_\Sigma \neq p'_\Sigma$. Trial and error quickly finds one of two possibilities, but having a theoretical reason for choosing the value of p_Σ for a diatomic state is clearly desirable.

Equation (G.5) shows a breaking of parity into two parts, p_Σ and $(-)^J$. In quantum field theory, parity is similarly broken into two parts, one depending upon parity and the other independent of parity and called the *intrinsic parity*. Intrinsic parity is believed to be a characteristic property of a fundamental particle. For example, the electron has been assigned the intrinsic parity -1. Parity is multiplicative, and the intrinsic parity of a composite particle is the product of the intrinsic parities of particles composing it, e.g., nucleons have an intrinsic parity of $+1$. An atom or molecule has an intrinsic parity of $(-)^N$, where N is the number of electrons. Ionization and electron attachment can obviously change the intrinsic parity of a molecule, but standard electromagnetic transitions cannot. The selection rule $p'_\Sigma \neq p_\Sigma$ used in the computation of the spectra shown in figures (G.1)–(G.3) eliminates consideration of p_Σ as intrinsic parity. However, the value of p_Σ appears to be correctly given as the product of the intrinsic parity eigenvalue p_0 and the *gerade/ungerade* parity eigenvalue p_{gu},

$$p_\Sigma = (-)^N p_{gu} \tag{G.6}$$

$$= p_0\,p_{gu}\,. \tag{G.7}$$

Based upon the very limited number of comparisons with experiment given in the attached figures, we tentatively conclude that equation (G.7) describes how intrinsic parity influences the spectrum of a homonuclear diatomic molecule.

As long as the weak force is excluded from consideration, none will doubt that the unitary transformation that diagonalizes the Hamiltonian matrix will also diagonalize the parity matrix, but equation (G.5) is new and its derivation is required. The derivation of the Hund's case (a) matrix representation of parity is based upon the Wigner–Witmer diatomic eigenfunction, not the Born–Oppenheimer approximation, and thus represents a radical departure from current diatomic theory. The considered Hamiltonian matrices are not effective Hamiltonians but are simply composed of case (a) matrix elements. Furthermore, the Hamiltonians are not separated into positive and negative parity submatrices, and Van Vleck reductions of the matrix dimensions are not performed.

References

[1] Amiot C 1983 *Astrophys. J. Suppl. Ser.* **52** 329
[2] Roux F, Michaud F and Vervloet M 1989 *Can. J. Phys.* **67** 143
[3] Lazdinis S S and Carpenter R F 1973 *J. Chem. Phys.* **59** 5203
[4] Michaud F, Roux F, Davis S P, Nguyen A-D and Laux C O 2000 *J. Mol. Spectrosc.* **203** 1

Christian G Parigger and James O Hornkohl

Appendix H

Review of diatomic laser-induced breakdown spectroscopy

This appendix summarizes diatomic molecular spectroscopy applications in laser-induced breakdown spectroscopy (LIBS) that were recently communicated in a line shape conference [1]. Moreover, this appendix serves as well as a summary on how to apply the diatomic spectroscopy computations in part I for analysis of measured emission spectra.

In principle, an atomic or molecular spectrum would be computed as follows: upper and lower Hamiltonians would be enumerated in a complete basis, and numerically diagonalized to give the upper and lower energy eigenvalues and eigenvectors. The transition moments for the appropriate operator, e.g., the electric dipole transition moments, would be evaluated from the eigenvectors. The vacuum wave numbers $\tilde{\nu}$, i.e., energy eigenvalue differences, would be found for all non-vanishing transition moments. The line strengths for each spectral line of wave number $\tilde{\nu}$ would be determined as sum of the squares of the transition moments over all transitions producing the same $\tilde{\nu}$. A line list that includes line strengths would be generated by repeating the above computations over the required range of upper and lower total angular momentum quantum numbers. The spectrum from $\tilde{\nu}_{\text{min}}$ to $\tilde{\nu}_{\text{max}}$ would be separated into a number of pixels, and subsequently the contribution of each line to each pixel is calculated using the line list. This appendix reviews how this algorithm can be implemented for a diatomic spectrum if the required molecular parameters are available.

H.1 Introduction

This work reports our algorithm used to compute the synthetic spectra of diatomic molecules. Figure H.1 shows a measured spectrum of interest in LIBS [1–14], and it also shows two computed C_2 Swan spectra fitted to the experimental spectrum. With one caveat, the algorithm is the textbook method for computation of a spectrum. The caveat is that a comprehensive set of accurate diatomic parameters is required for its implementation. We describe our algorithm and give an explanation of why it

doi:10.1088/978-0-7503-6204-7ch38

is based on the Wigner–Witmer diatomic eigenfunction [15] and not the Born–Oppenheimer approximation [16].

H.2 Computation of a diatomic spectrum

Except for the baseline corrections, the synthetic spontaneous emission spectra in figure (H.1) were computed using the following algorithm.

- For given upper total angular momentum quantum number J' and lower quantum number J, the upper and lower Hamiltonian matrices H' and H are computed in the Hund's case (a) basis and numerically diagonalized to obtain the eigenvalues F' and F and eigenvectors U' and U,

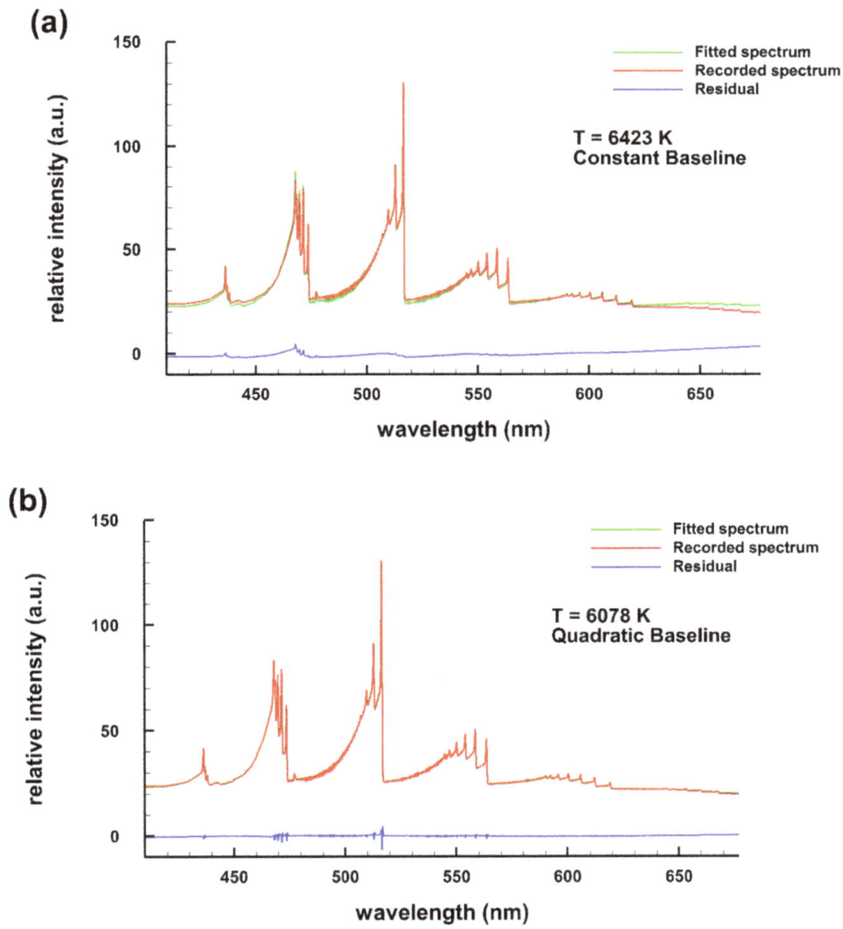

Figure H.1. A recorded C_2 Swan spectrum [17] compared with two computed Swan spectra containing (a) a small constant and (b) a quadratic baseline. The temperature and baseline values were obtained by trial and error implemented with the Nelder–Mead extremum algorithm by varying the temperature for here a thermal distribution of excited states [1].

$$F' = U'^{\dagger}\, H'\, U', \tag{H.1}$$

$$F = U^{\dagger}\, H\, U. \tag{H.2}$$

- The Hönl–London line-strength factors $S_{J'J}$ are computed [18],

$$S_{J'J} = (2J+1)\left| \sum_{a'}\sum_{a} \langle J\, |a\rangle\, \langle a|\, C^{J'\Omega'}_{J\Omega q\kappa} |a'\rangle\, \langle a'\, |J'\rangle \right|^2 \tag{H.3}$$

  and the term difference, i.e., line position, is computed for each nonvanishing Hönl–London line-strength factor,

$$\tilde{\nu} = F' - F \qquad \text{if } S_{J'J} \neq 0. \tag{H.4}$$

- A line list which includes line strengths is created by repeating the above for all required J' and J.
- A thermal distribution of excited states and a Gaussian spectrometer slit function having a specified full width at half maximum are assumed.
- The spectrum is separated into a number of pixels, and the the contribution of each line in the computed line list to each pixel is computed.

Use of the above algorithm requires the Wigner–Witmer diatomic eigenfunction,

$$\langle \mathbf{r}_1, \mathbf{r}_2, \ldots, \mathbf{r}_N, \mathbf{r}\, |nvJM\rangle = \sum_{\Omega=-J}^{J} \langle \rho, \zeta, \mathbf{r}'_2, \ldots, \mathbf{r}'_N, r\, |nv\rangle\, D^{J*}_{M\Omega}(\phi, \theta, \chi), \tag{H.5}$$

in which the total angular momentum states are *exactly* separated instead of a Born–Oppenheimer eigenfunction in which segregation of electronic and nuclear coordinates is enforced, thereby preventing separation of the total angular momentum. The quantum numbers in equation (H.5) are the vibrational quantum number v, the total angular momentum quantum number J, the magnetic quantum number M for the z-component of $\mathbf{J}$, the magnetic quantum number Ω for the z' component of $\mathbf{J}$, and n, which is a symbol representing all other required quantum numbers and labels for continuum indices. In equation (H.5), primed and unprimed coordinates are related by coordinate rotation,

$$\begin{bmatrix} x' \\ y' \\ z' \end{bmatrix} = \mathcal{D}(\phi, \theta, \chi) \begin{bmatrix} x \\ y \\ z \end{bmatrix}. \tag{H.6}$$

The Euler angles ϕ, θ, and χ and the coordinate rotation matrix $\mathcal{D}(\phi, \theta, \chi)$ used here,

$$\mathcal{D}(\phi, \theta, \chi) = \begin{bmatrix} \cos\phi\cos\theta\cos\chi - \sin\phi\sin\chi & \sin\phi\cos\theta\cos\chi + \cos\phi\sin\chi & -\sin\theta\cos\chi \\ -\cos\phi\cos\theta\sin\chi - \sin\phi\cos\chi & -\sin\phi\cos\theta\sin\chi + \cos\phi\cos\chi & \sin\theta\sin\chi \\ \cos\phi\sin\theta & \sin\phi\sin\theta & \cos\theta \end{bmatrix}. \tag{H.7}$$

are those used almost universally in the quantum theory of angular momentum [19–25]. Kovács [26] uses a different set of Euler angles. Primes do not appear on ρ (the distance of one of the electrons from the internuclear axis), ζ (the distance of this electron above or below the plane perpendicular to the internuclear axis and passing through the origin at the center of mass of the nuclei), and r (the internuclear distance) because they are scalars whose value is the same in all coordinate systems. The polar coordinates r, θ, and ϕ are those of the fictitious particle of reduced mass of the two nuclei who motion has replaced their motion. There are N electrons and their coordinates are labeled 1, 2, ⋯, N. The cylindrical coordinates ρ, χ, and ζ are those of one of the electrons. In accord with standard practice in angular momentum theory, the quantum numbers J, M, and Ω refer to the *total* angular momentum. The spectroscopic quantum numbers F, M_F, and Ω_F replace J, M, and Ω in equation (H.5) when it is written in spectroscopic notation.

In the Wigner–Witmer diatomic eigenfunction, the Euler angles ϕ, θ, and χ are both the parameters of coordinate rotation and the angles of physical rotation. The diatomic molecule is perhaps the most complicated actual system in which a single set of Euler angles can serve both purposes. For example, in polyatomic theory the Euler angles ϕ, θ, and χ are physical rotations describing a frame to which the nuclei are attached, and are not the parameters of coordinate rotation α, β, and γ that one would use to demonstrate rotational invariance and conservation of the total angular momentum in a polyatomic model which included vibrational angular momentum.

The Wigner–Eckart theorem breaks the transition moment into two parts: the so-called reduced matrix element, whose value is controlled by the initial and final total angular quantum numbers and the degree of transition tensor operator, and the Clebsch–Gordan coefficient, whose value is controlled by the magnetic quantum numbers and the indices of the components of the tensor operator. The exact separation of the total angular momentum in the Wigner–Witmer eigenfunction simplifies and improves the accuracy of computation of the Hönl–London strengths. Line positions $\tilde{\nu}$ and Hönl–London factors are normally independently evaluated. In the present algorithm, they are simultaneously computed. Of the myriad upper and lower term differences, only those for which the Hönl–London line-strength factor is nonvanishing become spectral lines.

The Born–Oppenheimer approximation is not totally eliminated from computation of a diatomic spectrum. It is needed to break the electronic–vibrational eigenfunction into the product of electronic and vibrational eigenfunction. The Wigner–Witmer diatomic eigenfunction exactly separates the diatomic line strength into the product of the electronic–vibrational strength $S_{n'v',nv}$ and the Hönl–London line-strength factor $S_{J',J}$,

$$S_{n'v'J',nvJ} = S_{n'v',nv}\, S_{J',J}, \tag{H.8}$$

$$S_{n'v',nv} = \left| a_0 + a_1\bar{r}_{v'v} + a_2\bar{r}^2_{v'v} + \cdots \right|^2 q_{v'v}. \tag{H.9}$$

The Born–Oppenheimer approximation separates the electronic–vibrational strength into the product of the square of the electronic transition moment,

$$R_e(r) = a_0 + a_1 r + a_2 r^2 + \cdots, \tag{H.10}$$

evaluated at the r-centroids $\bar{r}^n_{v'v}$,

$$\bar{r}^n_{v'v} = \frac{\langle v' \,| r^n | v \rangle}{\langle v' \,| v \rangle}, \tag{H.11}$$

and with the Franck–Condon factors $q_{v'v}$,

$$q_{v'v} = \langle v' \,| v \rangle^2. \tag{H.12}$$

Whereas experimental and calculated line positions are often known with an accuracy of 1: 10^6 or better, spectral line intensity is rarely recorded with an accuracy of better than 1%. Thus, use of the Born–Oppenheimer approximation for calculation of the electronic–vibrational strength, but not for calculation of line positions, does not introduce significant error.

H.3 Determination of the molecular parameters

Implementation of the algorithm described above requires a comprehensive set of accurate molecular parameters. This is not a minor proviso, and the situation is further complicated by the fact the use of the Wigner–Witmer eigenfunction modifies the manner in which diatomic parameters are determined from a accurate line position measurements such as those provided by Fourier transform spectroscopy. The basic idea in both current practice [27, 28] and our determination of diatomic parameters is the calculation of matrix elements of the form

$$\langle a_i \,| H_k | a_j \rangle = \langle n_i v_i J_i M_i \Omega_i \Lambda_i S_i \Sigma_i \,| H_k | n_j v_j J_j M_j \Omega_j \Lambda_j S_j \Sigma_j \rangle \tag{H.13}$$

in which $|a\rangle$ is a Hund's case (a) basis function,

$$|a\rangle = |nvJM\Omega\Lambda S\Sigma\rangle = \sqrt{\frac{2J+1}{8\pi^2}}\, \langle \rho, \zeta, \mathbf{r}'_2, \ldots, \mathbf{r}'_N, r \,| nv \rangle \, D^{J*}_{M\Omega}(\phi, \theta, \chi) \,| S\Sigma \rangle \tag{H.14}$$

and H_k is a term from the diatomic Hamiltonian

$$H = \sum_k H_k. \tag{H.15}$$

Use of the Wigner–Witmer eigenfunction breaks required matrix elements of the Hamiltonian into two parts: the angular momentum part, which can be calculated exactly, and the electronic–vibrational part, which, except for the very simplest molecules, cannot be calculated with spectroscopic accuracy. Use of a Born–Oppenheimer eigenfunction breaks calculation into three parts: an electronic part, a part consisting of a sum over an infinite number of Born–Oppenheimer vibrational states, and a rotational part. In general, none of the three can be done exactly. Van Vleck transformations, parity partitioning, and the concept of an 'effective

Hamiltonian" reduce the dimensions of the Hamiltonian matrix. We simply compute a Hamiltonian matrix composed of case (a) matrix elements, and let the dimension of the matrix be determined by the range of Ω in equation (H.5) be that required make computed line positions $\tilde{nu}_{\text{cal}}$ equal the experimental positions $\tilde{\nu}_{\text{exp}}$ to within the estimated accuracy of the $\tilde{\nu}_{\text{exp}}$. With the exceptions that (1) our Hamiltonian of unmodified case (a) matrix elements replaces the effective Hamiltonian used by others and (2) instead of using coded selection rules we use nonvanishing of the Hönl–London line-strength factor as the only selection rule, our determination of molecular parameters is identical to that described by [29]. Trial values of upper and lower parameters are assumed, the line positions are computed, corrections to the parameters are computed from the errors in the computed line positions,

$$\Delta\tilde{\nu} = \tilde{\nu}_{\text{cal}} - \tilde{\nu}_{\text{exp}}, \tag{H.16}$$

and the process is repeated until the corrections to the parameters become negligibly small. The algorithm for finding molecular parameters by fitting computed line positions to measured line positions is described in more detail in [30].

H.4 Discussion

The Wigner–Witmer paper [15] appeared about a year after the Born–Oppenheimer paper [16]. The Born and Oppenheimer work treats all molecules, but the Wigner–Witmer paper is strictly limited to the diatomic molecule. Although Born's formulation[16, 31] in terms of

$$\kappa = \left(\frac{m}{M_0}\right)^{1/4} \tag{H.17}$$

in which m is the electronic mass and 'where M_0 can taken as any one of the nuclear masses or their mean' [31] is rarely used, the Born–Oppenheimer approximation became the foundation of molecular theory. Although the Born–Oppenheimer approximation is applicable to all molecules, in their final section Born and Oppenheimer show that the polar angle and azimuthal angle of the polar coordinates r, θ, and ϕ of the internuclear vector in a diatomic molecule are exactly separable in the spherical harmonic $Y_{\ell m}(\theta, \phi)$. The Wigner complex conjugate of the D-symbol, $D^{j*}_{m\omega}(\phi, \theta, \chi)$ is the mathematical extension of the spherical harmonics,

$$Y_{\ell m}(\theta, \phi) = \sqrt{\frac{2\ell + 1}{4\pi}}\, D^{\ell *}_{m\,0}(\phi, \theta, 0) \tag{H.18}$$

which allows one to deal with the total angular momentum $|jm\rangle$ rather that just the orbital angular momentum $|\ell m\rangle$. The first half of the Wigner–Witmer paper is the logical extension of the Born–Oppenheimer $Y_{\ell m}(\theta, \phi)$ result to include all three Euler angles in diatomic eigenfunction. The two-part Wigner–Witmer paper became famous for its second part, which gives correlation rules relating the electronic states of a diatomic molecule to the $L - S$ coupled states of the separated atoms.

Oddly, the exact diatomic eigenfunction with which Wigner and Witmer determined their correlation rules has been ignored. About 40 years passed between publication the Wigner–Witmer paper and entry of the Wigner *D*-function into the literature of diatomic theory. By then, the Wigner–Witmer diatomic eigenfunction had apparently been forgotten.

References

[1] Hornkohl J O and Woods A C 2014 *J. Phys.: Conf. Ser.* **548** 012033

[2] Singh J P and Thakur S N (ed) 2020 *Laser-Induced Breakdown Spectroscopy* (New York: Elsevier Science)

[3] Musazzi S and Perini U (ed) 2014 *Laser-Induced Breakdown Spectroscopy* (Heidelberg: Springer)

[4] Cremers D E and Radziemski L J 2006 *Handbook of Laser-Induced Breakdown Spectroscopy* (New York: Wiley)

[5] Noll R 2011 *Laser-Induced Breakdown Spectroscopy.* (Heidelberg: Springer)

[6] Singh J P and Thakur S N (ed) 2007 *Laser-Induced Breakdown Spectroscopy* (Amsterdam: Elsevier Science)

[7] Miziolek A W and Palleschi V (ed) 2006 *Laser-Induced Breakdown Spectroscopy (LIBS)—Fundamentals and Applications* (New York: Cambridge University Press)

[8] Parigger C G 2006 Laser-induced breakdown in gases: Experiments and simulation *Laser-Induced Breakdown Spectroscopy (LIBS)—Fundamentals and Applications* ed A W Miziolek, V Palleschi and I Schechter (New York: Cambridge University Press)

[9] Parigger C G, Surmick D M, Helstern C M, Gautam G, Bol'shakov A A and Russo R 2020 Molecular laser-induced breakdown spectrosocpy *Laser-Induced Breakdown Spectroscopy* ed J P Singh and S N Thakur (New York: Elsevier Science) ch 7

[10] Merten J, Jones M, Hoke S and Allen S A 2014 *J. Phys.: Conf. Ser.* 548 012042

[11] Parigger C G, Helstern C M and Gautam G 2019 *Atoms* **7** 7030074

[12] Tiwari P K, Rai N K, Kumar R, Parigger C G and Rai A K 2019 *Atoms* **7** 7030071

[13] Surmick D M, Dagel D J and Parigger C G 2019 *Atoms* **7** accepted

[14] Parigger C G, Sherbini A M E L and Splinter R 2019 *J. Phys.: Conf. Ser.* accepted

[15] Wigner E and Witmer E E 1928 *Z. Phys.* **51** 859
Hettema H 2000 *Quantum Chemistry: Classic Scientific Papers* p 287 (Singapore: World Scientific)

[16] Born M and Oppenheimer R 1927 *Ann. Phys.* **84** 457
Hettema H 2000 *Quantum Chemistry: Classic Scientific Papers* p 1 (Singapore: World Scientific)

[17] Nemes L, Keszler A M, Hornkohl J O and Parigger C G 2005 *Appl. Opt.* **44** 3661

[18] Nemes L, Hornkohl J O and Parigger C G 2005 *Appl. Opt.* **44** 3686

[19] Rose M E 1995 *Elementary Theory of Angular Momentum* (Mineola, NY: Dover)

[20] Edmonds A R 1974 *Angular Momentum in Quantum Mechanics* 2nd edn (Princeton, NJ: Princeton University Press)

[21] Brink D M and Satchler G R 1993 *Angular Momentum* 3rd edn (Oxford: Oxford University Press)

[22] Biedenharn L C and Louck J D 2009 *Angular Momentum in Quantum Physics* (Cambridge: Cambridge University Press)

[23] Zare R N 1988 *Angular Momentum* (New York: Wiley)

[24] Thompson W J 1994 *Angular Momentum* (New York: Wiley)
[25] Varshalovich D A, Moskalev A N and Khersonskii V K 1988 *Quantum Theory of Angular Momentum* (Singapore: World Scientific)
[26] Kovács I 1969 *Rotational Structure in the Spectra of Diatomic Molecules* (New York: American Elsevier)
[27] Lefebvre-Brion H and Field R W 2004 *The Spectra and Dynamics of Diatomic Molecules* (New York: Elsevier/Academic)
[28] Brown J M and Carrington A 2003 *Rotational Spectroscopy of Diatomic Molecules* (Cambridge: Cambridge University Press)
[29] Zare R N, Schmeltekopf A L, Harrop W J and Albritton D L 1973 *J. Mol. Spectrosc.* **46** 37
[30] Hornkohl J O, Nemes L and Parigger C G 2009 Spectroscopy, dynamics and molecular theory of carbon plasmas and vapors *Advances in the Understanding of the Most Complex High-Temperature Elemental System* ed L Nemes and S Irle (Singapore: World Scientific) ch 4
[31] Huang K and Born M 1954 *Dynamical Theory of Crystal Lattices* (Oxford: Clarendon)

IOP Publishing

Quantum Mechanics of the Diatomic Molecule (Second Edition)

Christian G Parigger and James O Hornkohl

Appendix I

Program MorseFCF.for

This appendix lists the program 'MorseFCF.for' and the subroutines 'MorseSubs. for.' This program has been utilized in investigations of titanium monoxide (TiO) spectra that were measured following laser-induced breakdown spectroscopy with 1 to 100 TW cm $^{-2}$ irradiance [1]. Franck–Condon factors and *r*-centroids were computed and listed for selected TiO transitions [1] using Morse potentials chosen to best fit the low lying vibrational levels. The electronic transition moments were taken from the most recently reported *initio* computations for A–X, B–X, and E–X transitions.

I.1 MorseFCF.for

This program computes Franck–Condon factors and first three *r*-centroids for a Morse potentials and sets of molecular parameters.

```
!       February 12, 2012
!       James O. Hornkohl
!       Program to compute Morse Franck-Condon factors.
        program morfcf
        use DeclaredStuff
        real(8) :: qs, r1s, r2s, r3s
!        real(8) q, r1, r2, r3
!        dimension q(0:100,0:100), r1(0:100,0:100), r2(0:100,0:100), r3(0:100,0:100)
        real(8) :: weu, wexeu, Beu, muu
        real(8) :: wel, wexel, Bel, mul
        real(8) :: au, betau, reu, gammau, xu, alpha
        real(8) :: al, betal, rel, gammal, xl
        real(8) :: rmin, rmax, rsave, tstu, tstl, dr, r, psi, r0
        real(8) :: lgnrmu(0:100), lgnrml(0:100)
        real(8) :: lgnorm
        real(8) :: alphavu(0:100), alphavl(0:100)
        real(8) :: yu, yl, x, y
        integer*4 :: vmaxu, vmaxl, npts, i, j, k, lun
```



```
      integer*4 :: nv, mv
      character*130 :: prnfil, fcffil

      write (*,1000)
 1000 format (15x, 'Franck–Condon␣Factors␣from␣Morse␣Eigenfunctions')
      write (*,1005)
 1005 format (20x, 'Version␣for␣Microsoft␣32–bit␣Compiler', /)
      write (*,1010)
 1010 format ('␣␣␣␣␣␣␣␣␣␣␣␣␣␣␣␣␣␣␣Print␣file␣name␣=␣',$)
      read (*,'(a)') prnfil
      write (*,1016)
 1016 format ('␣␣␣␣␣␣␣␣␣␣␣␣␣␣␣␣␣␣␣␣␣FCF␣file␣name␣=␣',$)
      read (*,'(a)') fcffil
!      write (*,1020)
! 1020 format (5x, 'Enter constants for upper electronic state.')
!   20 write (*,1030)
! 1030 format ('                             we = ',$)
!      read (*,*,err=20) weu
!   30 write (*,1040)
! 1040 format ('                           wexe = ',$)
!      read (*,*,err=30) wexeu
!   40 write (*,1050)
! 1050 format ('                             Be = ',$)
!      read (*,*,err=40) Beu
!   50 write (*,1060)
! 1060 format ('                   reduced mass = ',$)
!      read (*,*,err=50) muu
!   60 write (*,1070)
! 1070 format (' max vibrational quantum number = ',$)
!      read (*,*,err=60) vmaxu
      weu = 373.152d0
      wexeu = 1.5112d0
      Beu =  0.1542393d0
      muu = 18.64966011d0
      vmaxu = 9
      write(*,1075)  weu, wexeu, Beu, muu
1075  Format(4(1pg20.13))
      call const (au, alphavu, lgnrmu, betau, weu, wexeu, Beu, reu, muu, gammau, vmaxu)
!      write (*,1080)
! 1080 format (5x, 'Enter constants for lower electronic state.')
!  100 write (*,1030)
!      read (*,*,err=100) wel
!  110 write (*,1040)
!     read (*,*,err=110) wexel
!  120 write (*,1050)
!      read (*,*,err=120) Bel
!  130 write (*,1060)
!      read (*,*,err=130) mul!
```

```
!  140 write (*,1070)
!      read (*,*,err=140) vmaxl
      wel = 370.201d0
      wexel = 1.3732d0
      Bel = 0.152233d0
      mul = muu
      vmaxl = 9
      call const (al, alphavl, lgnrml, betal, wel, wexel, Bel, rel, mul, gammal, vmaxl)
  150 write (*,1170)
 1170 format ('␣␣␣␣␣␣␣␣␣␣␣␣r-min␣=␣', $)
      read (*,*,err=150) rmin
      if (rmin.eq.0.0) go to 160
      xu = au * dexp(-betau*(rmin-reu))
      alpha = alphavu(vmaxu)
      lgnorm = lgnrmu(vmaxu)
      tstu = psi (vmaxu, xu, alpha, lgnorm)
      xl = al * dexp(-betal*(rmin-rel))
      alpha = alphavl(vmaxl)
      lgnorm = lgnrml(vmaxl)
      tstl = psi (vmaxl, xl, alpha, lgnorm)
      write (*,1180) tstu, tstl
 1180 format ('␣At␣trial␣value␣of␣r␣the␣eigenfunction␣=', 1pg11.4, 3x, 1pg11.4)
      rsave = rmin
      go to 150
  160 rmin = rsave
  170 write (*,1190)
 1190 format ('␣␣␣␣␣␣␣␣␣␣␣␣r-max␣=␣', $)
      read (*,*,err=170) rmax
      if (rmax.eq.0.0) go to 180
      xu = au * dexp(-betau*(rmax-reu))
      alpha = alphavu(vmaxu)
      lgnorm = lgnrmu(vmaxu)
      tstu = psi (vmaxu, xu, alpha, lgnorm)
      xl = al * dexp(-betal*(rmax-rel))
      alpha = alphavl(vmaxl)
      lgnorm = lgnrml(vmaxl)
      tstl = psi (vmaxl, xl, alpha, lgnorm)
      write (*,1180) tstu, tstl
      rsave = rmax
      go to 170
  180 rmax = rsave
  190 write (*,1200)
 1200 format ('␣increment␣in␣r␣=␣', $)
      read (*,*,err=190) dr
      npts = idnint((rmax-rmin)/dr) + 1
      if (prnfil.ne.'␣') then
         open (10, file=prnfil, status='unknown')
         lun = 10
```

```
      else
         lun = 0
      endif
      write (lun,1000)
      write (lun,1090) reu, rel
 1090 format (6x, '              re =', 1pg17.10, 3x, 1pg17.10)
      write (lun,1100) betau, betal
 1100 format (6x, '            beta =', 1pg17.10, 3x, 1pg17.10)
      write (lun,1110) gammau, gammal
 1110 format (6x, '           gamma =', 1pg17.10, 3x, 1pg17.10)
      write (lun,1120) weu, wel
 1120 format (6x, '              we =', 1pg17.10, 3x, 1pg17.10)
      write (lun,1130) wexeu, wexel
 1130 format (6x, '            wexe =', 1pg17.10, 3x, 1pg17.10)
      write (lun,1140) Beu, Bel
 1140 format (6x, '              Be =', 1pg17.10, 3x, 1pg17.10)
      write (lun,1150) muu, mul
 1150 format (6x, '   reduced mass =', 1pg17.10, 3x, 1pg17.10)
      write (lun,1210) rmin, rmax
 1210 format (6x, '           r-min =', 1pg11.4, 2x, 'r-max =', 1pg11.4)
      write (lun,1220) dr
 1220 format (6x,'          delta r ='1pg11.4)
      write (lun,1230) npts
 1230 format (6x,'number of points =',i4)
      allocate(q(0:vmaxu,0:vmaxl), r1(0:vmaxu,0:vmaxl), r2(0:vmaxu,0:vmaxl), r3(0:vmaxu,0:vmaxl))
      r0 = dble(idint(50.0d0*(reu+rel))) / 100.0d0
      do 220 i=0, vmaxu
         do 210 j=0, vmaxl
            qs = 0d0
            r1s = 0d0
            r2s = 0d0
            r3s = 0d0
            do k=1, npts
               r = rmin + (k-1)*dr
               xu = au * dexp(-betau*(r-reu))
               alpha = alphavu(i)
               lgnorm = lgnrmu(i)
               yu = psi (i, xu, alpha, lgnorm)
               xl = al * dexp(-betal*(r-rel))
               alpha = alphavl(j)
               lgnorm = lgnrml(j)
               yl = psi (j, xl, alpha, lgnorm)
               y = yu * yl
               qs = qs + y
               x = r - r0
               y = y * x
               r1s = r1s + y
               y = y * x
```

```
            r2s = r2s + y
            y = y * x
            r3s = r3s + y
         enddo
         qs = qs * dr
         r1(i,j) = r1s * dr / qs
         r2(i,j) = r2s * dr / qs
         r3(i,j) = r3s * dr / qs
         q(i,j) = qs * qs
         nv = i - 1
         mv = j - 1
 210   continue
 220 continue
     call prnfcf (vmaxu, vmaxl, r0, lun)
     if (lun.ne.0)  close (lun)
!     De = weu*weu / (4d0*wexeu)
     if (fcffil.ne.' ')  then
        open (unit=11, file=fcffil)
        do i=0, vmaxu
           do j=0, vmaxl
              write (11,2000) i, j, q(i,j), r1(i,j), r2(i,j), r3(i,j)
           end do
        end do
        close (11)
     endif
2000 format (1x, i3, ',', i3,  4(',', 1x,1pg14.7))
     end

     subroutine prnfcf (vmaxu, vmaxl, r0, lun)
     use DeclaredStuff
     implicit none
     integer*4 vmaxu, vmaxl, lun
!     real(8) q, r1, r2, r3
!     common / block1 / q(0:100,0:100), r1(0:100,0:100), r2(0:100,0:100),
!     c r3(0:100,0:100)
     real(8) r0, sum
     integer*4 i, j, jmin, jmax
     write (lun,1000) r0
1000 format (/, 6x, 'r0 =', 1pg12.5,/)
     jmin = 0
     jmax = 4
  10 if (jmax.gt.vmaxl)  jmax = vmaxl
     write (lun,1010) (j, j=jmin, jmax)
     do 30 i=0, vmaxu
        write (lun,1020) i, (q(i,j), j=jmin, jmax)
        write (lun,1030) (r1(i,j), j=jmin, jmax)
        write (lun,1030) (r2(i,j), j=jmin, jmax)
        write (lun,1030) (r3(i,j), j=jmin, jmax)
```

```
  30 continue
     if (jmax.eq.vmaxl) goto 40
     jmin = jmin + 5
     jmax = jmax + 5
     goto 10
1010 format (/, 5x, 'v''␣', 4x, 'v"␣=', i3, 7x, 'v"␣=', i3, 7x,&
         & 'v"␣=', i3, 7x, 'v"␣=', i3, 7x, 'v"␣=', i3)
1020 format (2x, i4, 5(3x,1pg11.4))
1030 format (6x, 5(3x,1pg11.4))
  40 write (lun,1040)
1040 format (/, 6x, 'Emission␣sums')
     do 60 i=0, vmaxu
        sum = 0d0
        do 50 j=0, vmaxl
           sum = sum + q(i,j)
  50    continue
        write (lun,1050) i, sum
  60 continue
1050 format (6x, i5, 2x, 1pg22.15)
     write (lun,1060)
1060 format (/, 6x, 'Absorption␣sums')
     do 80 j=0, vmaxl
        sum = 0d0
        do 70 i=0, vmaxu
           sum = sum + q(i,j)
  70    continue
        write (lun,1050) j, sum
  80 continue
     return
     end
```

I.2 MorseSubs.for

This section list subroutines for the main program MorseFCF that computes Franck–Condon factors and the first three r-centroids.

```
      subroutine const(a, b, lgnorm, beta, we, wexe, Be, re, mu, gamma,
     c vmax)
!     July 8, 2011
!     James O. Hornkohl
!     Compute some constants needed for Morse eigenfunctions.
      implicit real*8 (a-h, o-z)
      real*8 h, c, Avog, pi, hbar, x
      real*8 a, b, lgnorm, beta, we, wexe, be, re, mu, gamma
      integer k, v, vmax
      dimension b(0:100), lgnorm(0:100)
      real*8 dlgama
```

```
      h = 6.6260693d-27
      c = 2.99792458d+10
      Avog = 6.0221415d+23
      pi = dacos(-1d0)
      hbar = h / (2d0*pi)
      beta = 1d-08 * sqrt(4d0 * pi * c * (mu/Avog) * wexe
     c / hbar)
!     Compute equilibrium internuclear distance
      gamma = 1d+16 * hbar * Avog / (4d0 * pi * c * mu)
      if (Be.ne.0d0) then
         re = sqrt(gamma/Be)
      else
         Be = gamma / (re*re)
      end if
      a = we / wexe
!     Compute logarithm of normalization factor.
      do v=0, vmax
!     b(v) is the constant alpha in the Laguerre polynomial.
         b(v) = a - 2d0*v - 1d0
!     lgnorm(v) is the logarithm of the normalization factor for the
!     Morse eigenfunctions.
         lgnorm(v) = dlog(beta)
         if (v.lt.2) go to 10
         do k=2, v
            lgnorm(v) = lgnorm(v) + dlog(1d0*k)
         enddo
   10    lgnorm(v) = lgnorm(v) + dlog(b(v))
         x = b(v) + v + 1d0
         lgnorm(v) = lgnorm(v) - dlgama(x)
         lgnorm(v) = lgnorm(v) / 2d0
      enddo
      return
      end

      real*8 function psi (v, x, alpha, lgnorm)
!     December 28, 2011
!     Compute eigenfunction for Morse oscillator.  Subroutine const
!     must be called once before calls to here are made.
!     The vibrational quantum number is n-1.
      implicit none
      real*8 x, alpha, lgnorm, laguer
      integer k, kmax, v
      psi = dexp(lgnorm+alpha*dlog(x)/2d0-x/2d0) * laguer(v, x, alpha)
      return
      end

      real*8 function laguer (n, x, alpha)
!     December 28, 2011
```

```
!       The degree of the Laguerre polynomial is n-1.
        implicit none
        real*8 x, alpha
        integer n, k, kmax
        real*8 poly(0:100)
        poly(0) = 1d0
        poly(1) = alpha + 1d0 - x
        if (n.lt.2) go to 20
        kmax = n - 1
        do 10 k=1, kmax
           poly(k+1) = ((2*k+alpha+1d0-x) * poly(k)  -  (k+alpha) *
     c      poly(k-1) ) / (k + 1)
   10   continue
   20   laguer = poly(n)
        return
        end

        real*8 FUNCTION DLGAMA (X)
!                                  SPECIFICATIONS FOR ARGUMENTS
        real*8   X
!                                  SPECIFICATIONS FOR LOCAL VARIABLES
        real*8   P1(9),Q1(8),P2(9),Q2(8),P3(9),Q3(8),P4(7)
        INTEGER            IER,J
        real*8   BIG1,XINF,PI,
     *                     Y,T,R,SIGN,A,B,TOP,DEN,EPS
        LOGICAL            MFLAG
!                                  COEFFICIENTS FOR MINIMAX
!                                  APPROXIMATION TO LN(GAMMA(X)),
!                                  0.5 .LE. X .LE. 1.5
        DATA               P1(1)/6.304933722864032D02/,
     *                     P1(2)/1.389482659233250D02/,
     *                     P1(3)/-2.331861065739548D03/,
     *                     P1(4)/-2.651470392943388D03/,
     *                     P1(5)/-8.953073589022869D02/,
     *                     P1(6)/-9.229503102917111D01/,
     *                     P1(7)/-1.940352203312667D00/,
     *                     P1(8)/4.368019694395194D00/,
     *                     P1(9)/1.279153645893113D02/
        DATA               Q1(1)/6.689575153359349D02/,
     *                     Q1(2)/2.419887329355996D03/,
     *                     Q1(3)/3.354196974608081D03/,
     *                     Q1(4)/1.860416170944268D03/,
     *                     Q1(5)/3.944307810159532D02/,
     *                     Q1(6)/2.682132440551618D01/,
     *                     Q1(7)/3.440812622259858D-01/,
     *                     Q1(8)/5.948212550303777D01/
!                                  COEFFICIENTS FOR MINIMAX
!                                  APPROXIMATION TO LN(GAMMA(X)),
```

```
!                               1.5  .LE.  X  .LE.  4.0
      DATA                P2(1)/1.071722590306920D04/,
     *                    P2(2)/6.527047912184606D04/,
     *                    P2(3)/1.176398389569621D05/,
     *                    P2(4)/-9.726314581896472D03/,
     *                    P2(5)/-1.266094622188023D05/,
     *                    P2(6)/-5.393468741199669D04/,
     *                    P2(7)/-3.895916159676326D03/,
     *                    P2(8)/5.397392180667399D00/,
     *                    P2(9)/5.334390026324024D02/
      DATA                Q2(1)/5.314589562326176D03/,
     *                    Q2(2)/5.493654949398033D04/,
     *                    Q2(3)/2.205757574602192D05/,
     *                    Q2(4)/3.602313576600391D05/,
     *                    Q2(5)/2.273446951911101D05/,
     *                    Q2(6)/4.560612434396495D04/,
     *                    Q2(7)/1.702062439974796D03/,
     *                    Q2(8)/1.670328399370593D02/
!                               COEFFICIENTS FOR MINIMAX
!                               APPROXIMATION TO LN(GAMMA(X)),
!                               4.0  .LE.  X  .LE.  12.0
      DATA                P3(1)/-6.114039864945718D07/,
     *                    P3(2)/-5.588132821261888D08/,
     *                    P3(3)/-9.078970022444525D08/,
     *                    P3(4)/3.662935130796460D09/,
     *                    P3(5)/4.658700336821218D09/,
     *                    P3(6)/-4.570725249206307D09/,
     *                    P3(7)/-2.220833171087439D09/,
     *                    P3(8)/-1.474322990113017D04/,
     *                    P3(9)/-1.959850795570400D06/
      DATA                Q3(1)/-5.636057205056241D05/,
     *                    Q3(2)/-2.691827587118628D07/,
     *                    Q3(3)/-4.411606716771217D08/,
     *                    Q3(4)/-2.774890551941383D09/,
     *                    Q3(5)/-6.579874397740792D09/,
     *                    Q3(6)/-4.980644951174248D09/,
     *                    Q3(7)/-6.677373781427094D08/,
     *                    Q3(8)/-2.722530175870899D03/
!                               COEFFICIENTS FOR MINIMAX
!                               APPROXIMATION TO LN(GAMMA(X)),
!                               12.0  .LE.  X
      DATA                P4(1)/8.40596949829D-04/,
     1                    P4(2)/-5.9523334141881D-04/,
     2                    P4(3)/7.9365078409227D-04/,
     *                    P4(4)/-2.777777777769526D-03/,
     *                    P4(5)/8.333333333333333D-02/,
     *                    P4(6)/9.189385332046727D-01/,
     6                    P4(7)/-1.7816839846D-03/
```

```
      DATA                    XINF/.4494232D+308/
      DATA                    EPS/.2220446050D-15/
      DATA                    PI/3.141592653589793D0/
      DATA                    BIG1/1.28118D305/
!                                       FIRST EXECUTABLE STATEMENT
      IER = 0
      MFLAG = .FALSE.
      T = X
      IF (DABS(T).LT.BIG1) GO TO 5
      IER = 129
      DLGAMA = XINF
      GO TO 9000
    5 IF (T.GT.0.0D0) GO TO 20
      IF (T.LT.0.0D0) GO TO 10
      IER = 130
      DLGAMA = XINF
      GO TO 9000
!                                       ARGUMENT IS NEGATIVE
   10 MFLAG = .TRUE.
      T = -T
      R = DINT(T)
      SIGN = 1.0D0
      IF (DMOD(T,2.0D0).EQ.0.0D0) SIGN = -1.0D0
      R = T-R
      IF (R.NE.0.0D0) GO TO 15
      IER = 130
      DLGAMA = XINF
      GO TO 9000
!                                       ARGUMENT IS NOT A NEGATIVE INTEGER
   15 R = PI/DSIN(R*PI)*SIGN
      T = T+1.0D0
      R = DLOG(DABS(R))
!                                       EVALUATE APPROXIMATION FOR
!                                         LN(GAMMA(T)), T .GT. 0.0
   20 IF (T.GT.12.0D0) GO TO 60
      IF (T.GT.4.0D0) GO TO 50
      IF (T.GE.1.5D0) GO TO 40
      IF (T.GE.0.5D0) GO TO 25
!                                       0.0 .LT. T .LT. 0.5
      B = -DLOG(T)
      A = T
      T = T+1.0D0
      IF (A.GE.EPS) GO TO 30
      DLGAMA = B
      GO TO 9005
!                                       0.5 .LE. T .LT. 1.5
   25 TOP = T-0.5D0
      B = 0.0D0
```

```
      A = TOP-0.5D0
   30 TOP = P1(8)*T+P1(9)
      DEN = T+Q1(8)
      DO 35 J=1,7
         TOP = TOP*T+P1(J)
         DEN = DEN*T+Q1(J)
   35 CONTINUE
      Y = (TOP/DEN)*A+B
      IF (MFLAG) Y = R-Y
      DLGAMA = Y
      GO TO 9005
!                                1.5 .LE. T .LE. 4.0
   40 B = T-1.0D0
      TOP = P2(8)*T+P2(9)
      DEN = T+Q2(8)
      A = B-1.0D0
      DO 45 J=1,7
         TOP = TOP*T+P2(J)
         DEN = DEN*T+Q2(J)
   45 CONTINUE
      Y = (TOP/DEN)*A
      IF (MFLAG) Y = R-Y
      DLGAMA = Y
      GO TO 9005
!                                4.0 .LT. T .LE. 12.0
   50 TOP = P3(8)*T+P3(9)
      DEN = T+Q3(8)
      DO 55 J=1,7
         TOP = TOP*T+P3(J)
         DEN = DEN*T+Q3(J)
   55 CONTINUE
      Y = TOP/DEN
      IF (MFLAG) Y = R-Y
      DLGAMA = Y
      GO TO 9005
!                                12.0 .LT. X .LT. BIG1
   60 TOP = DLOG(T)
      TOP = T*(TOP-1.0D0)-.5D0*TOP
      T = 1.0D0/T
      Y = TOP
      IF (T.LT.EPS) GO TO 70
      B = T*T
      A = P4(7)
      DO 65 J = 1,5
         A = A*B+P4(J)
   65 CONTINUE
      Y = A*T+P4(6)+TOP
   70 IF (MFLAG) Y = R-Y
```

```
      DLGAMA = Y
      GO TO 9005
9000  CONTINUE
      write (*,'("␣Error␣in␣DLGAMA")')
9005  RETURN
      END
```

Reference

[1] Parigger C G, Woods A C, Keszler A, Nemes L and Hornkohl J O 2012 *AIP Conf. Proc.* **1464** 628

IOP Publishing

Quantum Mechanics of the Diatomic Molecule (Second Edition)

Christian G Parigger and James O Hornkohl

Appendix J

Boltzmann equilibrium spectrum (BESP) and Nelder–Mead temperature (NMT) scripts

J.1 BESP.m

The script BESP.m is designed following the FORTRAN/Windows 7 version [1]. The individual diatomic molecular data files for selected transitions are concatenated to only show wave numbers, upper-term values, and line strengths; see table 15.5. Adjustments of input parameters for MATLAB [2] are rather straightforward, equally, for generalizing the script for automatic input by converting the script to a function. Individual lines are computed using Gaussian profiles [1]. For the generation of a spectrum, only one temperature is needed for equilibrium computation. Conversely, as one infers temperature from a measured spectrum, a modified Boltzmann plot [3] is constructed for the determination of the equilibrium temperature. A Gaussian line shape is selected to model the spectrometer/intensifier transfer function profile. However, one usually considers a natural linewidth for electronic state-to-state transitions, and a Gaussian line shape (equation (J.1)) for Doppler broadening [4], viz.

$$\Delta\lambda = 7.16 \times 10^{-7}\lambda \sqrt{\frac{\mathrm{T}}{\mathrm{M}}}, \tag{J.1}$$

leading to Voigt line shapes. Here, $\Delta\lambda$ is the full width at half maximum, λ the wavelength, T is the temperature, and M is the molecular weight. For example, with $\lambda = 306$ nm, $\mathrm{T} = 3.5$ kK, and $\mathrm{M} = 17$ (OH), $\Delta\lambda = 0.0031$ nm. The spectral resolution, $\delta\lambda$, for the OH emission spectra-fitting, discussed in this appendix, amounts to $\delta\lambda = 0.33$ nm. Consequently, a Gaussian line shape is considered instead of a Voigt line shape for fitting of the OH data in the appendix, but the communicated MATLAB scripts can be adjusted for Voigt profiles, which is important for cases when individual electronic state-to-state molecular transitions/ resonances are investigated. Equally, when investigating individual transitions/

doi:10.1088/978-0-7503-6204-7ch40

resonances, asymmetric molecular line shapes can be implemented in the scripts. There is usually a volley of lines for electronic transitions of a diatomic molecules, e.g. OH [6] in excess of 3kK, within a wavelength bin and for an experimental spectral resolution of the order of 0.33 nm.

The program BESP.m receives input from the LSFs that contain relative line strengths. The output is generated in graphical format, and the program is slightly adjusted for the generation of the spectra illustrated in figures 15.1–15.9. However, figure 15.6 is generated with the BESP.m script given below.

```
% BESP.m
%
% Calculates diatomic specta using line strength data files constructed for
    selected transitions.
% The program is designed using a previous FORTRAN/Windows7 implementation
    including private communications
% with James O. Hornkohl and David M Surmick.
%
% David M. Surmick, 04-27-2016; edited by Christian G. Parigger 11-27-2022.

% input paramters, output: WL_exp (N-1 x 1 array), I (intensity)
wl_min=300; wl_max=325; T=3530; FWHM=0.35; N=10001; norm=1; x='OH-lsf.txt';

% generate wavelengths/wavelength-bins for computation akin to an experiment
nSpec=N-1; delWL=(wl_max-wl_min)/(nSpec); WL_exp=linspace(wl_min,wl_max,nSpec)
    ; WL_exp=WL_exp';

% constants in MKS units (Boltzmann factor bfac in cgs units)
h=6.62606957e-34; c=2.99792458e8; kb=1.3806488e-23; bFac=(100*h*c)/kb; gFac=2*
    sqrt(log(2));

% read line strength file
[p]=load(x); WN=p(:,1); Tu=p(:,2); S=p(:,3);

% convert vacuum wavenumber to air wavelength: CGP 11-27-2022
%a0=2.72643e-4; a1=1.2288; a2=3.555e4; r=1+a0+(a1./(WN.*WN))+(a2./(WN.*WN.*WN
    .*WN));
k0=238.0185;k1=5792105;k2=57.362;k3=167917;r=(1+k1./(1d8*k0-(WN.*WN))+k3./(1d8
    *k2-(WN.*WN)));WL=1.e7./(r.*WN);

% get LSF table wavelengths that most closely match the wavelength-bins
A=find(WL>wl_min & WL<wl_max); WLk=WL(A);
```

```
% get term values and line strengths at WLk in the range wl_min to wl_max
Sk=S(A); Tuk=Tu(A); TuMin=min(Tuk);

% calculate peak intensities and initialize peak_k calculation
peak=-4*log(WLk)+log(Sk)-(bFac/T)*(Tuk-TuMin); peak_k=zeros(nSpec,1); peakMax
    =-1;
for i=1:length(peak);
    if peak(i) > peakMax; peakMax=peak(i); end;
    if peak(i) ~= 0; peak_k(i)=peak(i)-peakMax; end;
end; peak_k=exp(peak_k);

% get wavelength-bin positions that most closely matches line strength table
    wavelengths
n0=zeros(length(WLk),1); for i=1:length(WLk); [~,n0(i)]=min(abs(WL_exp-WLk(i))
    ); end;

% calculate spectrum using Gaussian profiles for peaks, and for wavelength
    dependent FWHM
I=zeros(nSpec,1); FWHMk=(FWHM*WLk)/wl_max;
for i=1:length(WLk); deln=round(FWHM/delWL,0); nMin=n0(i)-deln;
    if nMin < 1; nMin=1; end; nMax=n0(i)+deln;
    if nMax > nSpec; nMax=nSpec; end;
    for j=nMin:nMax;  u=abs(gFac*(WL_exp(j)-WLk(i))/FWHMk(i));
        if u <=9.21;  I(j)=I(j)+peak_k(i)*exp(-u*u); end;
    end;
end; I=norm*I/max(I);

%Display graphical output
figure; plot(WL_exp,I,'LineWidth',1.5); set(gca,'FontWeight','bold','FontSize'
    ,20,'TickLength',[0.02, 0.02]);
LimitsX=xlim; LimitsY=ylim; title('    ','HorizontalAlignment','left','Position
    ', [LimitsX(1)-4, LimitsY(2)]);
xlabel('wavelength (nm)','FontSize',24,'FontWeight','bold');
ylabel('intensity (a.u.)','FontSize',24,'FontWeight','bold');
```

J.2 NMT.m

```
% NMT.m
%
% Fits measured diatomic specta using line strength data files constructed for
    selected transitions.
% The program is designed using a previous FORTRAN/Windows7 implementation
   including private communications
% with James O. Hornkohl and David M Surmick.
%
% inputs: WL_exp - exerimental wavelengths (n x 1 array)
%         Dat    - experimental spextrum (n x 1 array)
%         FWHM   - measured spectral resolution, seed for varried FWHM or
%                  fixed
%         T      - temperature seed for fitting
%         tol    - tolerance of Nelder-Mead fit
%         x      - name of line strength file for calculating theory spectra
%         FIT    - enter 1 for fitting linear offset and temperature
%                  enter 2 for fitting linear offset, temperature, and FWHM
%
% outputs: profile - matrix containing experimental wavelengths, measured
%                    spectrum, fitted spectrum, fitted baseline offset
%                    (n x 4 matrix)
%          vals    - array containing fitted paramters (3x1 or 4x1 array),
%                    temperature is always last entry
%
% sub-functions: FitSpec, FitSpec1, SynthSpec
%
% Example call: [I,v]=NMT(x,y1,0.15,3000,1e-8,'OH-LSF.txt',2);
%
% David M. Surmick, 04-28-2016, edited by Christian G Parigger 11-27-2022

function [profile,vals] = NMT (WL_exp,Dat,FWHM,T,tol,x,FIT)
tic % start code timer

% global variables
global bFac gFac WLk Tuk TuMin Sk n0 nSpec fwhm delWL temp wl_max;

% constants in MKS units (Boltzmann factor bfac in cgs units)
h=6.62606957e-34; c=2.99792458e8; kb=1.3806488e-23; bFac=(100*h*c)/kb; gFac=2*
    sqrt(log(2));

%load experimental data, here an OH spectrum 100 microsecond time delay in air
     breakdown.
xexp='OH100micros.dat';data=load(xexp);WL_exp=data(:,1);Dat=data(:,2);nSpec=
    length(Dat);

% input paramters
T=2000; FWHM=0.3; x='OH-LSF.txt'; temp=T; fwhm=FWHM; wl_min=min(WL_exp);
    wl_max=max(WL_exp); delWL=(wl_max-wl_min)/(nSpec);

% read rpovided LSF file
ZZ=readtable(x); WN=ZZ.Var1; Tu=ZZ.Var2; S=ZZ.Var3;

% convert vacuum wavenumber to air wavelength: CGP 11-27-2022
% a0=2.72643e-4; a1=1.2288; a2=3.555e4; r=1+a0+(a1./(WN.*WN))+(a2./(WN.*WN.*WN
```

```
    .*WN));
k0=238.0185;k1=5792105;k2=57.362;k3=167917;WLoffset=0;r=(1+k1./(1e8*k0-(WN.*WN
    ))+k3./(1e8*k2-(WN.*WN)));
WLoffset=0;if(xexp=='OH100micros.dat'); WLoffset=-0.05;end;WL=1.e7./(r.*WN)+
    WLoffset;

% get LSF table wavelengths in experimental range
A=find(WL>wl_min & WL<wl_max); WLk=WL(A);

% get Term Values and LineStrengths at WLk
Sk=S(A); Tuk=Tu(A); TuMin=min(Tuk);

% get exerpimenal wavelength positions that most closely matches line strength
     table wavelengths
n0=zeros(length(WLk),1); for i=1:length(WLk);  [~,n0(i)]=min(abs(WL_exp-WLk(i)
    )); end;

% normalize data
%Dat=Dat/max(Dat);

% Fitting with Nelder-Mead parameters including two cases options
tol=1.e-6; FIT=2; options=optimset('TolX',tol,'MaxIter',1e8,'MaxFunEvals',1e8)
    ;
switch FIT
    case 1 % fit offset, temperature
        theta=ones(3,1);
        theta(3)=T; % temperature seed
        vals=fminsearch(@(x) FitSpec(x,WL_exp,Dat),theta,options);
        bkg=vals(1)+vals(2)*WL_exp; % calculate fitted offset
        [I,bkg1]=SynthSpec(WL_exp,vals(3),FWHM,Dat,bkg); % calculate fit
    case 2 % fit offset, fwhm, temperature
        theta=ones(4,1);
        theta(3)=FWHM; % fwhm seed
        theta(4)=T; % temperature seed
        vals=fminsearch(@(x) FitSpec1(x,WL_exp,Dat),theta,options);
        bkg=vals(1)+vals(2)*WL_exp; % calculate fitted offset
        [I,bkg1]=SynthSpec(WL_exp,vals(4),vals(3),Dat,bkg); % calculate fit
end

% Visualize Fit
fname=regexprep(x,'-LSF.txt','-fit:');
figure
switch FIT
    case 1
        plot(WL_exp,Dat,'o',WL_exp,I,WL_exp,bkg1,'LineWidth',1.5)
        legend('experiment','fit','base line')
        set(gca,'FontWeight','bold','FontSize',16,'TickLength',[0.02, 0.02]);
        val3=round(vals(3),3, 'significant')
```

```
        title([num2str(fname),'T=',num2str(vals(3)),'K ,FWHM=',num2str(FWHM),'
            nm'])
        xlabel('wavelength (nm)')
        ylabel('intensity (a.u.)')
    case 2
        plot(WL_exp,Dat,'o',WL_exp,I,WL_exp,bkg1,'LineWidth',1.5)
        legend('experiment','fit','base line')
        set(gca,'FontWeight','bold','FontSize',20,'TickLength',[0.02, 0.02]);
        round(vals(4),3,'significant'); round(vals(3),2,'significant');
        val4=round(vals(4),3, 'significant'); val3=round(vals(3),2, '
            significant');
        title([num2str(fname),' T=',num2str(val4),' K, FWHM=',num2str(val3),'
            nm'])
        xlabel('wavelength (nm)','Fontsize',24,'FontWeight','bold')
        ylabel('intensity (a.u.)','Fontsize',24,'FontWeight','bold')
end

toc % end code timer

end % main function

% temperature, offset fit function
function [err] = FitSpec (p,WL_exp,Dat);
global fwhm;
bkg=p(1)+p(2)*WL_exp; [F,~]=SynthSpec(WL_exp,p(3),fwhm,Dat,bkg); c=F\Dat; z=F*
    c; err=norm(z-Dat);
end % fit spec

% temperature, fwhm, offset fit function
function [err] = FitSpec1 (p,WL_exp,Dat);
bkg=p(1)+p(2)*WL_exp; [F,~]=SynthSpec(WL_exp,p(4),p(3),Dat,bkg); c=F\Dat; z=F*
    c; err=norm(z-Dat);
end % fit spec 1

% calculate synthetic spectrum for fit
function [I1,bkg1] = SynthSpec (WL_exp,T,FWHM,Dat,bkg);
global bFac gFac WLk Tuk TuMin Sk n0 nSpec delWL wl_max;
FWHMk=(FWHM*WLk)/wl_max; % wavelength dependent FWHM

% Calculate Peak Intensities
peak=-4*log(WLk)+log(Sk)-(bFac/T)*(Tuk-TuMin); peak_k=exp(peak);

% calculate synthetic spectrum
I=zeros(nSpec,1); % initialize synthetic spectrum output
for i=1:length(WLk); deln=round(2.5*FWHMk(i)/delWL); nMin=n0(i)-deln;
    if nMin < 1; nMin=1; end;
    nMax=n0(i)+deln;
    if nMax > nSpec; nMax=nSpec; end;
```

```
    for j=nMin:nMax; u=abs(gFac*(WLk(i)-WL_exp(j))/FWHMk(i)); I(j)=I(j)+peak_k
        (i)*exp(-u*u); end;
end % synthetic spectrum loop

% normailze data to measured spectrum
I=I/max(I); I=I+bkg; sxy=sum(Dat.*I); syy=sum(I.*I); nf= sxy/syy; I1=I*nf;
    bkg1=bkg*nf;
end % SynthSpec
```

References

[1] Parigger C G, Woods A C, Surmick D M, Gautam G, Witte M J and Hornkohl J O 2015 *Spectrochim. Acta* B **107** 132

[2] *MATLAB Release R2022a Update 5* (MA: Natick)

[3] Parigger C G and Hornkohl J O 2020 *Quantum Mechanics of the Diatomic Molecule with Applications* (Bristol: IOP Publishing)

[4] Corney A 1977 *Atomic and Laser Spectroscopy* (Oxford: Clarendon)

[5] Parigger C G 2022 *Foundations* **2** 934

[6] Parigger C G 2023 *Foundations* **3** 1

IOP Publishing

Quantum Mechanics of the Diatomic Molecule (Second Edition)

Christian G Parigger and James O Hornkohl

Appendix K

Abel-inversion scripts

K.1 Abel-inversion programs

The Abel integral inversion algorithm utilizes function expansion techniques [1]. Chebyshev polynomials accomplish minimization of the maximum error [2]. The advantages of the expansion techniques include direct inversion of recorded, sensitivity corrected, and wavelength calibrated time-resolved data. A summary and detailed discussion of numerical inversion of the Abel integral is communicated in [3]. The published Matlab code [4, 5] is selected to accomplish analysis of spatially- (along the slit-height) and wavelength-resolved images. A typical adapted MATLAB [5] script, MixAnalysis.m, shows the implementation for analysis of CO_2: N_2 1:1 mixed gas. The Chebyshev expansion is computed using Expansion.m

The adaptation includes provision for correction of a slight asymmetry in the otherwise spherically symmetric expansion for specific time delays. The Matlab script CGPimage.m generates graphical output that is also included in this appendix. The MixAnalysis.m script includes lines for preparation of the plotting routines that have been available. However, the CGPimage.m script applies simple MATLAB display methods.

doi:10.1088/978-0-7503-6204-7ch41

K.1.1 MixAnalysis.m

```
function [ f_rec , X ] = abel_inversion(h,R,upf,plot_results,lsq_solve)
upfin=15;%8;%was 12 initially; nspectra=256; nwavel=1001;
midpoint=138;%8_8mix%145;%128;%139;%87;
npoints=80;%8_8_mix%73;%60; ninvertedspectra=2*npoints-1;
dr=1024./nspectra*13.6/1000.; %0.0544=4*0.0136
R=npoints*dr;                          % radius
X=(0:dr:R-dr)';          % spatial coordinates
ndim=length(X);
nsymmetric=1; %=1 for symmetric, =0 for asymmetric
nfilestart=1;%8_8mix%16;
nfileend=21;%8_8mix%21;
nlambda(1)=385;%660; nlambda(2)=490; nlambda(3)=438; nlambda(4)=414;
for i=1:1
nlam=nlambda(i)
for nfile=nfilestart:nfileend%:13
fnamein=sprintf('%02dabelc%03d.dat',nfile,nlam);

if nsymmetric == 1
    fnameout=sprintf('%02dabelc%03d_symmetric.dat',nfile,nlam);
    fnameout3D=sprintf('%02dabelc%03d_symmetric3D.dat',nfile,nlam);
else
    fnameout=sprintf('%02dabelc%03d_asymmetric.dat',nfile,nlam);
    fnameout3D=sprintf('%02dabelc%03d_asymmetric3D.dat',nfile,nlam);
end
fileID=fopen(fnamein,'r');%'08abelc.dat','r');
tline=fgetl(fileID); formatSpec='%f %f %f'; sizeDor=[3 Inf];
D=fscanf(fileID,formatSpec,sizeDor);
fclose(fileID);
D=D'; L=D(:,[1]); Warr=L(1:nwavel);%L(1:1001); H=D(:,[2]); Val =D(:,[3]);
for j=1:nspectra %256
    for i=1:nwavel %1001
       Arr(i,j)=Val(i+1001*(j-1));
    end
end
Arr=Arr';

yarr=zeros(length(X),1); garrR=zeros(length(X),1); garrL=zeros(length(X),1);
f_rec_l=zeros(length(X),1); f_rec_r=zeros(length(X),1);

%jwave=86;
for jwave=1:1001

    for k=1:ndim
        yarr0(k,1)=1./2.*(Arr(k-1+midpoint,jwave)+Arr(midpoint-k+1,jwave));
        if (yarr0(k,1) < 0.0001)
            yarr0(k,1)=0.0001;
        end
        garrR(k,1)=Arr(k-1+midpoint,jwave);%/yarr0(k,1);
        garrL(k,1)=Arr(midpoint-k+1,jwave);%/yarr0(k,1);
    end
garrL=garrL.\yarr0;
garrR=garrR.\yarr0;
for c=1:length(X)
 h(c,1)=yarr0(c,1);%Arr(c+66,jwave);
```

```
end
%
if ~exist('h', 'var') || isempty(h)
    [X,h,R]=generate_test_data;
    plot_results=1;
else
    plot_results=0;
    X=linspace(0,R-0.1,length(h))';
end

% default value for number of expansion elements
if ~exist('upf', 'var'); upf=upfin; end;

% avoiding problems if flags are not given in input
if ~exist('plot_results', 'var'); plot_results=0; end;
if ~exist('lsq_solve', 'var'); lsq_solve=0; end;

%% calculate series expansions fn and corresponding integrals hn

[fn,hn] = compute_expansion( X,upf,R );

%% solve equation system A*L=B for the amplitudes A
if lsq_solve ~= 1

    B = zeros(1,upf+1); L = zeros(upf+1,upf+1);  %create arrays

    for k=1:upf+1

        for l=1:upf+1
            L(l,k)=2.*sum(hn(:,k).*hn(:,l));
        end

        B(k)=sum(h(:).*hn(:,k));
    end

    A=B/L;

else

    x0=1*ones(upf+1,1);       % guess some initial values for optimisation
    A=solve_lsq(h,hn,x0);     % solve for amplitudes A
end

%% final stage: calculate the resulting distribution profile

% create vector for resulting reconstructed distribution
f_rec=zeros(length(h),1);
```

```
% special case for n=0 (where f_0(r) = 1)
f_rec = f_rec + A(1)*1;

% iterate eq. (1) for n=1:upf
for c=1:upf+1
    f_rec = f_rec + A(c).*fn(:,c);
end

for c=1:ndim
    f_rec_r(c,1)=f_rec(c,1)*garrR(c,1);
    f_rec_l(ndim-c+1,1)=f_rec(c,1)*garrL(c,1);
end

for c=1:ndim
     f_rec_all(c,1)=f_rec_l(c,1);
     h_all(c,1)=h(ndim-c+1,1);
     x_all(c,1)=-X(ndim-c+1);
     f_rec_all(ndim+c-1,1)=f_rec_r(c,1);
     h_all(ndim+c-1,1)=h(c,1);
     x_all(ndim+c-1,1)=X(c,1);
     g_all(ndim+c-1,1)=garrR(c,1);
     g_all(c,1)=garrL(ndim-c+1,1);
     inp_all(ndim-c+1,1)=garrR(c,1)*yarr0(c,1);
     inp_all(ndim+c-1,1)=garrL(c,1)*yarr0(c,1);
     f_rec_sym(ndim+c-1,1)=f_rec(c,1);
     f_rec_sym(ndim-c+1,1)=f_rec(c,1);
end

%jwave loop
fileID=fopen(fnameout,'a');%'08abelc_symmetric.dat','a');
if nsymmetric == 1
    fprintf(fileID,'%f %f ',Warr(jwave),f_rec_sym);
else
    fprintf(fileID,'%f %f ',Warr(jwave),f_rec_all);
end
fprintf(fileID,'\n');
fclose(fileID);

if plot_results==1
    figure; % normalized profiles for better comparison
    set(gca,'linewidth',1.5,'fontsize',16);
    plot(x_all,f_rec_all,'g','Linewidth',1.5);
    hold on; plot(x_all,inp_all,'r','Linewidth',1.5);
    grid on; box on;
    title(sprintf('number of cos expansions: %i',upf))
    legend('Abel reconstructed profile','measured profile','Location','South')

    figure; % normalized profiles for better comparison
```

```
    set(gca,'linewidth',1.5,'fontsize',16);
    plot(x_all,f_rec_sym,'g','Linewidth',1.5);
    hold on; plot(x_all,h_all,'k','Linewidth',1.5);
    grid on; box on;
    title(sprintf('wavelength position: %i',jwave))
    legend('Abel reconstructed profile','symmetrized profile','Location','
       South')

end

end
%prepare for 3D plot of Abel-inverted data
fileID=fopen(fnameout,'r'); formats='%f';
jspectra=(ndim-1)*2+1; sizeResult=[jspectra+1 nwavel];
Result=fscanf(fileID,formats,sizeResult);
fclose(fileID)
Result=Result';
fileID3D=fopen(fnameout3D,'a')
fprintf(fileID3D,'Zone f=point i=%d j=%d\n',nwavel,jspectra);
      for jdummy=2:jspectra+1
        wave=Result(:,[1]);
        waveout=wave(1:nwavel);
        waveout=waveout';
        spectrum=Result(:,[jdummy]);
        spectrumout=spectrum(1:nwavel);
        spectrumout=spectrumout';
        for iout=1:nwavel
            x=waveout(iout);
            y=(jdummy-(ndim+1))*dr;
            z=spectrumout(iout);
            fprintf(fileID3D,'%f %f %f \n',x,y,z);
            %fprintf(fileID3D,'\n');
        end
      end
fclose(fileID3D);
end
end
```

K.1.2 Expansion.m

```
function [ fn,hn ] = compute_expansion( X,upf,R )
% COMPUTE_EXPANSION calculates the Fourier series expansion terms, on which
% the Abel inversion algorithm [1] is based.
%
% Details: The unknown distribution f(r) is expanded as
%          f(r) = sum_{n=lof}^{upf} (A_n * f_n(r))                  (1)
% where the lower frequency is set to 1 and the upper frequency upf is
% important for noise-filtering. f_n(r) is a set of cos-functions:
%          f_n(r) = 1 - (-1)^n*cos(n*pi*r/R)  and f_0(r) = 1        (2)
% For the Abel inversion, the integrals h_n have to be calculated
%          h_n(x) = int_x^R f_n(r) * r / sqrt(r^2-x^2) dr           (3)
%
%   [1] G. Pretzler, Z. Naturforsch. 46a, 639 (1991)
%
%                                        written by C. Killer, Sept. 2013


% allocate matrices for f_n and h_n - rows are x-values,
% columns are the number of expansion elements (n+1 since we start with n=0)
fn=zeros(length(X),upf+1);
hn=zeros(length(X),upf+1);


% special case: first column for n=0, where f_0(r)=1;
fn(:,1)=1;
for c=1:length(X);
    x=X(c);

    % evaluation of (3)
    fun=@(r) r./sqrt(r.^2-x.^2);
    hn(c,1) = integral(fun,x,R);
end


% all the other columns
for n=1:upf
    for c=1:length(X)
        x=X(c);

        % evaluation of (2)
        fn(c,n+1) = (1 - (-1)^n*cos(n*pi*x/R));

        % evaluation of (3)
        fun=@(r) (1 - (-1)^n*cos(n*pi*r/R)).*r./sqrt(r.^2-x.^2);
        hn(c,n+1) = integral(fun,x,R);
    end
end

% remove the next comment to plot the integrals
% figure; plot(hn); title('cos-expansion integrals h_n(x)')
```

K.2 Display of wavelength calibrated and sensitivity corrected data

```
clear
pixelsize=0.0136; numberofspectra=256; initialdelay=200; delaystep=250;
    ngatestep=125;
for nfile=01:06
  fnamein=sprintf('%02dabelc385.dat',nfile);
  fileID=fopen(fnamein,'r');
  tline=fgetl(fileID); narrsize = sscanf(tline, 'Zone I=%d J= %d');
  narrlambda = narrsize(1); narrspectr = narrsize(2);
  nwavel=narrlambda; nspectra=narrspectr;
  formatSpec='%f %f %f'; sizeD=[3 Inf]; D=fscanf(fileID,formatSpec,sizeD);
  fclose(fileID);
  D=D';L=D(:,[1]);Warr=L(1:nwavel);%L(1:1001);H=D(:,[2]);
    for ii=1:numberofspectra
      Harr(ii)=ii*pixelsize*4.;
    end
  Val =D(:,[3]);
  for j=1:nspectra %256
    for i=1:nwavel %1001
      Arr(i,j)=Val(i+nwavel*(j-1));
    end
  end
  figure
  xdummylow=Warr(1); ximage=linspace(xdummylow,xdummyhig,nwavel);
  ydummy=(nspectra-1)*pixelsize*4;
  nlow=61; nhig=211; ylow=nlow/nspectra*ydummy; yhig=nhig/nspectra*ydummy;
  yimage=linspace(ylow,yhig,nspectra);
  clims=[0 500000];
  imagesc(ximage,yimage,Arr(1:nwavel,nlow:nhig)');%,clims);%, clims);
  colormap jet;
  axis([380 390 ylow yhig]);
  set(gca,'YDir','normal');
  set(gca,'TickDir','out');
  set(gca,'Fontsize',15,'Fontweight','bold');
  xlabel ('wavelength (nm)', 'Fontsize',20, 'Fontweight', 'bold');
  ylabel ('slit height (mm)', 'Fontsize',20,'Fontweight', 'bold');
  nfiledelay=initialdelay+(nfile-1)*delaystep;
  fname=sprintf('Filter #%d: %d ns delay, %d ns gate.',nfile,nfiledelay,
     ngatestep);
  title (fname, 'Fontsize',22,'Fontweight','bold');
end
```

Figure K.1 displays typical wavelength calibrated and detector sensitivity corrected data. The images illustrate the spectra recorded along the slit-direction.

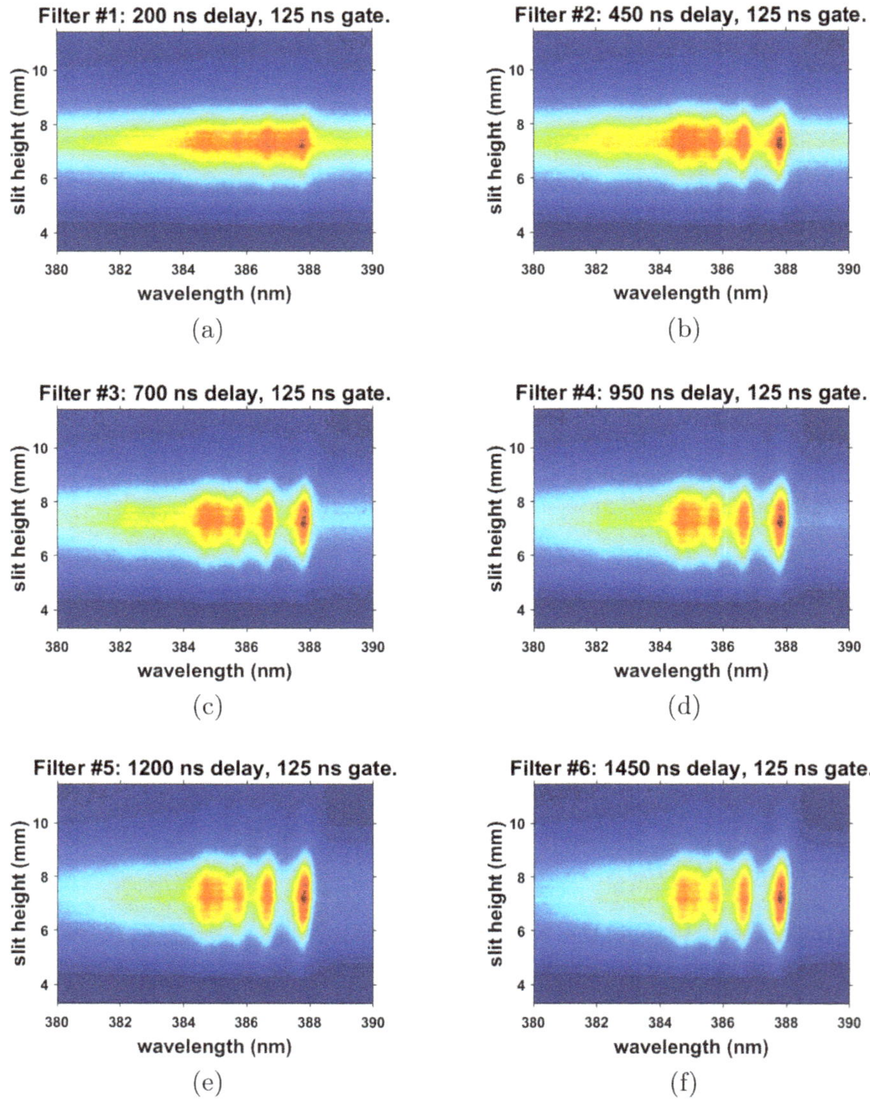

Figure K.1. Typical captured, wavelength calibrated, and sensitivity corrected CN spectra—1:1 molar CO_2:N_2 gaseous mixture held at atmospheric pressure and recorded with a CN spectra cut-on filter. Time delay: (a) 200 ns; (b) 450 ns; (c) 700 ns; (d) 950 ns; (e) 1200 ns; and (f) 1450 ns. Reprinted with permission from [6].

K.3 Display of Abel inverted data

```
clear;
pixelsize=0.0136; numberofspectra=256; initialdelay=200; delaystep=250;
    ngatestep=125;

for nfile=01:06
fnamein=sprintf('%02dabelc385_symmetric3D.dat',nfile);
  fileID=fopen(fnamein,'r');
    tline=fgetl(fileID); narrsize = sscanf(tline, 'Zone f=point i=%d j=%d');
    narrlambda = narrsize(1); narrspectr = narrsize(2);
    nwavel=narrlambda; nspectra=narrspectr;
    formatSpec='%f %f %f'; sizeD=[3 Inf]; D=fscanf(fileID,formatSpec,sizeD);
  fclose(fileID);
  D=D';L=D(:,[1]);Warr=L(1:nwavel);%L(1:1001); H=D(:,[2]);
  for ii=1:numberofspectra
    Harr(ii)=ii*pixelsize*4.;
  end
  Val =D(:,[3]);
    for j=1:nspectra %256
      for i=1:nwavel %1001
        Arr(i,j)=Val(i+nwavel*(j-1));
      end
    end
  figure
    xdummylow=Warr(1);%370.3000;xdummyhig=Warr(nwavel);%393.3547;
    ximage=linspace(xdummylow,xdummyhig,nwavel);
    ydummy=(nspectra-1)/2*pixelsize*4;%0.0544;%/2*0.0544/2;
    yimage=linspace(-ydummy,ydummy,nspectra);
    clims=[0, 250000];
    imagesc(ximage,yimage,Arr');%,clims); colormap jet;
    axis([380 390 -ydummy*0.8 ydummy*0.8]);
    set(gca,'YDir','normal');
    set(gca,'TickDir','out');
    set(gca,'Fontsize',15,'Fontweight','bold');
    xlabel ('wavelength (nm)', 'Fontsize',20, 'Fontweight', 'bold');
    ylabel ('radius (mm)', 'Fontsize',20,'Fontweight', 'bold');
    nfiledelay=initialdelay+(nfile-1)*delaystep;
    fname=sprintf('filter #%d: %d ns delay, %d ns gate.',nfile,nfiledelay,
       ngatestep);
    title (fname, 'Fontsize',22,'Fontweight','bold');
end
```

Figure K.2 displays typical output of the Abel inverted data.

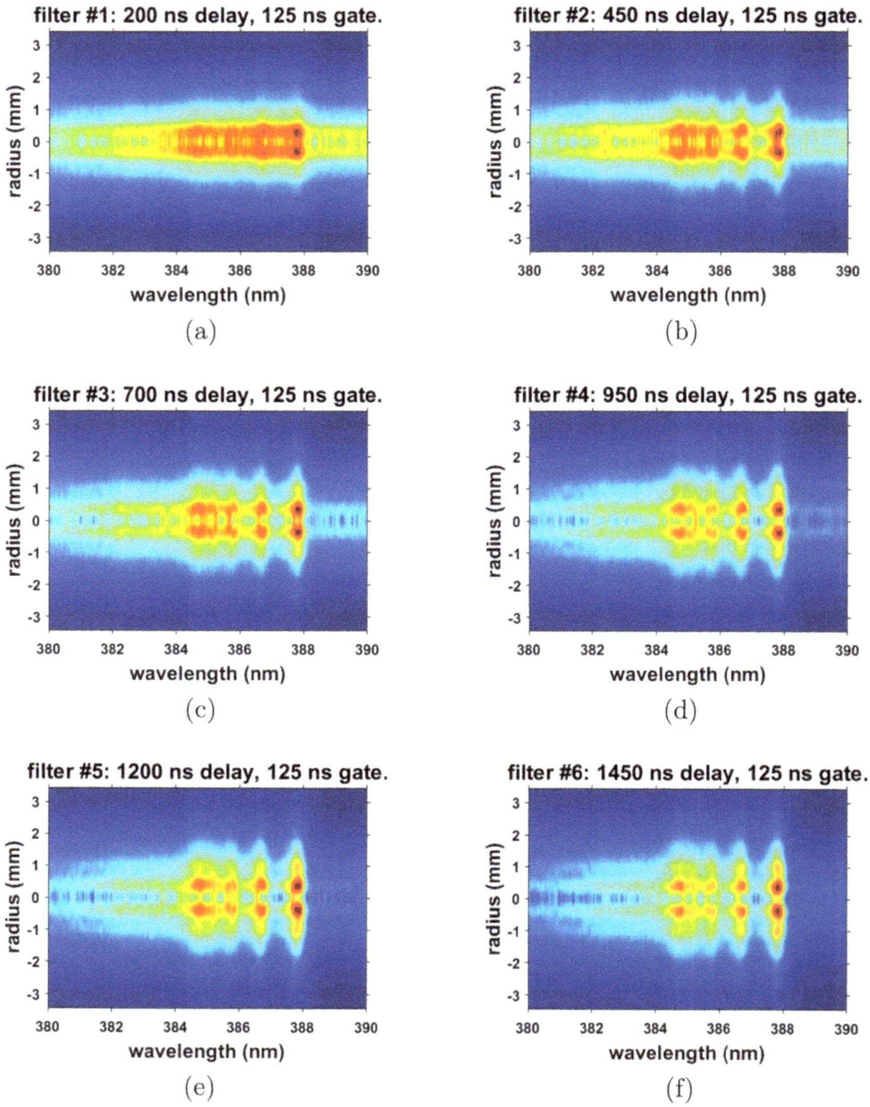

Figure K.2. Typical Abel inverted CN spectra—1:1 molar CO_2:N_2 gaseous mixture held at atmospheric pressure and recorded with a CN spectra cut-on filter. Time delay: (a) 200 ns; (b) 450 ns; (c) 700 ns; (d) 950 ns; (e) 1200 ns; and (f) 1450 ns. Reprinted with permission from [6].

References

[1] Pretzler G and Naturforsch Z 1991 *Z. Naturforsch* **46a** 639

[2] Arfken G B, Weber H J and Harris F E 2012 *Mathematical Methods for Physicists, A Comprehensive Guide* 7th edn (New York: Academic)

[3] Pretzler G, Jäger H, Neger T, Philipp H and Woisetschläger J 1992 *Z. Naturforsch.* A **47** 955

[4] Killer C 2014 http://www.mathworks.com/matlabcentral/fileexchange/43639-abel-inversion-algorithm.
[5] *MATLAB Release R2022a Update 5* (MA: Natick)
[6] Helstern C M 2020 Laser-induced breakdown spectroscopy and plasmas containing cyanide *PhD Dissertation* (University of Tennessee)

IOP Publishing

Quantum Mechanics of the Diatomic Molecule (Second Edition)

Christian G Parigger and James O Hornkohl

Appendix L

LIBS: 2018 to 2023 publications that include C.G.P.

L.1 Introduction

The physics activities and works by the author C.G.P. led to his inclusion in the most recent 2023 Stanford University career list of the World's Top 2% Scientists [1]. This communication summarizes research on the subject of laser-induced transient micro-plasma diagnoses and selected publications during the years 2018 to 2023. Time-resolved spectroscopy elucidates plasma dynamics and species distributions that are generally of value in analytical chemistry. The contents of the summarized work include aspects of electron density, and atomic and molecular distributions. Applications extend from analyses of laboratory to stellar plasma. Of particular interest is the spectroscopy of the hydrogen Balmer series and several diatomic molecules. In most of the publications, nominal nanosecond radiation from table-top laser devices is employed for the generation of the micro-plasma, and spatio-temporal experimental methods capture phenomena that occur at well-above hypersonic, supersonic, and subsonic plasma and gas expansion speeds.

This research-summary addresses recent 2018 to 2023 investigations [2] that were primarily conducted at the Center for Laser Applications at The University of Tennessee Space Institute. However, a few selected publications with international collaborators are also included. The author, Dr Christian Parigger, has been engaged in laser-plasma research at the University of Tennessee from 1987 to 2023. Recent publications in Multidisciplinary Digital Publishing Institute (MDPI) journals Atoms, Molecules, Foundations, and Symmetry encompass various research aspects. The 21 MDPI articles referenced in this summary reflect scientific, open-access, and peer-reviewed engagements. Various conference contributions, including in the *Journal of Physics: Conference Series*, further portray recent research associated with the biannual and well-established International Conferences on Spectral Line Shapes (ICSLS). The transition from previous archived journals such as Applied Optics, Optics Letters, Spectrochimica Acta

doi:10.1088/978-0-7503-6204-7ch42

Part A and/or Part B, and Journal of Quantitative Spectroscopy and Radiative Transfer to peer-reviewed open-access journals is in accord with worldwide transition to open-access viz. access-for-everyone. In addition, the moderated Cornell University https://arxiv.org and MDPI https://www.preprints.org preprint servers also convey aspects of research. And, of course, the Auburn University electronic International Review of Atomic and Molecular Physics (IRAMP) journal https://www.auburn.edu/cosam/departments/phyiscs/iramp/index.htm communicates peer-reviewed research activities.

L.2 Summary

L.2.1 Laser-plasma atomic and molecular spectroscopy

Hydrogen and selected diatomic emission spectroscopy includes analysis of laboratory and stellar astrophysical plasma, e.g., from white dwarfs [3–8]. These works include self-absorption assessments. Expansion dynamics at hypersonic, supersonic, and subsonic are usually measured with spatio-temporal spectroscopy [9–17]. Fundamental aspects of diatomic molecular spectroscopy [18] lead to consistent data analyses without invoking the concept of reversed angular momentum—the Nelder–Mead temperature (NMT) program and the Boltzmann equilibrium spectra program (BESP) are freely available [19] as clear-text scripts with data files. Plasma diagnosis is elaborated with selected diatomic molecules, including comparisons of the published database with other readily available databases for OH, CN, C_2, and AlO [19–25]. In addition, the collaboration with the University of Prayagraj (formerly Allahabad), India, on meteorite and gypsum laser-induced breakdown spectroscopy (LIBS) [26, 27] and on medical applications that include gallstone and pointed gourd leaves analyses has been elaborated [28, 29]. Collaborations with the University of Cairo include research on plasma involving silver nano-particles [30–32]. Recent collaborations with the Chemical Research Center in Hungary focus on microwave plasma methylidyne (CH) cavity ring-down spectroscopy [33].

L.2.2 Molecular spectroscopy chapter and e-book

The two fundamental works in 2020 are a book chapter [34] on molecular LIBS and an e-book on diatomic spectroscopy [35]. The former communicates molecular spectroscopy and applications to plasma, combustion, and astrophysics analyses. Primary interests include plasma in gases; however, the book chapter [34] includes laser ablation, including the coauthors' work on laser-ablation molecular isotopic spectrometry (LAMIS). Diatomic molecules include cyanide (CN), aluminum monoxide (AlO), titanium monoxide (TiO), Swan bands of C_2, and the hydroxyl radical (OH). Aspects of spherical aberrations from focusing with a single lens are elaborated, and Abel inversion techniques are discussed for determination of spatial molecular distribution. The latter derives diatomic spectroscopy transition strengths [35] employing the Wigner–Witmer diatomic eigenfunction. The diatomic line strength is composed of electronic, vibrational, and rotational transition terms, including Franck–Condon, Hönl–London and r-centroid factors.

L.3 Discussion

Both atomic and molecular species can be readily discerned from comparisons of measured and computed atomic line shapes and molecular band appearances. Several of the investigations elucidate experimental spatially- and temporally-resolved LIBS records' analysis details. The molecular emission spectroscopy comparisons require accurate databases. The established and well-tested databases for selected electronic, vibrational, and rotational diatomic transitions and the associated analysis programs are now published for applications in LIBS research [19]. The 2018 to 2023 summary shows a research focus in 2022 to 2023 on molecular diagnosis by comparing accurate line strengths predictions [19–23] with those from readily available other corresponding databases [36, 37], including ExoMol [37]. Future applications are envisioned to include laser-induced hydrogen-based combustion—analysis of hydrogen emission lines and OH molecular bands are expected to benefit from the research publications communicated in this summary.

References

[1] Ioannidis J P A 2023 *October 2023 data-update for: Updated science-wide author databases of standardized citation indicators,* Elsevier Data Repository, Electronic data https://ecebm.com/2023/10/04/stanford-university-names-worlds-top-2-scientists-2023/
[2] Parigger C G 2023 *Int. Rev. At. Mol. Phys.* **14** 89
[3] Parigger C G, Drake K A, Helstern C M and Gautam G 2018 *Atoms* **6** 36
[4] Parigger C G 2020 *Contrib. Astronom. Observat. Skalnaté Pleso* **50** 1
[5] Parigger C G, Helstern C M, Gautam G and Drake D A 2019 *J. Phys.: Conf. Ser.* **1289** 012001
[6] Parigger C G, Helstern C M and Gautam G 2019 *Atoms* **7** 63
[7] Surmick D M and Parigger C G 2019 *Atoms* **7** 101
[8] Parigger C G, Sherbini A M E L and Splinter R 2019 *J. Phys.: Conf. Ser.* **1253** 012001
[9] Gautam G and Parigger C G 2018 *Atoms* **6** 46
[10] Parigger C G 2019 *Atoms* **7** 61
[11] Parigger C G, Helstern C M and Gautam G 2020 *Symmetry* **12** 2116
[12] Parigger C G, Helstern C M and Gautam G 2019 *Atoms* **7** 74
[13] Helstern C M and Parigger C G 2019 *J. Phys.: Conf. Ser.* **1289** 012016
[14] Parigger C G and Helstern C M 2023 *J. Phys.: Conf. Ser.* **2439** 012003
[15] Surmick D M, Dagel D J and Parigger C G 2019 *Atoms* **7** 86
[16] Parigger C G, Helstern C M, Jordan B S, Surmick D M and Splinter R 2020 *Molecules* **25** 615
[17] Parigger C G 2020 *Spectrochim. Acta* B **179** 106122
[18] Parigger C G 2021 *Foundations* **1** 208
[19] Parigger C G 2023 *Foundations* **3** 1
[20] Parigger C G 2022 *Foundations* **2** 934
[21] Parigger C G 2023 *Atoms* **11** 62
[22] Parigger C G 2023 *Preprints* **2023** 2023050423
[23] Parigger C G 2023 *Preprints* **2023** 2023041258

[24] Parigger C G, Helstern C M, Jordan B S, Surmick D M and Splinter R 2020 *Molecules* **25** 988
[25] Parigger C G and Helstern C M 2023 *J. Phys.: Conf. Ser.* **2439** 012004
[26] Rai A K, Pati J K, Parigger C G, Dubey S, Rai A K, Bhagabaty B, Mazumdar A C and Duorah K 2020 *Molecules* **25** 984
[27] Rai A K, Pati J K, Parigger C G and Rai A K 2019 *Atoms* **7** 72
[28] Pathak A K, Rai N K, Kumar R, Rai P K, Rai A K and Parigger C G 2018 *Atoms* **6** 42
[29] Kumar T, Rai P K, Rai A K, Rai N K, Rai A K, Parigger C G, Watal G and Yadav S 2022 *Foundations* **2** 981
[30] Sherbini A M E L, Sherbini A E E L and Parigger C G 2018 *Atoms* **6** 44
[31] Sherbini1 A M E L, Sherbini A E E L, Parigger C G and Sherbini T M E L 2019 *J. Phys.: Conf. Ser.* **1289** 012002
[32] Sherbini A M E L, Farash A H E L, Sherbini T M E L and Parigger C G 2019 *Atoms* **7** 73
[33] Nemes L and Parigger C G 2023 *Foundations* **3** 16
[34] Parigger C G, Surmick D M, Helstern C M, Gautam G, Bol'shakov A A and Russo R 2020 Molecular laser-induced breakdown spectroscopy *Laser Induced Breakdown Spectroscopy* ed J P Singh and S N Thakur 2nd edn (New York: Elsevier)
[35] Parigger C G and Hornkohl J O 2020 *Quantum Mechanics of the Diatomic Molecule with Applications* (Bristol: IOP Publishing)
[36] McKemmish L K 2021 *WIREs Comput. Mol. Sci.* **11** e1520
[37] Tennyson J *et al* 2020 *J. Quant. Spectrosc. Radiat. Transf.* **255** 107228

www.ingramcontent.com/pod-product-compliance
Ingram Content Group UK Ltd.
Pitfield, Milton Keynes, MK11 3LW, UK
UKHW051905100926
8556IPUK00005B/775

9 780750 362